W9-BEX-466

INGENIOUS MECHANISMS
FOR DESIGNERS AND INVENTORS

VOLUME IV

INGENIOUS MECHANISMS

FOR DESIGNERS AND INVENTORS

VOLUME IV

Mechanisms and Mechanical Movements Selected from Automatic Machines and Various Other Forms of Mechanical Apparatus as Outstanding Examples of Ingenious Design Embodying Ideas or Principles Applicable in Designing Machines or Devices Requiring Automatic Features or Mechanical Control

Edited by
JOHN A. NEWELL
and
HOLBROOK L. HORTON

INDUSTRIAL PRESS INC.
200 MADISON AVENUE, NEW YORK 10016

Industrial Press Inc.
200 Madison Avenue
New York, New York 10016-4078

INGENIOUS MECHANISMS
FOR DESIGNERS AND INVENTORS—VOLUME IV

Copyright © 1967 by Industrial Press Inc., New York, N.Y. Printed in the United States of America. All rights reserved. This book or parts thereof may not be reproduced in any form without permission of the publishers.

12 14 13 11

FOURTH VOLUME OF INGENIOUS MECHANISMS

A considerable period of time has elapsed since the publication of the third volume of Ingenious Mechanisms. During this period we have received many inquiries about the possible publication of a fourth volume in the series, indicating a continuing interest in this area.

This fourth volume follows the same pattern as its predecessors. The mechanisms described have been developed for application in a wide variety of fields. Rather than classify them by application, however, they have been grouped by type of mechanical movement. Thus, the reader is quickly guided to those mechanisms which may provide a possible solution to his problem.

Furthermore, the grouping is closely similar to, if not exactly the same as, that in the previous volumes so that the entire set may be used as an integrated reference library on the subject of mechanisms.

CONTENTS

CHAPTER 1

Cam Applications and Special Cam Designs

In the design of mechanisms to obtain irregular movements of various kinds, cams are frequently employed. Those which are described or illustrated in connection with the mechanisms covered by this chapter are notable for some ingenious arrangement or design. Other applications of cams and cam-operated mechanisms will be found in Chapter 1, Volume I; Chapter 1, Volume II; and Chapter I, Volume III of "Ingenious Mechanisms for Designers and Inventors."

Cam Produces Motion on Alternate Revolutions

A conventional plate-cam was used on a machine producing a wire product to operate a forming press. The press had to be actuated once during each revolution of the driving shaft. A subsequent product change necessitated an alteration in the operating cycle of the cam — it was now required to operate the press twice during one revolution, then to remain at rest during the next revolution. Figures 1 and 2 show the design and operation of a cam which produced the desired movements with no alterations being required on the machine.

Cam body *A*, Fig. 1, is in the shape of a disc having an integral hub on its front face. The cam is keyed to shaft *B* and rotates in the direction indicated by the arrow. Two studs *C* pass through the disc and are free to rotate. Welded to them are curved bars *D* which act as cam lobes. Compression springs *E* apply sufficient frictional resistance to the studs to prevent movement by centrifugal force. Lever *F*, which operates the forming press, carries

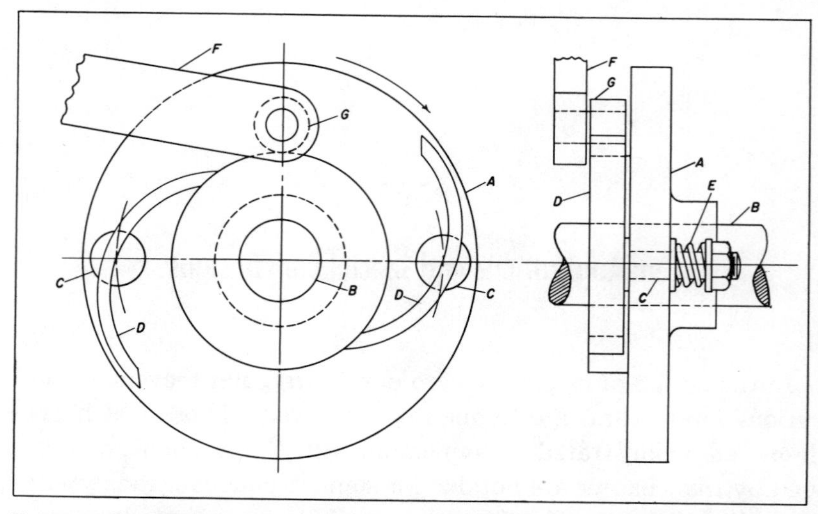

FIG. 1. Cam body *A* carries two moving cam-bars *D*. This design imparts two movements to lever *F* during one revolution, followed by one revolution at rest.

follower-roller *G* and is held against the cam by a spring (not shown).

Operation of the cam is illustrated in Fig. 2. At W, cam-bar *D* is in the same position as in Fig. 1, but the entire cam has rotated 90 degrees. Stud *C* and roller *G* are now on the same center line, bar *D* having caused the roller to rise and operate the press through lever *F*. The cam continues to rotate, as at X, and spring tension on lever *F* overcomes the frictional resistance of stud *C*, forcing bar *D* into the position shown. When the other lobe of the cam comes into position this action is repeated, so that two movements of the lever are produced, 180 degrees apart, in one revolution of shaft *B*.

No movement of lever *F* is produced during the next revolution of the shaft. This is because the leading ends of the cam-bars have been lifted from the hub of disc *A* and now pass over roller *G*, as shown at Y. As the cam rotates further, and the roller passes the center line of stud *C*, cam-bar *D* is forced to pivot as at Z — thus being returned to its original position (Fig. 1). In this manner, each two-revolution cycle of the cam produces two

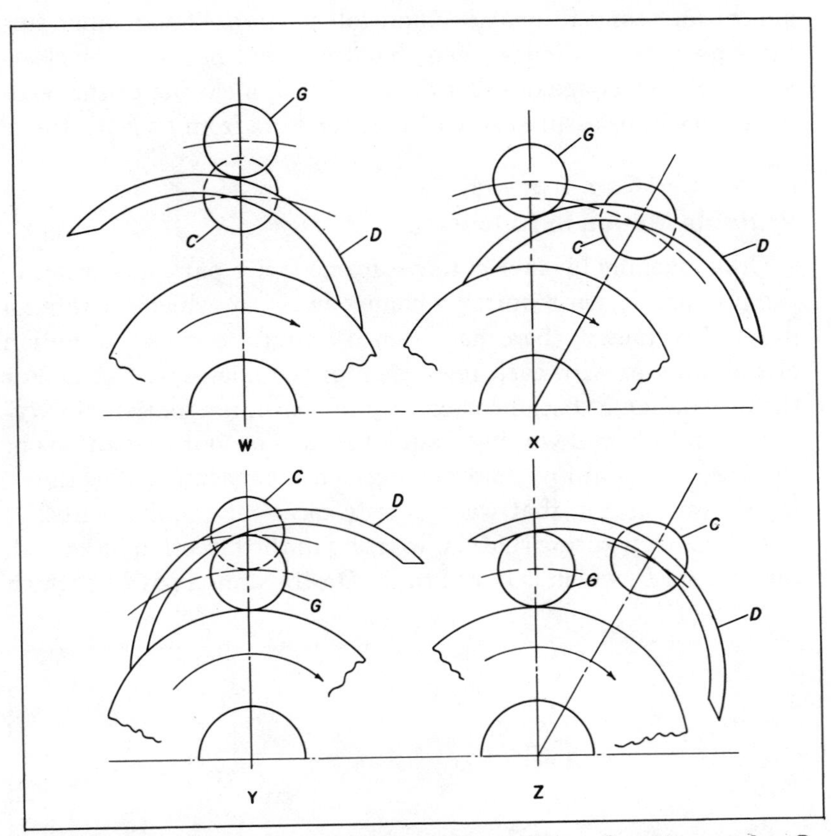

FIG. 2. During the active revolution, follower *G* rises along the cam-bar *D*, view W, then forces it to pivot for the downward movement, view X. On the next trip around, the upended cam-bar passes over the follower and is returned to its original position, views Y and Z.

movements of lever *F*, followed by a rest period of 540 degrees.

There may appear to be an undesirable feature in the design of this cam in that there would be a rapid drop of roller *G* on the falling side of bar *D* (view X, Fig. 2). This, however, does not occur, due to the fact that the outer surface of the cam-bar is on a rising angle (view W). Thus, downward movement takes place almost immediately after the center of stud *C* passes the center of roller *G*.

Outer surfaces of the cam-bars may be contoured to produce

almost any conventional rise and fall pattern. Their inner surfaces must be so dimensioned that there will be sufficient clearance for the passage of roller *G*, and that full closing of the leading ends will be assured when the roller exits from beneath them.

Four-Lobed Cam Transmits Variable Motion to Follower

On a machine for fabricating a formed wire part, a revision in the product design required a change in a cam which operated a press. Previously, there had been a uniform oscillating motion of the follower with each revolution of the operating shaft. For the new design, it became necessary to transmit a motion of varying magnitude and varying timing for each of four revolutions of the shaft, without any major changes in the machine. The drawing shows the cam that was made to meet the requirements.

In Fig. 3, operating shaft *A*, rotating in the direction indicated, carries arm *B*, which is keyed to it. On the arm is pawl *C*, which

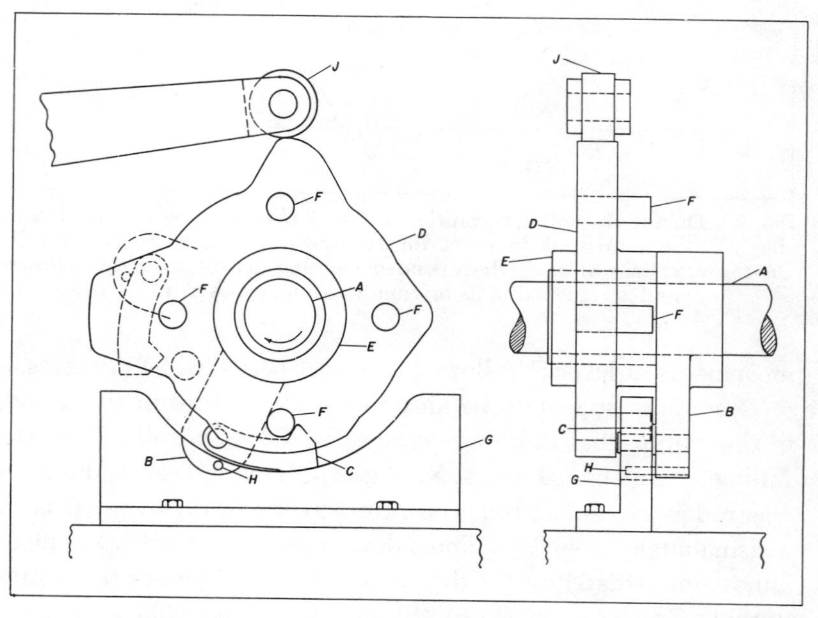

FIG. 3. As long as pawl *C* remains in contact with angle-plate *G*, cam *D* rotates. When the pawl leaves the angle-plate, the cam stops.

can swing on its stud. Four-lobed cam *D* is free on the extended hub of the arm, and is retained by collar *E*. Projecting from the face of the cam are four pins *F*, equally spaced around the center of rotation. Angle-plate *G* is machined to a true arc of a circle on its upper edge, and is attached to a stationary part of the machine.

The shaft, arm, and pawl rotate as a unit. No motion is transmitted to the cam until the pawl contacts the upper edge of the angle plate, when the pawl is brought into position to engage one of the pins. The drawing shows the position of the components at about the midpoint of the cam movement.

Having engaged one of the pins, the pawl carries the cam with it, until it no longer contacts the angle-plate. At this point, due to the angularity of the contact surface of the pawl, it disengages automatically, and the cam stops. (The position of the pawl at the time of disengagement is shown in broken lines on the left side of the drawing.) Stop *H* limits the swing of the pawl so that it will be in position to engage the next pin upon rotating to the right side of the angle-plate. Thus, the cam rotates 90 degrees for each revolution of the shaft, and the four lobes of the cam are brought consecutively into position to actuate cam follower *J* as required.

Intermittent Rotary Motion from a Uniform Reciprocating Drive

Two devices on a machine had to be rotated intermittently, with a rest period at each of eight stations in each cycle. Although the loading devices were widely separated, they could be placed in axial alignment and, be carried on the same shaft. The required movement, shown in Fig. 4, was obtained from a barrel-cam driven by a reciprocating part.

Shaft *A*, on which the loading devices are mounted, is supported in bearings and carries a barrel-cam *B* with an irregular groove. Roller *C* operates in this groove and is carried on a slide-bar *D* mounted on a stationary part of the machine by gibs *E*. Member *D* is given a uniform reciprocating motion by a cam (not shown). Since there are eight stations, eight axial follower

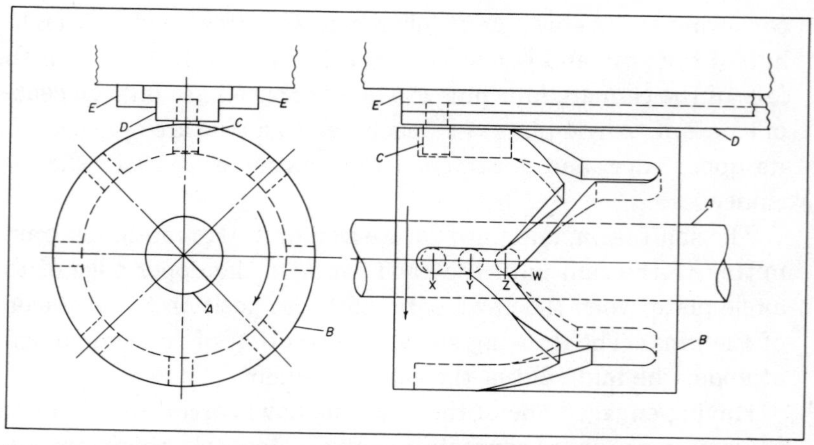

FIG. 4. Cam mechanism for converting uniform reciprocating motion to intermittent rotary motion.

grooves are machined on opposite sides of the cam barrel. These are connected by other grooves milled in the periphery at an angle of about 45 degrees with the axis of the cam. The vertex of the angle formed by any two of the angular grooves is approximately in line with one side of the axial grooves, as indicated by line *W*.

The assembly is shown with slide-bar *D* at the extreme left position. To demonstrate the action, three positions of the roller in the center groove are seen at *X*, *Y*, and *Z*. Roller *C*, in moving from the extreme right position, acts against the angular side of the groove, causing the cam *B* to rotate in the direction indicated by the arrow. When roller *C* reaches position *Y*, rotation of the cam ceases, and it remains at rest during the continued movement of the roller to the extreme left position *X*.

On the return movement, no rotation of cam *B* is produced until roller *C* again contacts the angular groove at position *Z*. Since the vertex of the angular groove surfaces are not aligned with the centers of the axial grooves, the roller cannot return to the groove previously traversed, but must enter the next one. Continued movement of roller *C* causes cam *B* and shaft *A* to rotate to the next station, and the cycle is repeated.

Cylindrical Cam Positions
Wire Guide

Figures 5 and 6 show two views of a mechanism designed to guide a strand of wire through an irregular path in a machine which produces a woven wire product. The purpose of this mechanism is to create a continuously varying pattern in the weave. Position of the wire strand *W* in the weave pattern must bear a given relationship to other parts of the weave over a required length of the fabric, and then repeat. Figure 5 is a plain view of the mechanism, and Fig. 6 is a front view.

The driving shaft *A* carries the worm *B*, which meshes with the worm-gear *C* on shaft *D*. Shaft *D* carries the cylindrical cam *E*, which imparts a transverse guiding movement to wire *W*. The two rounded grooves in cam *E* are identical, although the axes of the grooves are offset from the shaft axis and are 180 degrees apart.

Shaft *G* receives motion from worm *B* through worm-gear *F*, and carries the disc *H*, which is connected to block *J* by the pitman *I*. Block *J* is attached to the dovetailed slide *K*, which is

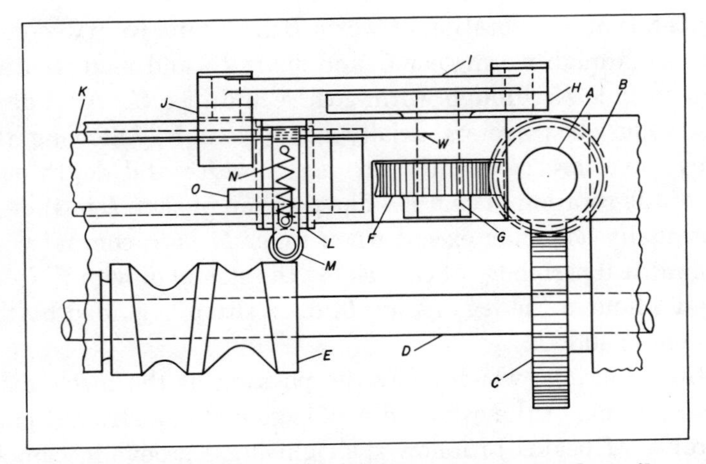

FIG. 5. Plan view of wire-guide mechanism for a metal textile weaving machine designed to impart an intricate transverse motion pattern to wire strand *W*.

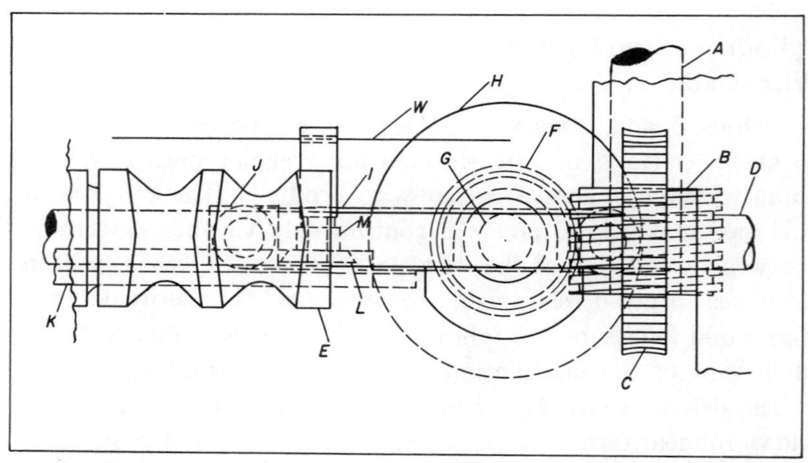

FIG. 6. Sixty turns of shaft *A* are required to completely cycle follower *M* through a single out-and-back traverse of rotating cam *E*.

given a reciprocating motion by the rotation of crank disc *H*. Slide *K* carries the dovetailed slide *L*, which in turn mounts the roller follower *M*. The follower roll is held in contact with cam *E* by the spring *N* attached to a pin in slide *L* and a pin in bracket *O* attached to slide *K*.

In operation, the rotation of worm *B* transmits rotary motion to cam *E* through worm-gear *C* and shaft *D*, and reciprocating motion to slide *K* through worm-gear *F* and disc *H*. As slide *K* moves, roller *M* traverses axially along cam *E*, following the grooves which are constantly varying in width and depth as a result of the rotation of cam *E*. The position of slide *L* in slide *K* is continually changing except when roller *M* is in contact with the cylindrical periphery of cam *E*. As the strand of wire *W* feeds through a hole in the leg of slide *L*, its position is guided by the movement of slide *L*.

In the diagrams, which show the position of the mechanism at the beginning of the cycle, wire *W* is guided in a straight path until roller *M* begins to follow the right-hand groove in cam *E*. Thus the wire is moved from start position. It returns to its starting position when roller *M* returns to the periphery of cam *E*. After a short period of rest, slide *L* again moves as roller *M* enters the second groove. This is followed by a period of rest as roller *M* again reaches the periphery of cam *E*. Worm-gears *C* and *F* are of different pitch diameters. *G* has thirty teeth and *F* has twenty teeth. Therefore, the rotation of cam *E* and the movement of slide *K* are not synchronized. Thus the path followed by roller *M* varies as the varying contours of the cam are presented to it. This action results in varying rest periods at the ends of the movement of slide *K*, as well as a varying timing pattern and positioning of slide *L* at different points of the cycle, setting up an intricate pattern in the positioning of wire *W*.

While the diagrams indicate the starting point of the cycle, the completion of the cycle is accomplished only when all of the moving members of the mechanism are returned to this start position. As stated, worm-gears *C* and *F* have thirty and twenty teeth, respectively, having a ratio of 3 to 2. Therefore, for a complete cycle, gear *F* must complete three revolutions, and gear *C* must complete two revolutions. The complete cycle, therefore, requires sixty revolutions of drive-shaft *A*.

Combination Cam Controls Stock Feed of Wire-Forming Machine

A combination end and radial cam is the heart of a stock feed for a multiple-slide wire-forming machine. Figure 7 shows the arrangement of the parts.

The combination cam *A* is carried by the machine's shaft system. It is basically a two-diameter plug, the shoulder having been modified to form an axial cam, and the small diameter to form a radial cam. Follower *B* in lever *C* rides on both cam surfaces.

This lever fulcrums at its center on stud *D*. By being joined to the stud by cross-pin *E*, the lever can swing both left to right (to follow the end cam surface) and in and out (to follow the radial cam surface).

Directly behind the lower end of the lever is a dovetail slide *F* containing quill *G*. The quill has two parts. One part fits the dovetail, and the other has a pin carrying bushing *J* which fits a slot in the lower end of the lever. There is a semicircular section

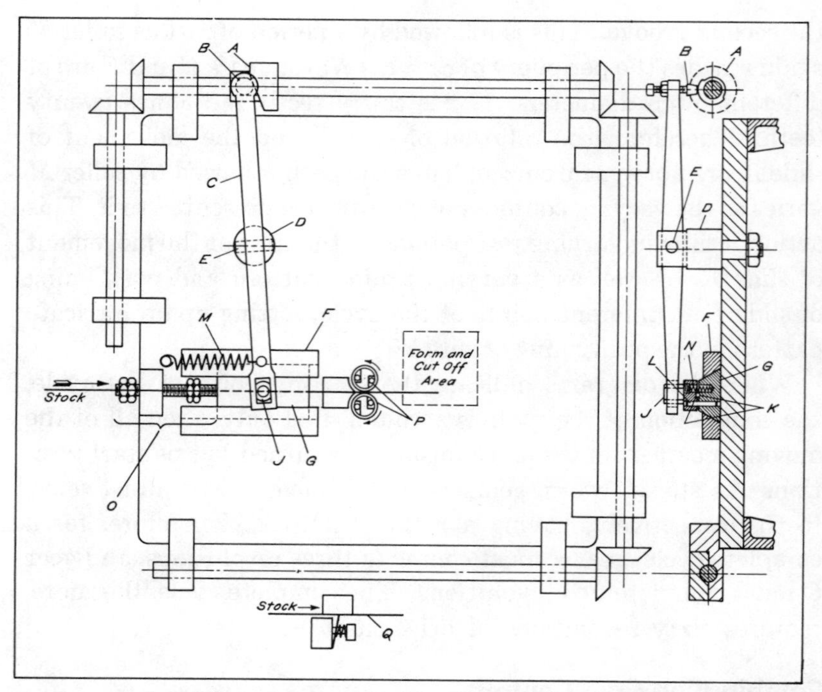

FIG. 7. Cam *A* controls the feed of the wire by having its motion translated to quill *G* through lever *C*.

K in the mating surfaces of the two parts of the quill, through which the stock advances. (The section can be modified to accept whatever stock size or shape is used.) The outer part of the quill is held in position by loose-fitting pin *L*.

The drawing shows the lever position at a point in the feed stroke. At the start of the stroke, spring *M* pulls the bottom of the lever to the left. As the cam starts to rotate, the lobe on its small diameter bears against the follower, causing the bottom of the lever to swing in and thus force the quill to close tightly over the wire. At the same time, the end cam forces the bottom of the lever to the right, thus advancing the wire.

When the lever reaches the end of the forward stroke, the follower no longer has any thrust on it from either cam surface. Spring *N*, contained in a hole extending into both parts of the quill, then allows the quill to release its grip on the wire. Next, spring *M* operates, pulling the lever to the left, and the feed cycle is completed.

Stop-block assembly *O* provides stroke-length adjustment by controlling the point to which the lever can return. A simple way to prevent the wire from tending to move back on the return stroke is to add a pair of non-reversing rolls *P*. Or, a spring check *Q* can be used.

Mechanism Loops and Twists Wire Ends

Completion of a certain job necessitated the production of wire components having a twisted loop at each end. The soft steel wire parts were required in lengths of 16 inches and longer. A typical twisted loop is shown at X in Fig. 8.

Hollow shaft *A* has a running fit on drive shaft *B*, and is coupled to it by means of a clutch, a portion of which may be

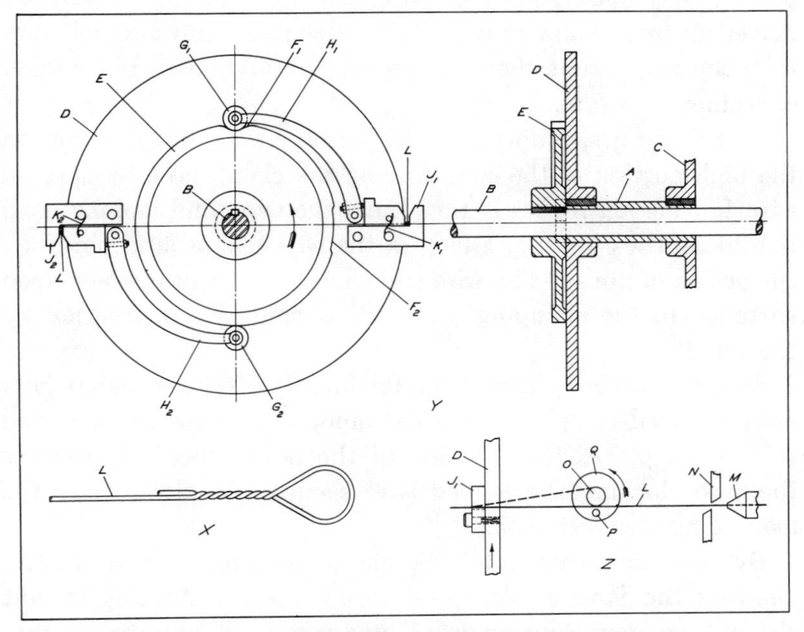

FIG. 8. Device loops and clamps wire component, then transfers it on a rotary disc to a point where the loop is twisted into the form shown at X.

seen at C. This clutch allows shaft A, shown at Y, to rotate one-half revolution for each complete revolution made by shaft B. During the remaining one-half revolution of the drive shaft, shaft A dwells. Keyed to the left-hand end of the hollow shaft is mounting disc D.

Disc cam E, having two notches, F_1 and F_2, is keyed directly to the drive shaft. The purpose of the cam is to actuate follower rollers G_1 and G_2 which, in turn, pivot levers H_1 and H_2. This action either causes the jaws of clamps J_1 and J_2 to close or permits them to open under the influence of springs K_1 and K_2. The two clamps, together with their actuating levers, are mounted on disc D. There is a second complete unit (not shown), identical to the one illustrated, at the left-hand end of shaft B to twist a loop in the opposite end of the wire part.

In operation, the wire L to be formed is fed through guide nozzle M, as shown at Z, between two cutting blades N and then between two pins O and P of looping head Q. When the correct developed length of wire has been fed out, clamp J_1 moves up from below to nest it. At this point clutch C releases, with the result that the clamp remains stationary while cam E continues to rotate.

As this occurs, follower G_1 leaves notch F_1 and rides up on the high portion of the cam, forcing the clamp jaws to close on wire L. Cutting blades N now shear the wire, and looping head Q rotates, thus forming a loop in the wire end around pin O by the action of pin P. The wire end comes to rest in the V-shaped entrance to the clamping jaws and is held in this position by the pin P.

As cam notch F_2 arrives under follower G_1, the clamp jaws are permitted to open a sufficient amount to allow the wire end to drop in place. The passing of the notch once again closes the jaws, locking the looped wire securely in place while the looping head is retracted.

By this time, cam notch F_1 has arrived under follower G_2, allowing the jaws of clamp J_2 to open wide. At this instant clutch C engages, causing disc D and cam E to once again rotate in unison for one-half revolution. During this movement,

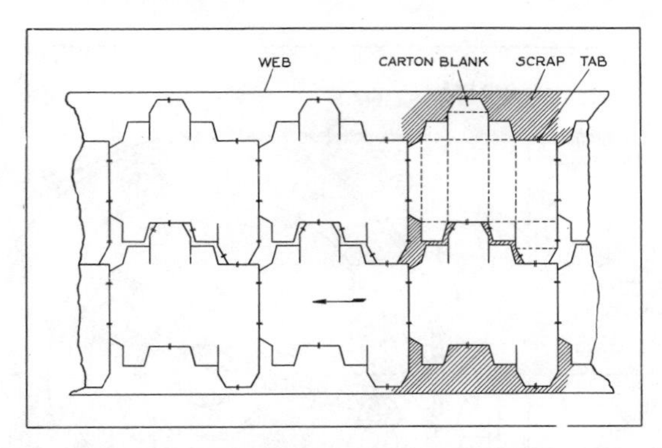

FIG. 9. Cardboard web layout showing die-cut carton blanks and the attached scrap material.

clamp J_2 moves up into contact with a newly fed length of wire, which begins the next cycle.

While in this position, with clamp J_1 dwelling at a location 180 degrees away from its starting point, a twisting head engages the clamped wire loop. This head, not shown, completes the twist as shown in the illustrated example of the workpiece. Following the twisting operation, cam notch F_1 moves beneath follower G_1, allowing the clamp jaws to open and to drop the finished wire components clear of the machine. The cycle of operation then continues.

Rotary Scrap-Stripping Device

One step in the manufacture of cardboard cartons is the die cutting of the developed form. Each printed carton blank is joined to the adjacent carton and to the scrap material surrounding the end flaps by means of small tabs. A typical layout, Fig. 9, shows the arrangement of the die-cut shapes on the web, or continuous cardboard strip; the scrap material, depicted by shading; and the tabs, indicated by two parallel lines.

Upon completion of the die-cutting operation, the scrap pieces must be removed from the web. This procedure, known as stripping, is frequently carried out by hand. However, the rotary scrap stripper shown in Fig. 10 has been designed to replace this manual operation.

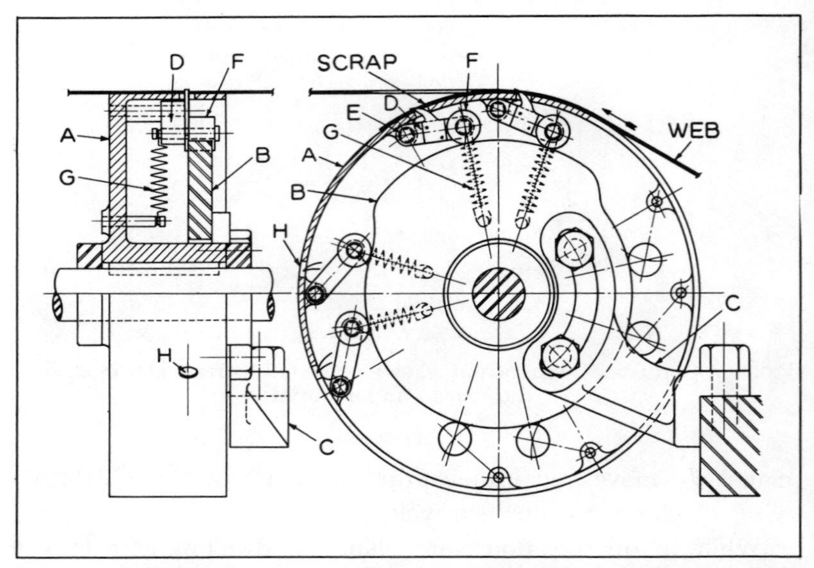

FIG. 10. A rotating unit strips scrap material from a die-cut web on a production line handling cardboard cartons.

The main member of the device is a cylindrical housing *A*. This housing is driven at the same surface speed and is the same diameter as the printing cylinders that are located ahead of the cutting dies. Cam *B* is stationary, being mounted on a fixed portion of the machine by means of bracket *C*. An arcuate slot in the bracket permits adjustment of the cam position.

Scrap pieces are picked out by a hook attached to follower arm *D*. This arm pivots on shaft *E* which fits into a bearing hole bored through an integral housing lug. On the other end of the arm is a cam-follower *F*. Because arm *D* and its associated parts rotate with the cylindrical housing, tension springs *G* are necessary to overcome the centrifugal force developed during normal operation and maintain the cam-followers in contact with the cam surface. A clearance opening *H* is provided for each hook.

It is apparent from the lay-out of the carton blanks on the web that there is a periodic repetition of the scrap pieces and, therefore, that they will always contact the same area on housing *A*. Because of this, one or more hooks are located within each contact area.

During normal operation, follower *G* rides up on the cam lobe as the leading edge of the scrap piece contacts the housing. Follower arm *D* is raised just enough to allow the hook to puncture the scrap piece and remain there while the follower is in contact with the cam lobe. As the web leaves housing *A* in a tangential path, the scrap material is retained by the curved hooks. This division in paths of travel causes the tabs to break, thus effecting a separation between the carton blanks and the unwanted trimmings. After the cam-followers leave the cam lobe, the hooks are withdrawn, and the scrap is free to fall into disposal cans.

Single Closed-Track Cam Drives Glue-Transfer Mechanism

One station of a large machine for packaging material in paper sacks is devoted to the application of glue before the final fold is made to seal the container. An interesting mechanical arrangement is employed to accomplish the various motions required.

At this station of the machine, the package carrier *A* is brought to a momentary halt (see Fig. 11). During this interval, an applicator *B* (carried by slide *C*) must pick up glue from cylinder *D*, turn through an arc of 180 degrees (by means of a stationary cam, not shown), and move to the package fold to deposit a strip of the adhesive.

All movements necessary to advance and retract the applicator slide, rotate the glue cylinder between packages, and synchronize these functions are furnished basically by just one closed-track cam *E*. Follower-roller *F* is carried by follower-lever *G*, which is integral with short lever *H*. The roller rides in cam track *J*.

Long lever *K* pivots freely about the same shaft used by *G* and *H*. However, it is coupled to them as shown in the auxiliary

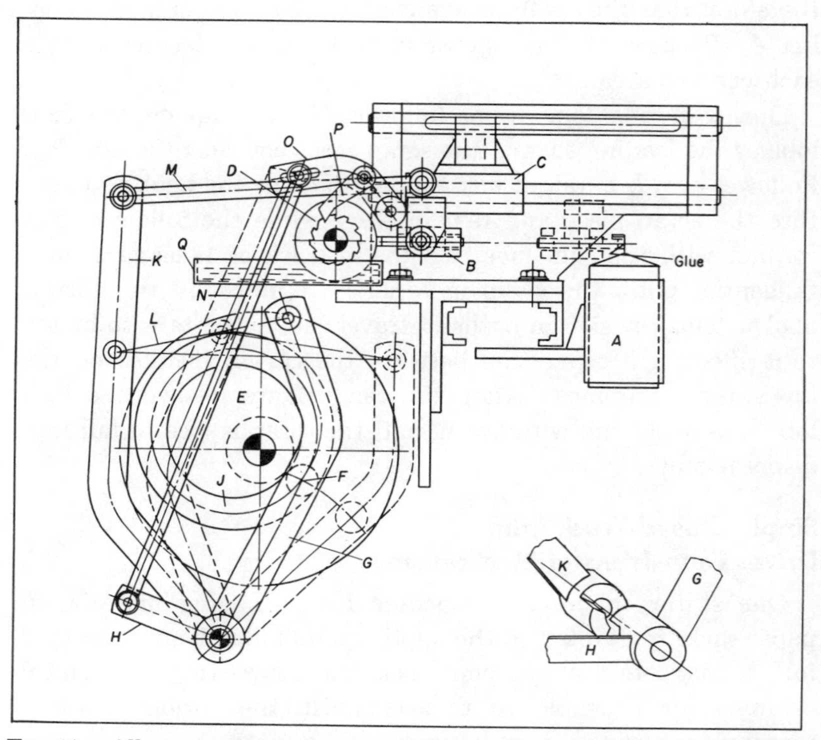

FIG. 11. All movements necessary to the proper functioning of this glue-transfer mechanism are under the control of one closed-track cam.

sketch so that there is over 90 degrees of play between them. The coupling is designed so that lever *G* will drive lever *K* to the left only. *G* and *K* are shown in their closest position.

As the cam rotates, follower-roller *F* is forced to the right. Because the coupling does not provide positive drive in this direction, a spring *L*, attached to levers *G* and *K*, causes the latter to follow the movement of the former.

In this way, both desired end motions are achieved. First, lever *K*, through connecting-rod *M*, advances and retracts slide *C* and, in turn, applicator *B*. Second, lever *H*, through connecting-rod *N*, raises and lowers the pivoting support arm *O* that permits ratchet and pawl *P* to function, thus rotating the glue cylinder in glue tray *Q* during each reciprocation of slide *C*. On the return movement of lever *G* the coupling engages and lever *K* is forced back to its original position under direct drive.

The cam is designed to yield a 180-degree rise and a 180-degree return motion. Although this high-speed mechanism is capable of transferring large forces, its capacity can be increased by cutting a close-tolerance cam groove and by replacing the line-contact follower-roller with a plane-contact sliding piece.

Indexing Attachment that Controls Ratchet Operation

On a machine for producing ornamental wire screening, it is necessary for the movement of the screening through the machine to be interrupted at certain times, depending on the screen pattern. During the idle period of the feeding mechanism, work is performed on the screening. In its movements, the screening is fed by a ratchet mechanism. How this mechanism was designed is shown in Fig. 12.

Drive-shaft *A* carries a worm *C* and disc *B* which rotate with the shaft in a clockwise direction. Disc *B* is provided with a slide-bar *M* which operates in a groove against light frictional resistance that is provided by two springs. The ends of bar *M* are shaped to serve as cam surfaces, and contact roller *L* which is carried on follower bar *K*. This latter bar is provided with a return spring, not shown.

Worm *C* meshes with worm-gear *D* on shaft *E*. This shaft also carries disc *F*, which has twelve holes into which cylindrical buttons are pressed, in various positions, as required by the specified screen pattern. These buttons periodically actuate the swinging lever *G* as disc *F* revolves. The worm-gear has twelve teeth, so that there is one rotation of shaft *E* to twelve turns of shaft *A*. A cylindrical plunger *H* mounted in block *N* carries a roller *I* at its upper end. A pin extends through vertical slots in plunger *H* and the wall of block *N*, and into a horizontal slot in lever *G*. This lever is mounted to swivel freely on a stud at its left-hand end.

In Fig. 12 slide-bar *M* is in a position to permit roller *L* to pass freely between disc *B* and the contoured offset end of bar *M*. At

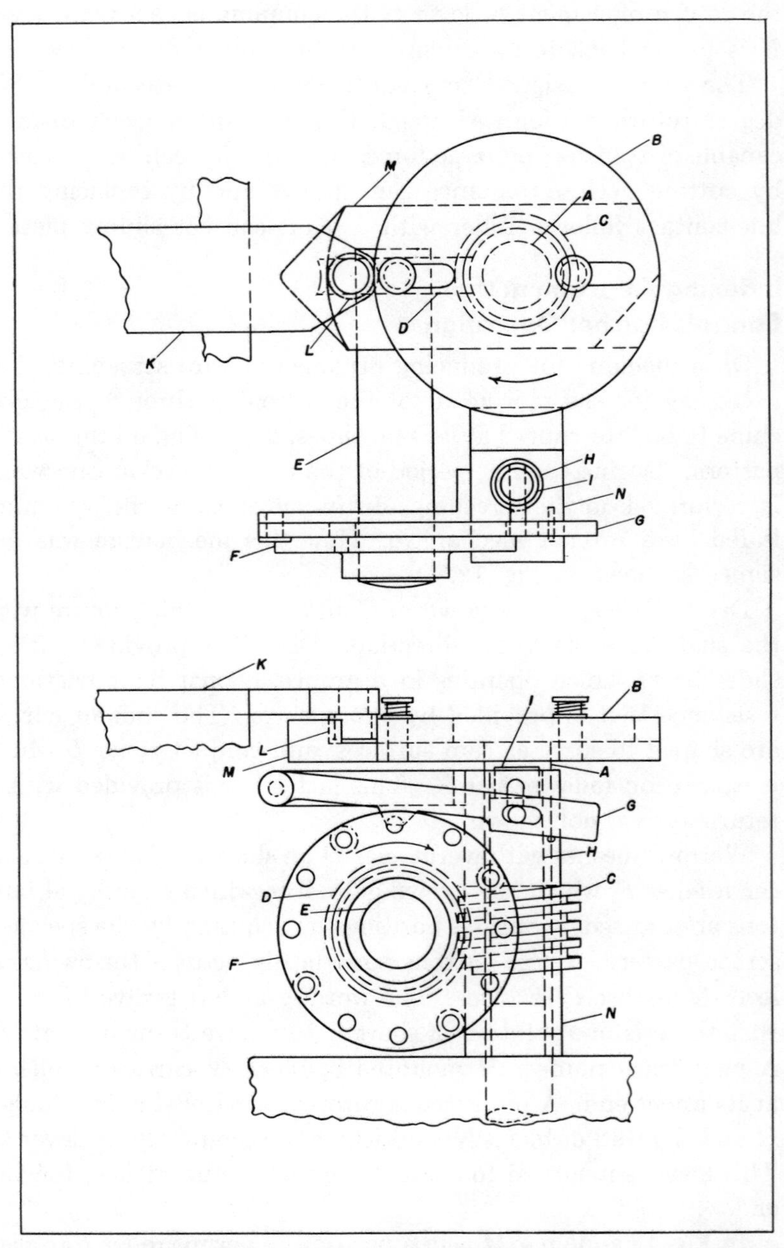

Fig. 12. Mechanism designed to periodically interrupt a feed movement as required for an operation.

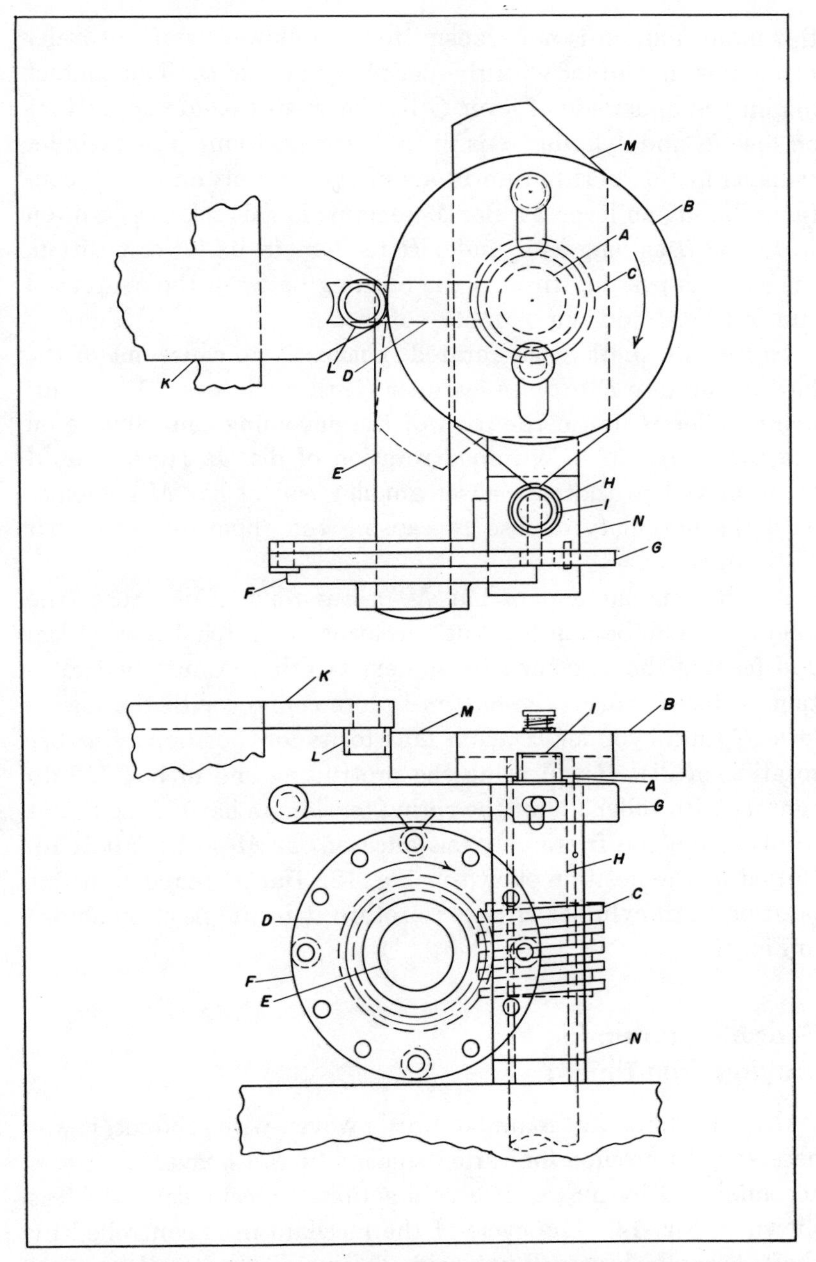

FIG. 13. Position of the various components of the mechanism when screening is being fed.

this point, motion is not transmitted to follower bar *K*, as roller *L* remains in contact with the periphery of disc *B*. The contact lug on the underside of lever *G* lies between two of the buttons on disc *F*, and cylinder *H* is in its lower position. The cylinder remains in this position until one of the buttons on disc *F* contacts the lug on lever *G*. Bar *M* remains in this relative position with disc *B* as long as cylinder *H* remains in its lower position, the roller *L* passing through the opening between the contoured end of bar *M* and the periphery of disc *B*.

In Fig. 13, shaft *E* has rotated sufficiently to cause one of the buttons on disc *F* to raise cylinder *H*, through lever *G*. At this point, roller *H* lies in the path of the oncoming cam surface on the end of bar *M*. Continued rotation of disc *B* causes bar *M* to be moved so that the offset angular end of bar *M* coincides with the periphery of disc *B*, causing the opposite end of bar *M* to protrude.

As the angular end of bar *M* passes roller *L* it causes the follower bar to be moved, thus actuating the ratchet mechanism and feeding the screening to its next position. Continued rotation of disc *F* causes the button to lose contact with the lug on lever *G*, and cylinder *H* again falls to its low position. Further rotation of disc *B* will bring the protruding end of bar *M* into contact with roller *L*, but the spring tension on bar *K* is sufficient to overcome the frictional resistance of bar *M* so that it is returned to the position shown in Fig. 12. Bar *M* remains in this position until cylinder *H* rises to return it to the position shown in Fig. 13.

Gear Mechanism for Varying Cam Timing

On a machine that manufactures a woven-wire product, it was necessary to provide for varied spacing in the weave. This was accomplished by means of a cam-actuated mechanism which is shown in Fig. 14. The cycle of the mechanism is controlled by shaft *A* to which gear *B* is keyed. Gear *C* is fixed to the hub of cam *D* which rotates freely on the shaft and is retained by collar

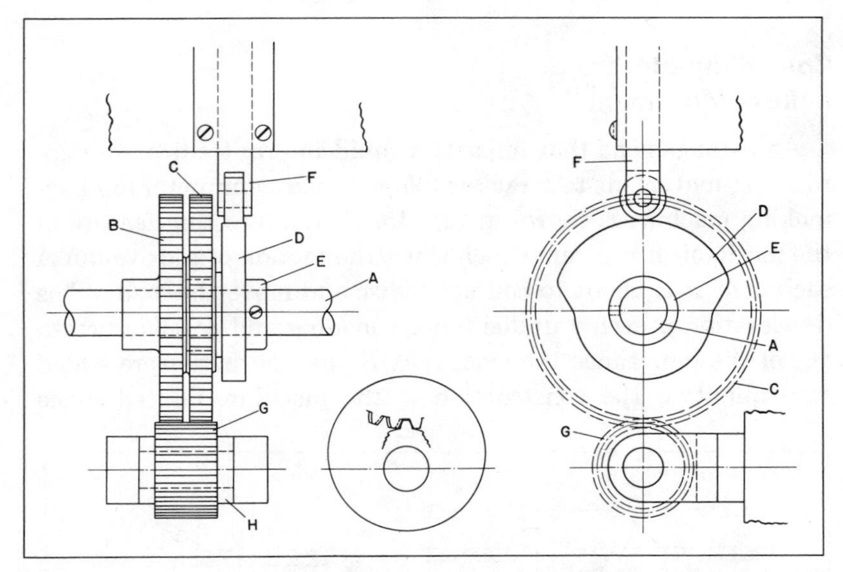

FIG. 14. A cam action which produces varied spacing in a woven product.

E. Cam *D* operates the follower bar *F* which actuates the spacing mechanism. Pinion *G* is supported to rotate freely on bearing bracket *H* and mesh with gears *B* and *C*. Gear *B* is of standard pitch and has fifty teeth. Gear *C* has fifty-one teeth but the same outside and pitch diameters as gear *B*. The additional tooth, however, decreases the circular pitch since the width of the teeth must be narrower than standard.

In operation, shaft *A* rotates gear *B*, the motion being transmitted to gear *C* through pinion *G*. Since gear *C* rotates slower than gear *B*, due to the difference in the number of teeth, cam *D*, actuated by gear *C*, will turn less than one revolution in relation to gear *B*. In the situation described, the loss in radial movement would be approximately 7 degrees, and shaft *A* would require fifty-one turns to produce a complete timing cycle of cam *E*. The modification of gear *C* is restricted by its relation to pinion *G*. Since the teeth of gears *B* and *C* must remain in alignment as they mesh with the pinion gear, the reduction in the thickness of the teeth of gear *C* must be sufficient to prevent any binding action.

Cam Eliminates Shock in Rack Movement

An arrangement that imparts a rapid intermittent and reciprocating movement to a rack employed in an aluminum foil bag-making machine is shown in Fig. 15. The interesting feature of the mechanism is a cam which slows the motion of a drive-pin *A* each time it is about to contact a disc and move the load. This deceleration prevents undue impact loading and possible breakage of the pin. Since the load, rack *B*, and its drive were added subsequent to the construction of the machine, limited space

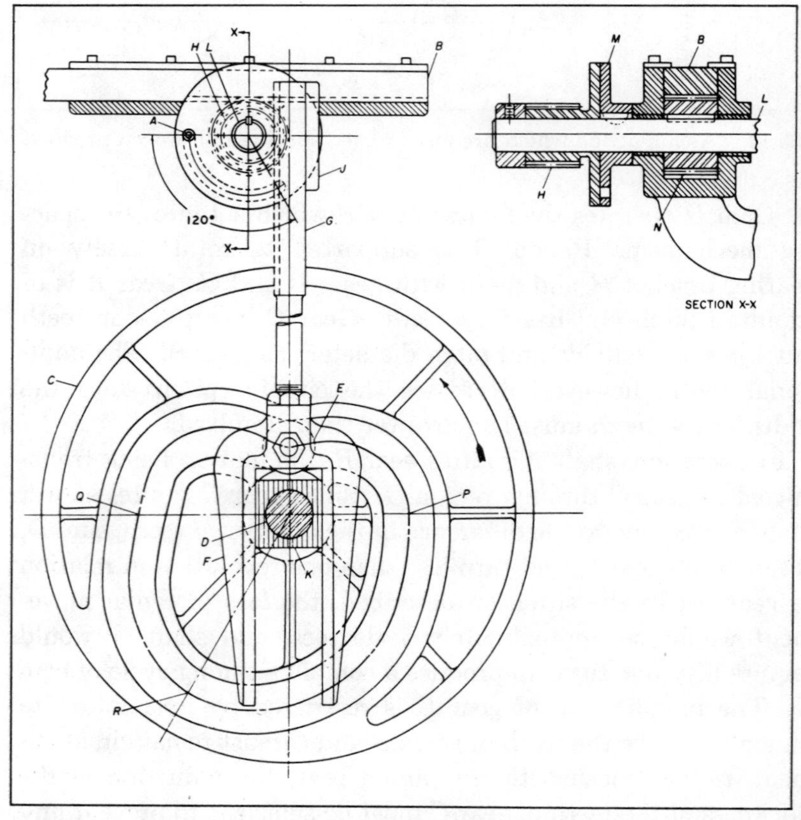

FIG. 15. Arrangement that incorporates a cam to prevent excessive shock in an intermittent, reciprocating rack movement.

and the position of existing members were major considerations governing design of the device.

A face-cam *C* is keyed to a drive-shaft *D* which existed prior to the modification of the machine. Roller follower *E* is free to rotate on a stud secured to a slider guide *F*. One end of the slider guide is turned and threaded for assembly to a rack *G*. A lock-nut secures the assembly in the proper position. Rack *G* is restrained to a vertical linear motion by a pinion *H*, guide *J*, and a stationary slider *K*, the drive-shaft *D* having a running fit in the slider. The pinion *H* is in mesh with rack *G*, and is free to rotate on a shaft *L*.

A disc integral with pinion *H* carries drive-pin *A* which is located in a 120-degree circular slot in an adjacent disc *M*. Disc *M* and a pinion *N*, which is in mesh with the guided horizontal rack *B*, are keyed to shaft *L*. Thus, if pinion *H* is rotated counterclockwise, pin *A* will traverse the circular slot and contact disc *M*. Further motion in the same direction will be transmitted through disc *M*, shaft *L*, and pinion *N* to drive rack *B*. Reversal of rotation of pinion *H* will again cause the drive-pin to traverse the circular slot before contacting disc *M* to drive the load in the other direction.

The rotational speed of pin *A* at impact with disc *M* is controlled by the path of the slot in cam *C*. If the rotation is not reduced to a minimum at the moment of contact, the resulting additional impact loading could cause breakage of the pin. Minimum impact is obtained by making the incremental radial rise of the cam-slot as small as possible in the areas of points *P* and *Q* which are traversed by the follower at the two moments of contact. In addition, cam *C* is designed to impart the movement necessary for rack *B* to complete its function in the machine.

In operation, cam *C* is rotated counterclockwise. The rise imparted to the vertical rack during the first 90 degrees of rotation of the cam traverses pin *A* through the 120-degree slot in disc *M*, rack *B* remaining stationary. When the center of the follower coincides with point *P* on the cam, pin *A* is in contact with disc *M*. For the next 120 degrees of cam rotation pin *A* moves to the horizontal rack to the left. Reversal in the rotation of the pin occurs when point *R* of the cam coincides with the center of the follower. Again, rack *B* is stationary as the pin traverses the slot in disc *M* clockwise during the following 60-degree movement of the cam. Pin *A* contacts the disc when the follower is at point *Q* and returns rack *B* to its initial position in the final 90 degrees of cam rotation. At both points *P* and *Q* the rate of rise or fall of the follower is at a minimum to prevent impact damage to the pin.

Wheel-Dressing Attachment for a Gear-Grinding Screw

A patented attachment for dressing a continuous helical rib on the abrasive wheel of a gear-grinding machine is shown in the accompanying drawing. The attachment, see Fig. 16, is mounted on the compound slide of the machine. During dressing, it is traversed parallel with the axis of the wheel-spindle by a screw driven from the spindle through pick-off gears.

Dressing is carried out simultaneously on both flanks of the helical rib by separate diamonds mounted in holders *A* and *B*. The holders are set to correspond with the pressure angle of the gear to be ground. At the end of the dressing stroke, the cross-slide and attachment are moved away from the spindle by hand, so that the diamonds clear the wheel.

Then, the longitudinal slide is returned to the starting position by power traverse. Next, the cross-slide is brought to the dressing position, and the diamond holders are adjusted lengthwise by means of screws *C* and *D* for applying another cut. In this way, the rib on the grinding wheel is dressed to its full depth in several passes.

The housing for the left-hand diamond holder *A* is fixed to base *E*, which is secured to the compound slide of the grinder. A slide *F* is provided for the right-hand diamond holder *B*. With this arrangement, the right-hand holder can be moved toward or away from the left-hand holder by means of a screw, for setting the distance between the diamonds in accordance with the required width of the helical rib to be dressed.

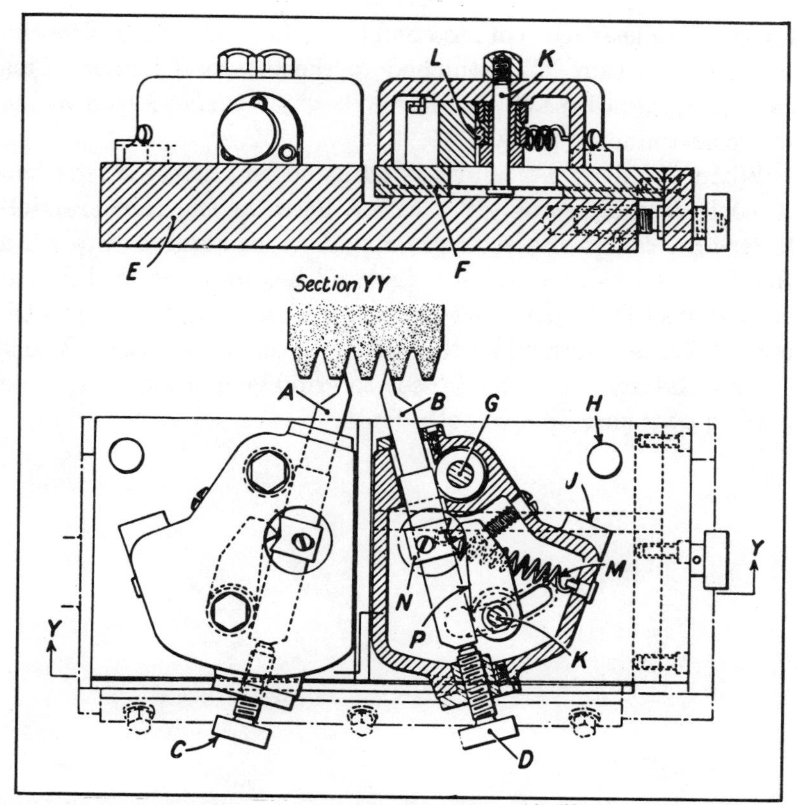

FIG. 16. Attachment for diamond dressing a continuous helical rib on the abrasive wheel of a gear-grinding machine. By changing cam *N*, the wheel may be dressed to various profiles.

Since both diamond-holder assemblies are of similar design, only the right-hand unit will be described. This unit can be pivoted about pin *G*. The diamond holder *B* is set to the required angle by means of gage-blocks, which are placed between the periphery of reference pin *H* and the machined surface *J*. The unit is then secured to slide *F* by nuts on the threaded upper ends of pins *G* and *K*. The enlarged-diameter lower end of pin *K* engages a T-slot in the slide.

An eccentric ring *L* is keyed to a sleeve surrounding the pin *K*. Diamond holder *B* is held in close contact with this ring and a sleeve surrounding pin *G* by a tension spring *M*. When a wheel for grinding gear teeth of modified involute form is to be dressed, a plate type cam *N* is attached to the diamond holder. This cam is engaged by a follower-arm *P*, which is also keyed to the sleeve surrounding pin *K*.

When the diamond holder is adjusted lengthwise at the end of each dressing stroke, the action between the cam and follower-arm causes the eccentric ring *L* to be rotated through a small angle. As a result, the diamond holder is set in different angular positions during the dressing operation so that a rib with curved flanks is formed on the wheel. By using cams of different shapes, the wheel may be dressed to grind gears having modified profiles over part or full tooth depth.

CHAPTER 2

Intermittent Motions from Gears and Cams

The term "intermittent motion" is applied to mechanisms for obtaining a "dwell" or possibly a series of dwells or moving and stationary periods of equal or unequal lengths. Many different designs of intermittent motions are in use because they are required on so many different types of automatic and semi-automatic machines. The intermittent motions illustrated and described in this and the following chapter supplement those presented in Volumes I, II and III of "Ingenious Mechanisms for Designers and Inventors."

Intermittent Worm-Gear Train

The worm-gear drive shown in Fig. 1 was designed to provide intermittent motion. It consists of a worm-wheel *A* having teeth in sectors *X*, *Y*, and *Z*; worms *B* and *C*, mounted on and keyed to shaft *D*; and a plunger *E*. Worm *B* is firmly attached to the shaft; Worm *C* is free to slide. The helices of these worms should be continuous.

In order to hold worm *C* to the right, out of engagement with the worm-wheel, a spring *F* is located in a longitudinal hole in shaft *D*. A cross-pin *G*, extending through a slot in the shaft and engaging a seat in worm *C*, transmits the spring pressure to this worm. A similar seat is provided in worm *B* for this pin. The spring is suitably secured at its left end.

The device operates as follows: as shaft *D* rotates in the direction shown by the arrow, there will be no rotation of the worm-wheel *A* in the position illustrated, as worm *B* is rotating in a plain sector of the worm-gear. In order to produce rotation of

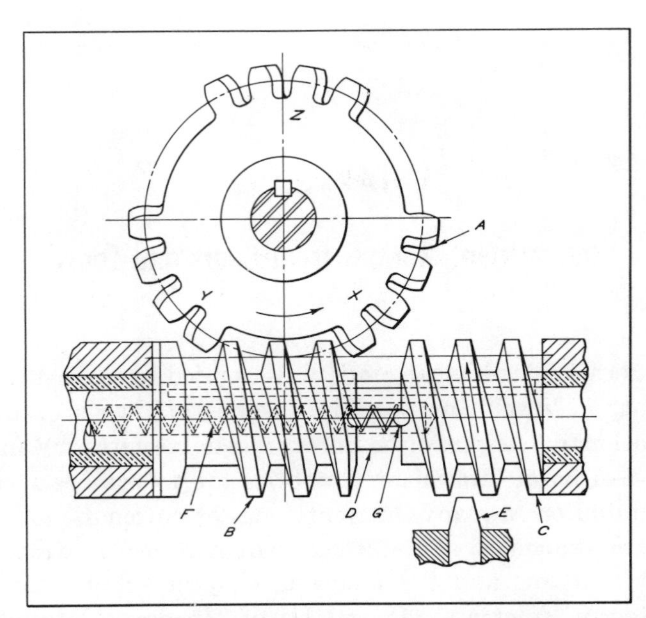

FIG. 1. Worm-gear train for producing intermittent motion. The worm-wheel *A* is rotated one-third of a revolution, followed by a stationary period, through the action of the two worm-gears *B* and *C*.

the worm gear — in this case one-third a complete revolution — plunger *E* (timed by other parts not shown) rises into engagement with the worm thread. Upon engagement, the rotating worm *C* moves to the left, as its thread moves past the plunger until the worm engages the last tooth in sector *X* of the worm-wheel. The plunger must be withdrawn before worm *C* reaches worm *B*.

When the end of worm *C* comes in contact with the end of worm *B*, the worm-wheel *A* is rotated in the direction of the arrow, both worms operating as a single unit. After worm-wheel *A* has rotated sufficiently to bring the last tooth of sector *X* out of engagement with worm *C*, spring *F* pushes the worm to the extreme right, where it cannot engage the teeth of sector *Y*. When the last tooth in sector *Y* is disengaged from worm *B*, worm-wheel *A* will stop, having made one-third of a revolution about its axis.

Indexing Movement that Starts without Shock

Most indexing mechanisms incorporate either cams, Geneva movements, or other components which present machining problems. An indexing mechanism made up of easily machined components and accurate gears that can be obtained from gear specialists is shown in Fig. 2.

The mechanism, illustrated in the top view of Fig. 2, is a planetary gear device incorporating two eccentrically located spur gears. The bore of each of these gears is machined off center by an amount equal to 20 per cent of its pitch radius. The desired dwell period is realized when the ratio of the number of turns of arm *C* to the number of turns of sun gear *H* is 3 to 1; that is, the arm must rotate three times faster than the sun gear, but in the opposite direction.

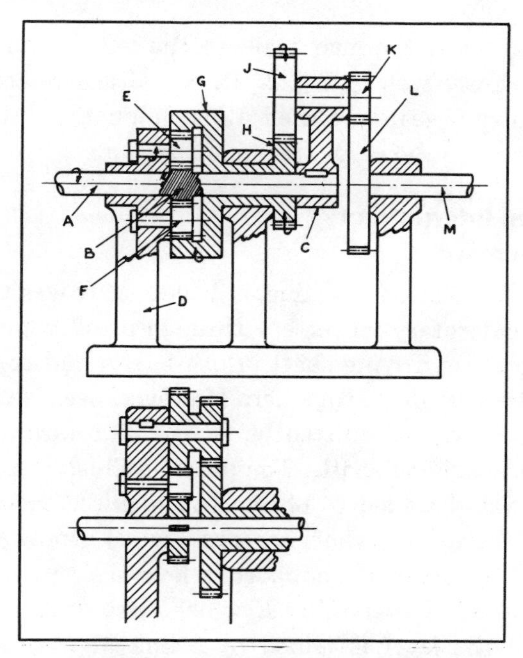

FIG. 2. (Top) Planetary gear type indexing device that provides fixed dwell periods in the movement of follower-shaft *M*. (Bottom) Alternate gear arrangement permits the elimination of internal gear *G*.

Drive-shaft *A*, which carries a 20-tooth pinion *B*, is keyed to arm *C*. Frame member *D* supports two pinions *E* and *F*, each having 20 teeth, which mesh with pinion *B* and also with internal gear *G*. This internal gear has 60 teeth and is integral with an eccentrically located, 40-tooth sun gear *H*. Gears *G* and *H* revolve around shaft *A*.

Meshing with the sun gear is another eccentrically located, 40-tooth gear *J*. Mounted on the same shaft with gear *J* is a 20-tooth pinion *K*. This gear meshes with a 60-tooth gear *L* that is mounted on follower-shaft *M*.

In operation, drive-shaft *A* turns clockwise as indicated by the arrow; pinions *E* and *F*, internal gear *G*, and sun gear *H* turn counterclockwise, while gear *J* and pinion *K* turn clockwise. Since gear *L* is driven in a counterclockwise direction, follower-shaft *M* receives this motion. One full revolution of the drive-shaft results in several fixed dwell periods in the follower-shaft rotation.

An adaptation of the gear train to the left of sun gear *H* is shown in the lower view of Fig. 2. In this alternate arrangement only external spur gears are used, thus eliminating internal gear *G*. The gear ratio, however, remains the same (3 to 1).

Chain Driven Intermittent Rotary Movement

On a wire fabricating machine, a driven shaft was to be given an intermittent rotary movement through a roller chain from a uniformly rotating driving shaft. Both shafts had to begin and complete each revolution together. However, because the driven shaft was to move intermittently, it had to rotate at a higher speed than the driving shaft. The drawing illustrates the mechanism that was designed to obtain the required motion.

Keyed to the driving shaft *A*, (see Fig. 3), are a sprocket *B* and a cam *C*. A bracket *L* supports a lever *J*. At its lower end, lever *J* carries a follower roller *K* which contacts the cam. The upper end of the lever is joined by a link *M* to a slide-bar *F* dovetailed to the frame of the machine. Four idler sprockets *D* are mounted on the machine, and two idler sprockets *E* are car-

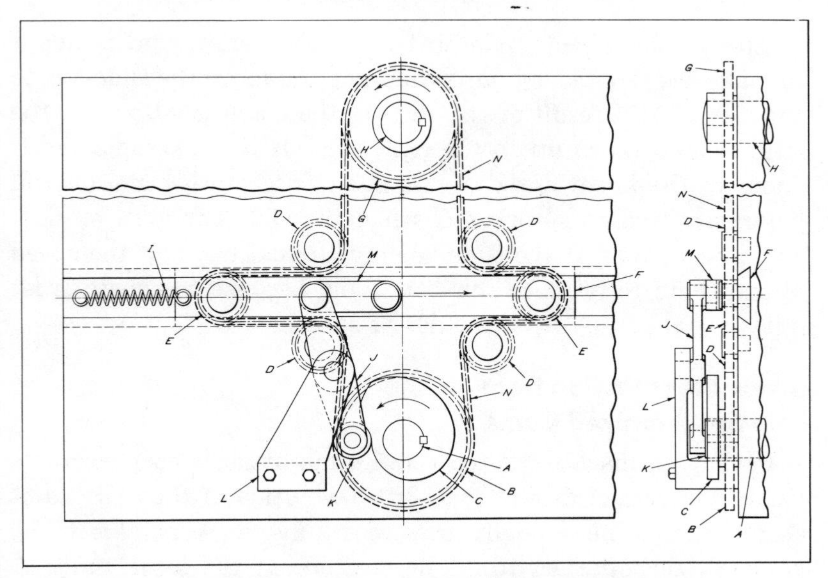

FIG. 3. Practical design for producing intermittent rotation in a driven shaft H from a uniformly rotating driving shaft A.

ried on the slide-bar. A spring I serves to resist any movement of the slide-bar to the right, and maintains the follower roller in contact with the cam. The driven shaft H has a sprocket G keyed to it which, like the sprocket on the driving shaft, is in mesh with a chain N.

In order to explain the operation of the mechanism, let it be assumed that the cam and lever were omitted. The rotation of shafts A and H would then be in the ratio of the number of teeth on sprockets B and G, the idler sprockets D serving merely to direct the chain over the required path. The two idler sprockets E carried on the slide-bar do not affect the motion of sprocket G, provided the slide-bar remains stationary. But if there is a change in the position of the slide-bar, there will also be a change in the relative positions of sprockets B and G. This is because the movement of the slide-bar causes the chain to be let out on one side and taken up on the other side, this action producing a partial rotation of sprocket G.

By referring to the drawing, it can be seen that the cam

rotates in the direction indicated by the arrow, and is about to make the lever swing on its fulcrum and move the slide-bar to the right. As a result of this action, the chain is let out on the left side and taken up on the right side. If the take-up speed is equal to the linear speed of the chain, no rotative motion will be transmitted to sprocket *G* while the slide-bar is in motion. The linear speed of the slide-bar must equal one-half the speed of the chain to produce this condition, because the chain is let out and taken up on both sides of sprockets *E*.

Intermittent Motion from Two Synchronized Cams

Packaging machines often require mechanisms to transmit a particular motion during each fifth revolution of the main camshaft. Such a need might arise where five packages are to be grouped, then pushed from the machine at the same time. A mechanism that has been arranged to satisfy these particular requirements is shown in Fig. 4.

The principal operating elements of this mechanism are two synchronized cams and one follower-lever. The upper end of single, L-shaped lever *A* drives the package-ejector unit (not shown). At the opposite end of the lever are two follower-rollers B_1 and B_2 which are held in contact with cams C_1 and C_2, respectively, by a spring *D* (attached to the upright lever arm).

Cam C_1 is pinned directly to the constantly rotating camshaft *E*, while the motion for cam C_2 is obtained indirectly from a gear *F*, also pinned to the camshaft. By means of gears *G* and *H* — the latter being keyed to cam C_2 — movement of gear *F* is reduced to one-fifth by the time it reaches the second cam. Thus, the speed of cam C_2 is only one-fifth that of the camshaft and C_1, although the rotational movement of both of the cams is in the same direction.

Bearing this in mind, and noting the cam configurations and positions in the right-hand view, it can be seen that in one revolution of cam C_1 cam C_2 will rotate a distance equal to the width of its cutout *J*. This cutout occupies approximately one-fifth of the otherwise circular cam.

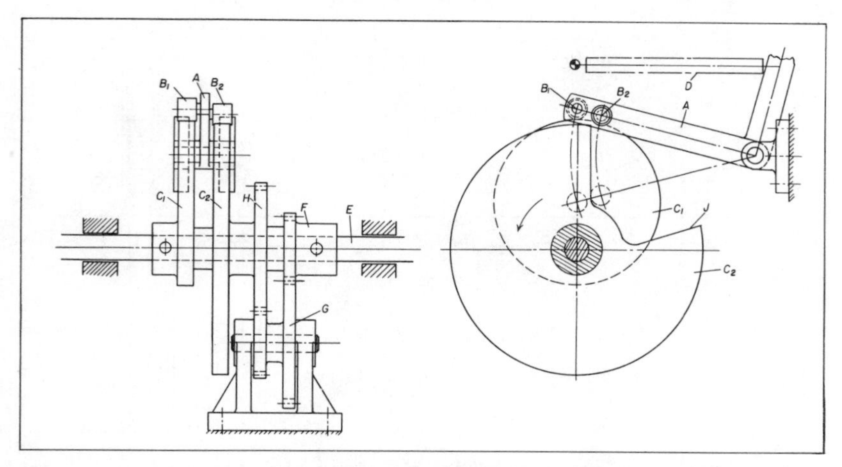

FIG. 4. Lever A is permitted to function only once for each five revolutions of camshaft E. This intermittent movement is controlled by the action of cams C_1 and C_2.

During this rotation of the camshaft, follower-roller B_2 is disengaged from the surface of cam C_2. This permits roller B_1 to track along the entire surface of cam C_1, thus causing lever A to pivot. For the next four rotations of the camshaft, follower-roller B_2 will ride along cam C_2, thereby preventing roller B_1 from being affected by the contour of cam C_1, and causing lever A to remain motionless.

Intermittent Rotary Movement with End-Cycle Reversal

On a machine for forming a product of flat wire, the material is fed intermittently through a twisting clamp. During the cycle, a shaft rotates one revolution, dwells, then rotates another revolution in the same direction. After several such revolutions, the shaft rotates in the reverse direction for a number of revolutions equal to the number of forward, separate revolutions, then stops to end the cycle.

In Fig. 5, tubular spindle shaft A, which carries the twisting clamp, is supported by bearing brackets B. Pinion gear C, brake-drum D, and counterweight drum E are keyed to shaft A. Friction is applied to the brake-drum by leather band F which is

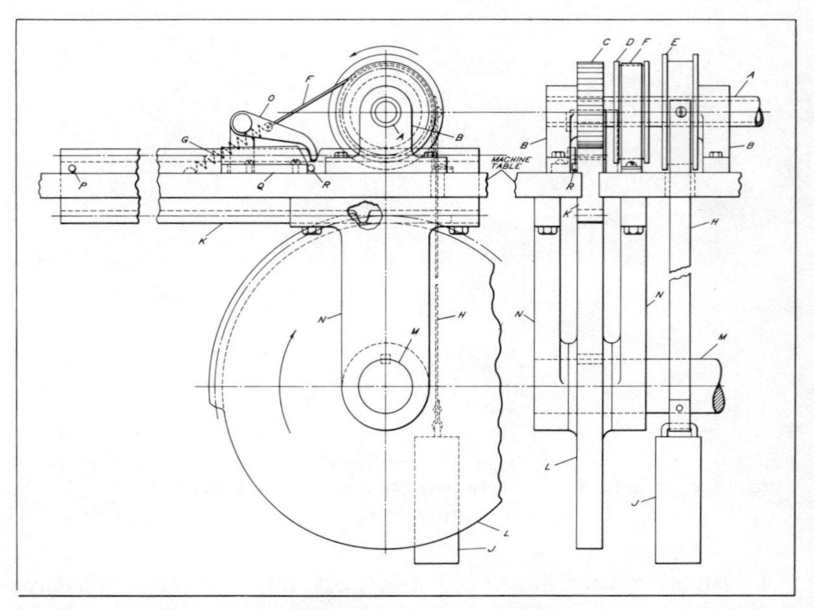

FIG. 5. Mechanism that provides intermittent rotary movement from a rotating drive-shaft, and which has provision for automatic reversal upon completion of operating cycle.

held under tension applied by spring *G*. The counterweight drum receives steel trap *H*, to which the counterweight *J* is fastened. This serves to return the tubular shaft *A* to its initial position at the end of each cycle. The purpose of the leather band is to control the speed with which the counterweight descends.

A rack *K*, with teeth cut on both upper and lower surfaces, meshes with pinion gear *C* and gear segment *L*. Gear segment *L* is keyed to the driveshaft *M*, which, in turn, is supported by bearing bracket *N*. The ratios of the pitch diameter of gears *L* and *C*, and consequently their speed ratios, are 4:1. The gear segment of *L* extends over 90 degrees and can rotate gear *C* one revolution.

The diagram shows the mechanism at the beginning of the cycle. Referring to the front elevation at the left, the first tooth in gear *L*, which is rotating in the direction indicated by the arrow, is in contact with the gear teeth on the under side of rack *K*. This causes the rack to move to the right, thereby rotating gear *C* in a counter-clockwise direction. Drum *E*, being keyed to spindle shaft *A* also rotates, thus causing counterweight *J* to rise as steel strap *H* is wrapped around the drum. After the last tooth in gear *L* has disengaged the teeth in the rack, gear *C* stops turning and the rack is held in position at this point by pawl *O*. The pawl drops between two of the upper rack teeth.

The spindle shaft then remains stationary until the first tooth of gear *L* again contacts the teeth of the under side of rack *K*. This phase is repeated until pin *P*, attached to the rack, contacts sliding dog *Q*, moving it to the right. This raises pawl *O* out of contact with the teeth in rack *K*. The action of counterweight *J* then causes a reverse rotation of pinion gear *C*, which, in turn, drives the rack to the left. Pawl *O* remains raised until pin *R* contacts sliding dog *Q*, moving it to the left, and allowing the pawl to drop back into contact with rack *K*, thus completing the cycle.

Fool-Proof Indexing Mechanism

Necessity for the design of an indexing mechanism that will not overshoot the desired position or otherwise inaccurately index, is often encountered. This is especially true when a weighty fixture is to be carried by a large index-table. A lever type indexing mechanism that is fool-proof in operation is shown in Fig. 6.

Lever *A* pivots about a spindle that is also common to both ratchet wheel *B* and spring-loaded dog *C*. The dog remains engaged in a slot milled across a boss integral with the lever. The ratchet-wheel teeth are cut to mesh with mating spaces in the periphery of index-plate *D*. Pawl *E*, resting against a flat spring, restricts the motion of the ratchet wheel to a clockwise direction only. The back end of the pawl rests against an adjustable cam-stop.

Slotted lever *F* pivots freely on the common spindle and is connected to locating plunger *G* by a headed pin. The locating

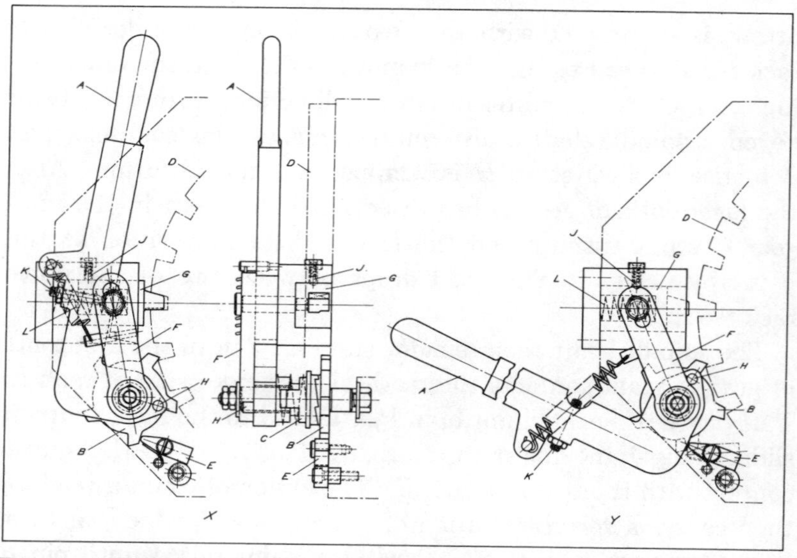

FIG. 6. Index-plate mechanism incorporates locating device and one-directional driving dog to provide fool-proof operation by eliminating the possibility of faulty indexing.

plunger slides in a guide block fastened to the base of the indexing table. A locating tongue on the plunger end is accurately ground to fit within the tooth spaces around index-plate *D*. This tongue is partially relieved on the face parallel to the horizontal center line to clear any burrs that may have been raised in the tooth spaces on index-plate *D* by the action of ratchet wheel *B*.

To move the index-plate from one position to the next, lever *A* is moved from the position shown at X to that shown at Y. During the initial part of this motion, pin *H*, pressed in the short leg of the lever, moves freely in an arc until it contacts slotted lever *F*. The pin, during the remainder of the stroke, forces the slotted lever to pivot around the common spindle, thereby disengaging locating plunger *G* from the index-plate. At the end of the stroke, spring-loaded ball *J* rides into a cone-shaped recess in the top plunger face. The plunger is thus held in the retracted position.

At the same time that these movements are taking place, spring-loaded dog *C* freewheels over the dog teeth that are integral with ratchet wheel *B* and drops into position after having moved a distance equal to one dog tooth. The ratchet wheel, which has the same number of dog teeth as ratchet teeth, is prevented from turning by pawl *E*.

Lever *A* is now moved back to its original position. In doing so, dog *C* drives ratchet wheel *B* one tooth which, in turn, moves the index-plate to the next position. During the first part of this stroke, spring-loaded ball *J* retains locating plunger *G*, allowing the index-plate to move unrestricted. Toward the end of the stroke an adjustable screw *K*, which is threaded through a pad on the lever, contacts the pin connecting slotted lever *F* with the plunger. This drives the plunger forward causing the locating tongue to enter a tooth slot on the index plate, thus locking it firmly in place. The plunger is held in this position by coil spring *L*.

Escapement Provides Regular Intermittent Drive

An escapement mechanism in which a pendulum is applied to control the timing of an intermittently rotating shaft was incorporated in a wire weaving machine to advance strands of wire a required distance at regulated time intervals. It resembles the pin-pallet, or Brocot, escapements used in French and pendulum clocks.

Drive-shaft *A*, see Fig. 7, has gear *B* keyed to it. This gear meshes with gear *C*, which is carried free on driven shaft *E*. Gear *C* carries a series of spring-loaded plungers which contact disc *D*, keyed to shaft *E*, see Fig. 8. Shaft *E* carries, at its outer end, a disc *G*, which is provided with a series of pins. A ring of friction material *F* is carried on the hub of disc *G* where it transmits rotary motion from gear *C* to disc *G* due to the pressure applied by the spring plungers. Bracket *H*, bolted to a stationary part of the machine, carries a pivot stud which supports a pallet *I* and a pendulum *J*, which are locked together by two screws in

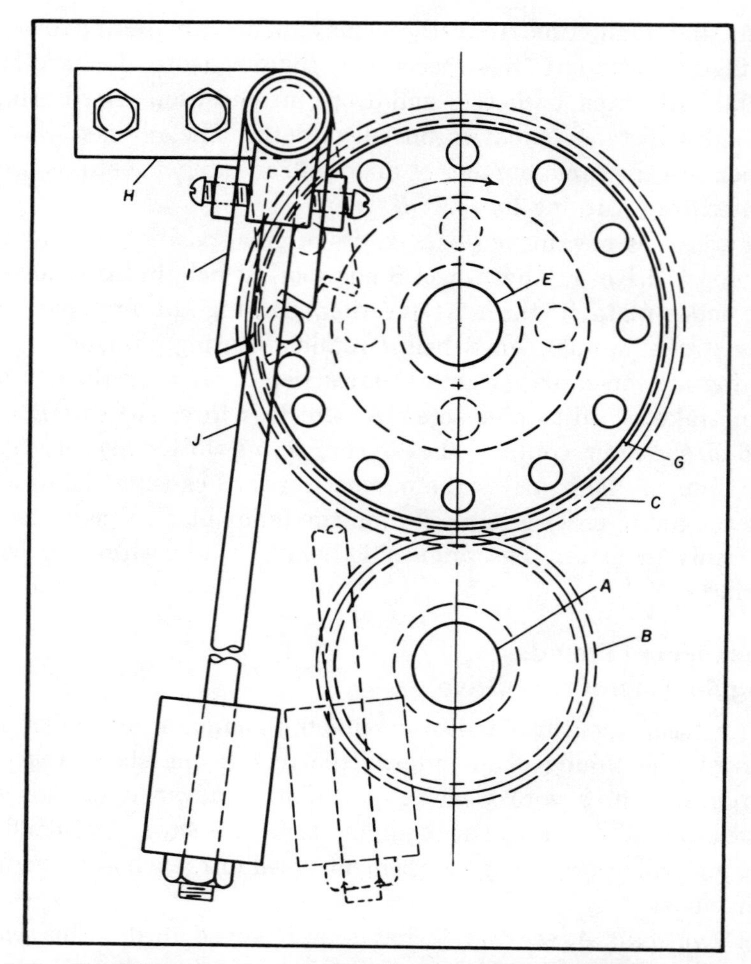

FIG. 7. Pendulum escapement gives disc *G* intermittent clockwise motion by means of the pin pallet *I* alternately releasing the pins in the disc as pendulum *J* swings from side to side.

the side plates attached to pallet *I*. These screws provide a means of adjusting the position of the pallet relative to the pendulum.

In operation, the continuously rotating shaft *A* transmits motion to the shaft *E* through gears *B* and *C*, friction ring *F*, and disc *G*, where there is no restriction to the movement of disc

G. In Fig. 7, the positions of pendulum *J* and pallet *I* are such that the upper foot of pallet *I* contacting a pin of disc *G* prevents its rotation. As pendulum *J* swings to the other end of its arc, as shown dotted, the contacting pallet foot slides off the pin, allowing disc *G* to rotate. But before the upper foot has completely lost contact with its pin, the lower pallet foot carries into position to catch the succeeding pin. Thus, at the end of the arc

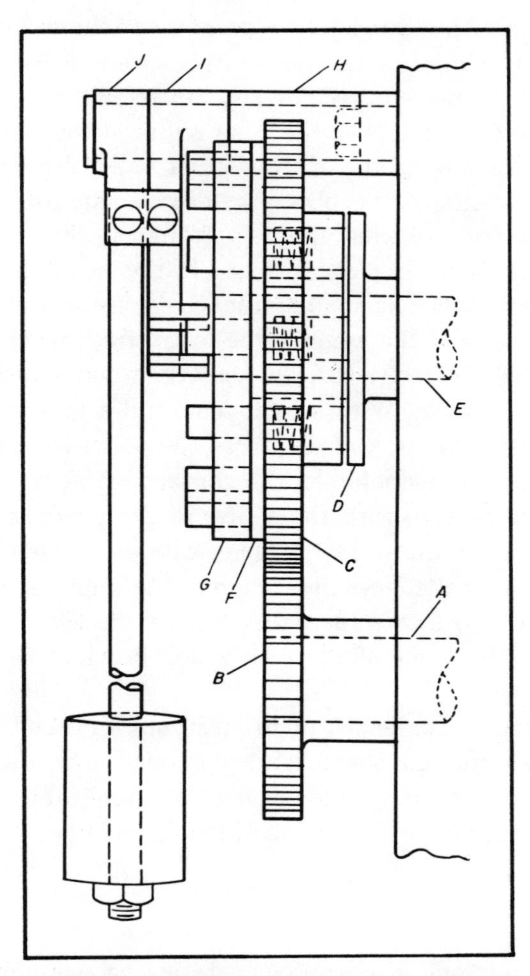

FIG. 8. Driven shaft *E* is turned intermittently under control of pendulum escapement, receiving power from shaft *A*.

to the right the pallet I will be in the position shown by the dotted outline, the succeeding pin having moved to contact the lower foot of pallet I. If the pallet feet are correctly located, the angular movement of shaft E will equal one-half the angular spacing between any two consecutive pins on disc G. Therefore, the number of angular movements of shaft E per revolution will equal twice the number of pins in disc G.

The number of movements per minute of shaft E is controlled by the swing of the pendulum regardless of the number of pins on disc G. But in all cases the rotative speed of disc G must be such that the time required for any pin to arrive at the locking position must be less than the time required for the pendulum to swing. The angularity of the feet of pallet I relative to the line of movement of the pins provides an impulse to produce continued motion of pendulum J. However, this angularity is regulated by operating conditions. If the angle is too small, there will be insufficient impulse applied to the pendulum to keep it swinging. But if the angle is too large, free movement of the pendulum will be restricted, particularly when a heavy load is applied by the pins. With a light load and a large angle, there will be a noticeable jerk of disc G as the pallet feet slide off the pins. If this is objectionable, the contact surfaces of the pallet feet must be curved, with the center of the pivot as the center of the arc of curvature. In the latter case, an angle on the leading side of the pallet feet must deliver the impulse. In general, it is advisable to locate the pallet feet at the lowest angle with the line of movement of the pins which provides the necessary impulse.

To determine the length of the pendulum needed to produce the required timing, the formulas applied to a free-swinging weight suspended on a length of cord are applicable. For determining the time of swing, the accepted formula is

$$t = \pi \left(\frac{l}{g}\right)^{\frac{1}{2}}$$

in which: t = time, in seconds; l = length of pendulum in feet; and g = the force of gravity in feet per second^2. The generally applied value of g is 32.2.

To determine the length of pendulum required to produce a specified rate of swing, the foregoing formula is transposed, thus:

$$l = g\left(\frac{t}{\pi}\right)^2$$

In the formulas, the symbol l is the distance from the pivot point to the center of the suspended weight. However, this formula is based on a weight suspended on a cord in which there is no friction influence and negligible weight of the cord. On the other hand, an accurate determination of the length of pendulum by this formula for the escapement application is not accurate because the combined weight of the rod and its fastening produce a distribution of weight which is not easily pinpointed. For all practical purposes, the calculated length is determined by the center of the suspended weight which is provided with a nut for adjustment, as shown. The pound value of the suspended weight in no way controls the timing of the swing. Its value lies in increasing momentum and providing the steadying influence of inertia when the impulse is applied. The lightest weight which will serve this purpose is recommended.

Rotary Work-Table with Mechanism for Automatic Indexing

An indexing work-table that can be used in conjunction with independent cutter-heads to form an automatic multiple-spindle machine is shown in Fig. 9. The table is intended to receive several work-holding fixtures according to the number of indexing stations provided. A variety of machining operations may be performed automatically while the work-pieces are located at these stations.

Referring to sectional view W-W, annular table *A* rotates on steel balls which surround fixed central disc *B*. Indexing is carried out by means of gear segment *C* (section X-X) which is secured to spindle *D*. The latter component is mounted in ball bearings which are housed in disc *B* and in the base.

Motion from gear segment *C* is transmitted by pinion *E*, which engages gear teeth in the bore of the table. The indexing action is controlled by compound cam *F* (section W-W) which

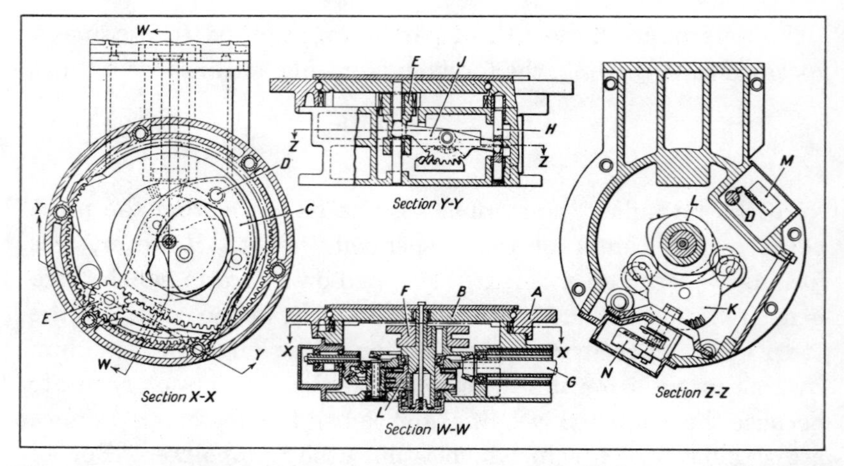

FIG. 9. Sectional views of work-table that can be set up for automatic indexing. Limit switch M stops indexing cycle and switch N starts machining cycle.

engages follower rollers housed in recesses in the gear segment C and is driven through bevel gears by shaft G. This shaft is driven by a motor, through V-belts, an electromagnetic clutch, and brake units. This driving equipment is not shown in the illustration.

At the beginning of the cycle, segment C dwells for a period, and the indexing motion is then completed during a 210-degree angular movement of cam F. Subsequently, segment C is again caused to dwell before it is returned to its original position. This is done in preparation for the next indexing cycle in the course of the final 90-degree angular movement of cam F.

During the dwell periods of the segment C before and after the indexing movement, pinion E and plunger H (section Y-Y) are moved vertically in opposite directions by lever J. At the beginning of the cycle, the plunger is withdrawn from one of a number of holes provided in the under side of the table at the indexing positions. Simultaneously pinion E is brought into engagement with the gear teeth in the table for the indexing movement. After indexing has been completed, the plunger is inserted into the next hole in the table. The latter is, therefore, positively located while the machining operations are being carried out on the work-piece. At the same time, the pinion is withdrawn from the gear teeth in the table in preparation for the return movement of the segment. At certain points in the cycle, the pinion and the plunger are in simultaneous engagement with the table, so that the latter is positively located during the entire indexing operation.

Movement is transmitted to the pinion and the plunger by bevel gear teeth on lever *J* and on pivoted segment *K* (section Z-Z). The segment carries two follower rollers which engage simultaneously with compound *L*. This cam is keyed to the lower end of the shaft which carries cam *F*. The arrangement may be seen in section W-W of the illustration.

The indexing cycle is started by means of a switch (not shown) which activates the electromagnetic clutch to engage the drive with shaft *G*. At the end of the indexing cycle, limit switch *M* (section Z-Z) is operated by means of a detent on the lower end of spindle *D*, with the result that the clutch and consequently the drive to the shaft *G* are disengaged. Concurrently, an arm attached to the pivot spindle for the lever *J* actuates the limit switch *N* to start the cycle of the cutterheads.

Intermittent and Pressure Applying Mechanism

The gluing of paper watch dials to their metal backings originally required the use of four presses and four operators. In order to reduce labor and equipment costs, a gluing device incorporating an ingenious operating mechanism was developed to do this work. The new device required only one operator, eliminated scrap, and enabled the work to be done with greater safety.

In designing this device, it was necessary to incorporate means for holding and pressing the paper dials and their metal backings together for a sufficient period of time to allow the glue to set. This requirement was met by providing the fixture with eight cam-operated pressing spindles mounted in an intermittently indexed eight-position spindle-carrier, with one position reserved for loading and unloading the work. The spindle-carrier is operated slowly enough to permit proper setting of the glue in one complete revolution of the carrier. A glued dial and its backing is removed from the loading and unloading station designated "0" and replaced by new pieces during the dwell period following each indexing of the spindle-carrier. Thus eight glued dials are completed in one revolution of the spindle-carrier, each dial and its backing being under pressure during one revolution of the carrier.

The essential features of the mechanism designed to operate the gluing device are shown in Fig. 10. The mechanism is driven by a motor through a belt passing over the friction pulley *H* and a clutch operated by lever *J*. When the clutch is engaged, worm *K* on shaft *F* turns the worm-wheel *L*. The upper portion of the worm-wheel carries the indexing pin of a Geneva mechanism. At each revolution of the worm-wheel, the indexing disc *M*, which is fastened rigidly to vertical shaft *N*, is rotated one-eighth revolution. Since the spindle-carrier *A* is keyed to the vertical shaft, it also rotates.

The eight equally spaced pressure spindles *C* are mounted in the carrier *A* as shown. Directly above, and in axial alignment with each pressure spindle *C*, is a spring-backed spindle *B*. It is between the work supporting disc *P* on the upper end of spindle *C* and the disc at the end of spindle *B* that the watch dial and its backing are pressed together to complete the gluing operation. Both spindles *C* and *B* are provided with keys that slide in keyways to prevent them from turning in the carrier *A*. Slots at the bottom ends of spindles *C* accommodate rollers, which are in constant contact with a circular cam-ring *D*.

The cam-ring has a raised portion throughout 246 degrees of its circumference, sloping portions throughout 47 and 44 degrees, and a flat portion throughout 23 degrees, which is located between the sloping portions. The flat portion at the position marked "0" is directly in front of the operator, and it is at this location that a pair of spindles is loaded and unloaded.

Bed *E*, besides acting as a support for cam-ring *D*, houses the Geneva motion and worm-wheel and serves as a support bracket for the worm-shaft *F*. The driving pulley *G* engages the spring-

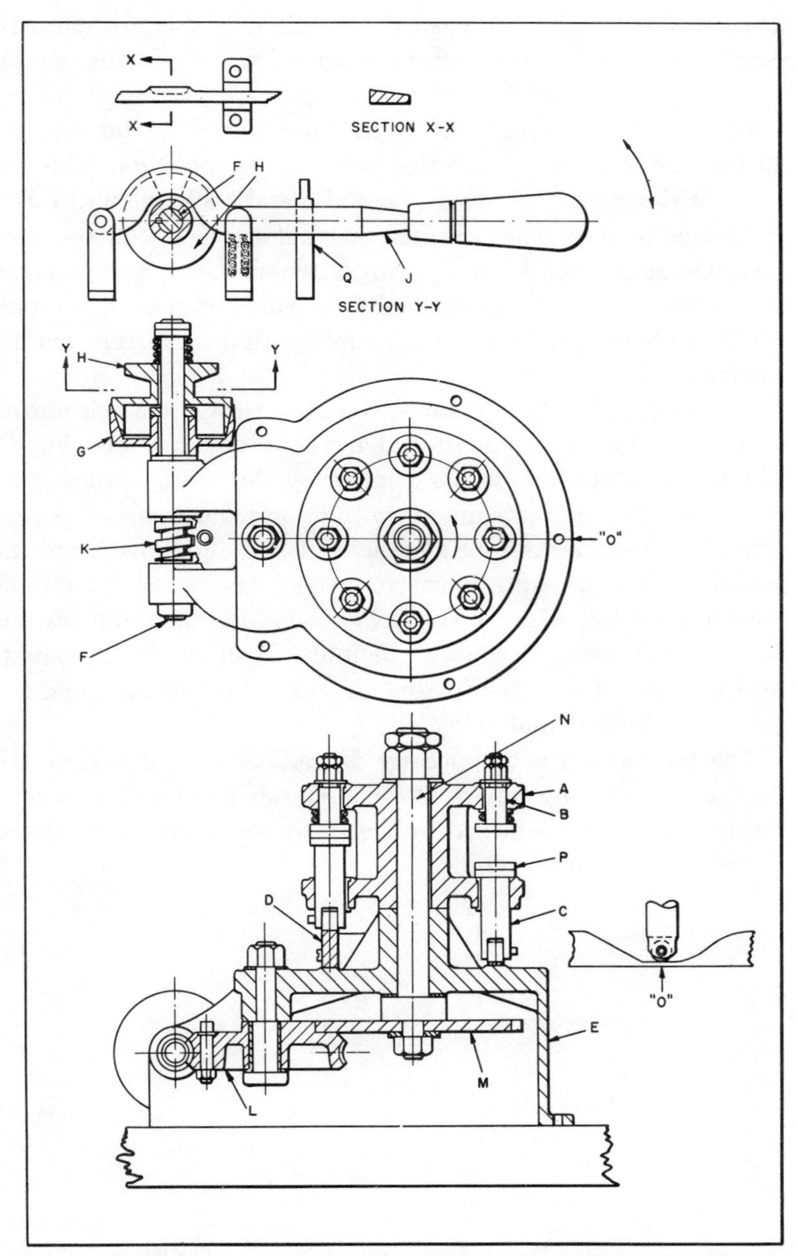

FIG. 10. Mechanism used to glue paper dials to metal discs by applying sufficient pressure for a predetermined length of time.

loaded friction pulley *H* when the clutch lever *J* is in the raised position, imparting rotation to shaft *F*, worm *K*, and worm-wheel *L*.

In operation, the parts to be glued are placed by the operator on the disc *P* of the spindle that is in the "0" position. The correct positioning of the paper dial and metal backing on disc *P* is facilitated by means of locating pins (not shown). The operator then releases the flat-hooked spring *Q*, which allows spring-loaded clutch lever *J* to move upward. This, in turn, releases driven pulley *H* and engages the train of members that drive the spindle-carrier *A*.

As the spindle-carrier rotates, the pressure spindle *C* is moved upward by the action of its roller on the circular cam-ring *D*. The pressure exerted on this member by the spring-loaded spindle *B* is sufficient, both in magnitude and duration, to permit setting of the glue. As one spindle is loaded and moved into the position where pressure is exerted, the pressure on the spindle immediately following is relieved. Its roller then follows the descent in the cam-ring under the influence of gravity in moving into position "0." Once in this position, the bottom spindle is quickly unloaded and reloaded.

The mechanism is stopped by depressing lever *J*, which disengages the driving pulley *G*. The clutch lever also acts as a friction brake. It is locked in the depressed position by means of flat hooked spring *Q*.

CHAPTER 3

Intermittent Motions from Ratchet and Geneva Mechanisms

Two methods of producing intermittent motion in which the periods of rest are evenly spaced and of equal length are by means of ratchet gearing and by using some modification of the Geneva motion. In its basic form this motion is obtained by means of a Geneva wheel, acting as a driven member, which has four radial slots located 90 degrees apart that successively engage a roller or pin on the driving member. The Geneva wheel thus turns with the driving member through one-quarter of a revolution and is idle for the remainder of the revolution of the driving member.

A number of ingenious mechanisms in which a ratchet arrangement or a Geneva motion play a prominent part are described in this chapter. For other mechanisms of a similar type, the reader is referred to Volumes I, II, and III of "Ingenious Mechanisms for Designers and Inventors."

Adjustable Intermittent Ratchet Mechanism

The device shown in Fig. 1 was used to give intermittent drive to a mechanism by means of a ratchet. The number of teeth per cycle was to be adjustable, as well as the location of the teeth in the cycle.

The device itself consists of two ratchets of the same diameter and number of teeth. Ratchet *A* is keyed to shaft *B*, and transmits the desired motion to this shaft. Ratchet *C* is free to turn on shaft *B*. On the extended hub of ratchet *C* is carried, on one side the pawl arm *D*, and on the other side the masks *E* and *F*.

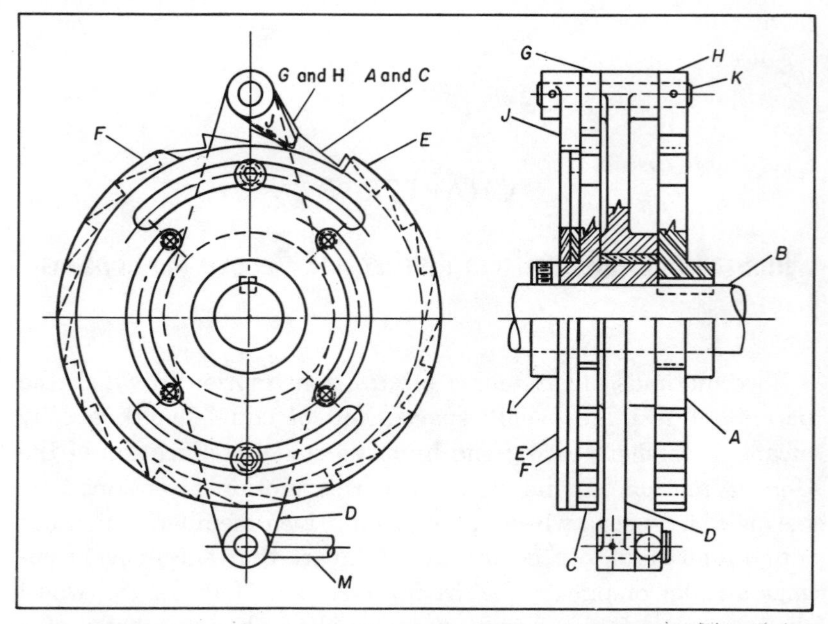

FIG. 1. Pawl G rotates ratchet C and attached masks E and F with every stroke. Masks E and F are capable of lifting pawl J and H, through pin K, thus controlling the rotation of ratchet A.

These masks have stepped diameters, the major equal to that of the ratchets, and the minor slightly less than the root diameter of the ratchet teeth. There is a series of arcuate slots in the masks, and tapped holes in ratchets C are so arranged that a variable number of teeth can be uncovered, and the position of these teeth located anywhere on the circumference of the ratchet. It is obvious that a modification of the profile of these masks will provide for an infinite number of conditions.

On the upper end of the pawl arm are carried two pawls, G, and H, and a pawl-like lever, J. These are carried on a pivot pin K, pawl H, and lever J being pinned on pin K. Pawl G is free on pin K.

Motion is transmitted to the pawl arm D by the connecting rod M by means not shown.

The operation of this device is as follows: As shown, the mechanism is set up to move four teeth per cycle, one tooth having been moved already. The next three movements of the pawl arm will move a tooth each, the whole mechanism rotating as a unit. The fourth backward movement of the pawl arm will cause lever *J* to ride up on the major diameter of mask *F*. Both lever *J* and pawl *H* being pinned to pivot pin *K*, this movement outward of lever *J* will lift pawl *H* out of engagement with ratchet *A*. Pawl *H* will remain out of engagement as long as *J* remains on the major periphery of the masks *F* and *E*. Pawl *G*, however, will engage ratchet *C*, moving it forward with its attached masks. This will continue until the lever *J* will be permitted to move down to the minor radius of the masks, when pawl *H* will re-engage ratchet *A* and the next movements of the pawl arm will carry the driven mechanism forward, in this case, four teeth.

Ratchet-Tripping Mechanism Controls Cut-off Length of Sheets

A mechanism employing a feed-ratchet tripped indirectly by a roller chain for pre-setting cut-off lengths was designed for an expanded-metal fabricating machine. The ram type machine has toothed blades attached to a slide that reciprocates across the sheet between strokes. The blades punch and expand openings in a solid steel sheet.

During the first stroke, the slide is laterally situated in one extreme position. Then, after the material has been fed forward a distance equal to one row of perforations, the blade slide moves to the opposite extreme position for a second stroke. Thus the rows of expanded openings are staggered on the sheet. The expanded metal is cut off from the solid sheet each time that the feed mechanism is prevented from functioning by the ratchet-tripping device here described.

The adjustable ratchet arrangement employed for feeding the metal sheets is shown in Fig. 2. Ratchet wheel *A* is intermittently rotated by feed-pawl *B*. The feed rate may be controlled by the adjustment of knob *C* to change the radial location of the driver that imparts motion to the feed pawl.

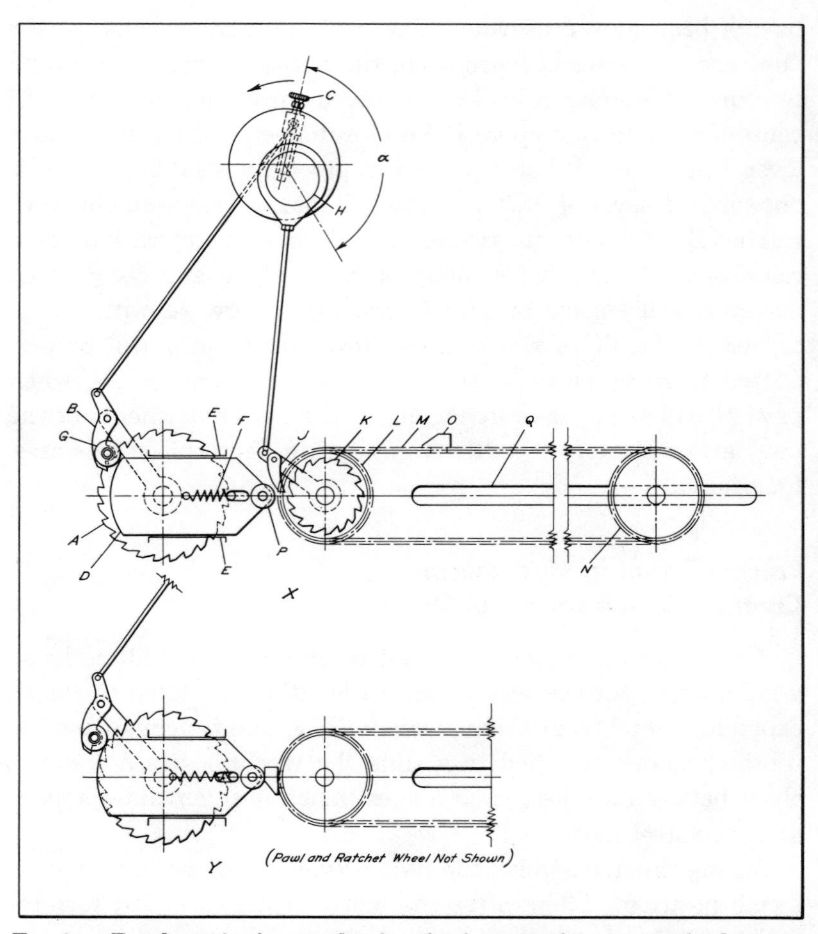

FIG. 2. Ratchet-tripping mechanism for interrupting a press feed at preset points, shown in feeding position at X, and non-feeding position at Y.

Slide *D*, which is retained between two guides *E*, trips the feed-pawl *B* directly. The slide is shown in its extreme right-hand position at X, held there by spring *F*. The left-hand end of the slide is an arc of approximately the same radius as ratchet wheel *A*. Roller *G*, attached to the feed-pawl, rides on this rounded end when the slide is shifted to the left.

Slide *D* is actuated by the ratchet-driven roller chain mechanism at the right in view X. A nonadjustable eccentric *H*, mounted on the drive-shaft, actuates pawl *J*. In turn, pawl *J* imparts intermittent motion to the ratchet wheel *K*. The eccentricity of member *H* was calculated to advance the ratchet wheel one tooth for each revolution of the drive-shaft.

A chain sprocket *L* is mounted on the same shaft as the ratchet wheel. The sprocket carries a roller chain *M* which is held in tension by idler sprocket *N*. Ratchet wheel *K* and sprocket *L* have the same number of teeth so that each feed movement of the pawl will advance the chain a distance of one link. Cam *O*, bolted to one of the chain links, pushes slide *D* against roller *G* once for each complete cycle of chain movement.

The arrangement of the mechanism during the feeding phase of the cycle is shown in view X. The slide is located at the right, allowing feed-pawl *B* to function. As the operation continues, the roller chain moves counterclockwise until the tapered front surface of cam *O* engages roller *P*. This forces slide *D* to the left, against spring pressure, to displace roller *G* and disengage the feed-pawl *B* from ratchet wheel *A*. The mechanism is then in the non-feeding phase of its cycle, as seen at Y. The next revolution of the drive-shaft will index the sprocket *K* through a distance of one tooth and move cam *O* past roller *P*. Spring *F* then returns slide *D* to the right-hand position so that feeding of the work will continue.

During the non-feeding portion of the cycle, the fabricated metal is cut off from the solid sheet. For various cut-off lengths, chains of different lengths are used. A slot *Q* is provided for the purpose of adjusting the idler sprocket either backward or forward to accommodate the various lengths of chain required for different jobs.

Angle α should be selected to make the two pawls operate in opposite directions. This will assure that either a feeding movement will occur or that there will be absolutely no movement; therefore the condition in which a feed movement is partially made before feed-pawl *B* is disengaged will never arise.

Positive Ratchet Mechanisms Designed for Silent Operation

Silent-operating positive-drive ratchet mechanisms are not too well known. By substituting a brake for the conventional spring, the pawl or finger member is lifted off the ratchet teeth on the idle stroke and made to engage the teeth again on the return stroke. Thus, although still being a positive intermittent mechanism, it works without the usual clicking noise made by the finger in riding or jumping over the teeth on the idle stroke, and therefore reduces the wear on the finger as well as on the teeth.

Referring to Fig. 3, the finger *F* is pivoted on an arm *A* which is, in turn, pivoted on the shaft *S*. The connecting-rod *C* pivots on the finger *F*. A spring-loaded brake *B* — prevented from rotating by a stud in the body of the machine — acts as a brake on a drum which is part of arm *A*. The toothed ratchet wheel *W* is keyed to the shaft *S*.

On the idle stroke (indicated by dotted-line arrow), the connecting-rod *C* will first pivot the finger *F*, thus lifting its point off the tooth on the ratchet wheel *W*. The arm *A* will not turn on the shaft at that time, as it is being restrained by the brake *B* and so offers more resistance to movement than the finger. This finger will pivot only through a certain angle until its short finger hits the stop *T* on the arm. It will then force the arm *A* to turn on the shaft, overcoming the friction of the brake and causing the finger and the arm to pivot on the shaft as one part.

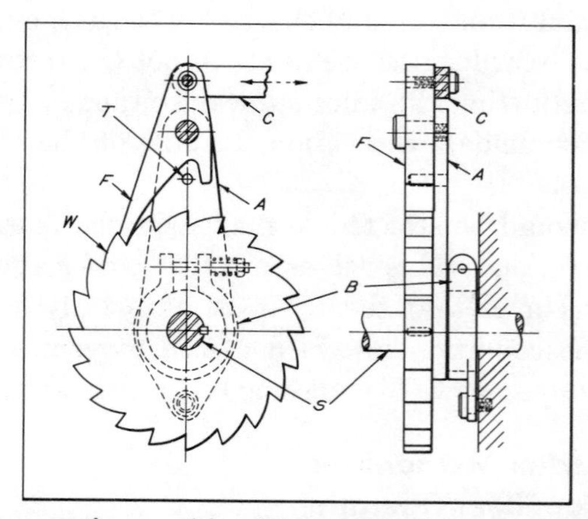

FIG. 3. Silent-operating, positive-drive mechanism with the ratchet pawl pivoted on the oscillating driving arm.

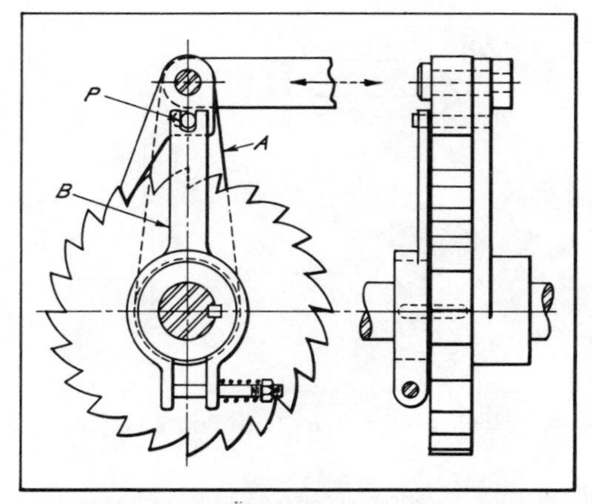

FIG. 4. Alternate design of silent, positive ratchet mechanism with connecting-rod and pawl pivoted on same pin.

On the return stroke, the finger will first pivot to engage a new tooth, and then the whole mechanism will turn as one piece, including the ratchet wheel and the shaft.

Another design of silent ratchet, in which the connecting-rod and the finger are both pivoted on the same pin, is shown in Fig. 4. The arm *A* and the brake arm *B* are arranged on opposite sides of the ratchet wheel. A pin *P* fixed in the finger provides

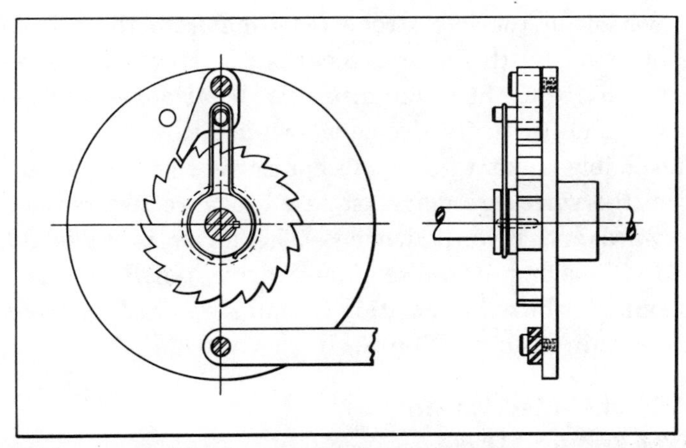

FIG. 5. Front and side views of ratchet designed for use on small-sized mechanisms.

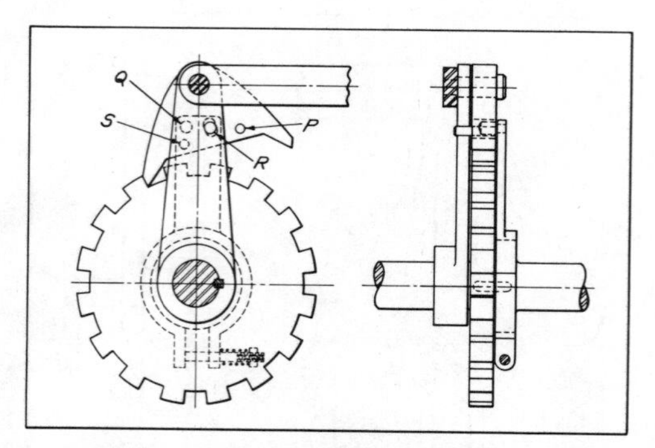

FIG. 6. Reversible ratchet with double-ended pawl.

the necessary stop by engaging an elongated slot in the brake arm on one side and a circular slot in the arm *A* on the other side.

A design that is suitable for small-size mechanisms, and which is similar to the one illustrated in Fig. 4, is shown in Fig. 5. The brake is made of a piece of spring wire. The big disc replaces the arm. The designs shown in Figs. 4 and 5 have the brakes operating on a drum which is attached to the shaft and ratchet wheel. This arrangement calls for a stationary finger to prevent the shaft from reversing on the idle stroke, or else the shaft with all the elements driven by it should offer enough resistance to prevent reversal on the idle stroke resulting from the grip of the brake. Obviously, this arrangement is not absolutely necessary for the design, and the brake drum can be attached to the body of the machine, as in the design shown in Fig. 3.

A reversible silent ratchet mechanism is seen in Fig. 6. The teeth on the wheel are made "square" to have two radial sides, for forward and reverse driving. The finger is made double-pointed. To change the direction of drive, pin *R* on the brake arm should be shifted to position *Q*, and stop-pin *P* on the finger should be shifted to position *S*.

Silent Ratchet Mechanism for Over-Running Drive

Ratchet mechanisms used on over-run drives frequently present problems of noise and wear. Shown in Fig. 7 is a ratchet mechanism designed to operate silently, with a minimum of wear on its working parts.

Consisting principally of a gear *A* and a ratchet *B*, the over-drive assembly is driven either by shaft *C*, to which the ratchet is keyed, or by the gear. The driving gear is mounted on the hub of the ratchet, and is free to turn on the hub, being retained by collar *D*. A recess is provided in the gear member to accommodate pawl *E*. Although the pawl pivots on pin *F*, it fits the pin loosely, and the actual pressure transmitted by the pawl is borne by the right-hand end of the recess in the gear.

When the shaft is driven by the gear, which revolves counter-clockwise, the pawl drives the ratchet in the usual manner. But when the shaft, which also rotates counter-clockwise, becomes the driver, the gear is stationary, and the pawl over-rides the ratchet. One of the functions of the mechanism at this time is to prevent the pawl from sliding over the ratchet teeth.

This is accomplished in the following manner: When the ratchet rotates counter-clockwise, a brass cam-plate *G* moves with it due to the friction developed by four cork-tipped spring plungers *H* in the ratchet as they ride on the cam-plate. The movement of the cam-plate lifts the pawl from the ratchet as pin *J*, which projects from the pawl, slides up slot *K* in the cam-plate.

When the gear drives the shaft, the ratchet remains stationary until the pawl is engaged. Since the friction generated by the

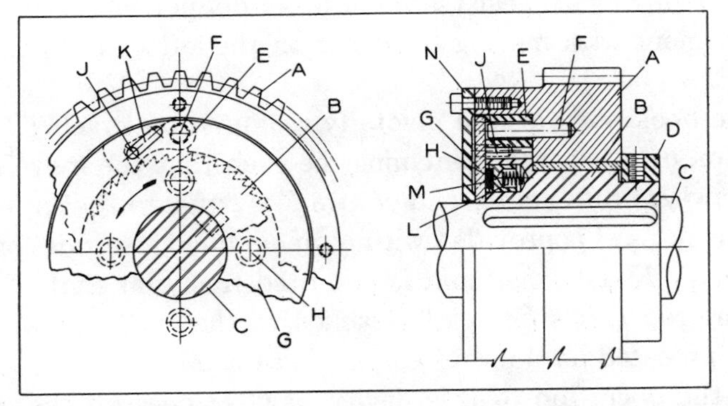

FIG. 7. Ratchet mechanism on an over-run assembly which operates silently and with minimum wear on the parts.

spring plungers in this case will retard the rotation of the cam-plate, pawl pin *J* is forced down to the left in slot *K*. Therefore, the pawl engages the ratchet teeth and the entire assembly revolves as a unit.

Spring plungers *H* are contained in four blind holes bored in the side of the ratchet. Outward pressure of each plunger is exerted by a spring *L* against a cork friction button *M*, fitted into a hole bored in the plunger. Covering the ratchet mechanism is a protective plate *N*, which is relieved to avoid a large area of contact with the cam-plate. To assure the frictional movement of the cam-plate only under the desired circumstances, the area of contact is reduced to a minimum. There is also clearance between the cover plate and the shaft for the same reason.

Additive and Subtractive Ratchet Mechanism

In the operation of a ratchet-driven device it was found desirable to automatically, and frequently, add extra tooth movements. Occasionally, all movement must stop.

The device illustrated in Fig. 8 includes ratchet wheel *A* which is keyed to the driven shaft *B*. Mounted on an extended hub of the ratchet is a bushed push-pawl arm *C*. Push pawl *D* is pivotally mounted on arm *C*. The pawl is provided with a projecting pin *E* that rides on the periphery of mask *K*. Arm *C* is provided with gear teeth cut around a portion of its hub for engagement with mating teeth cut on the left end of hook-pawl arm *F*.

The hook-pawl arm is pivotally mounted on bracket *G*. The location of this pivot point must be such that the movement of the pawl *H* on the outer end of arm *F* is equal to that of pawl *D*. The hook pawl is provided with a pin *J* which also rides on mask *K*. Mask *K* is bushed and is mounted freely on shaft *B*. Connecting-rod *L* has one end attached to the mask. Pawl arm *C* is reciprocated by another connecting-rod *M*.

In the operation of this device pawl *D* does all the driving, ordinarily moving a distance of one tooth on ratchet wheel *A*

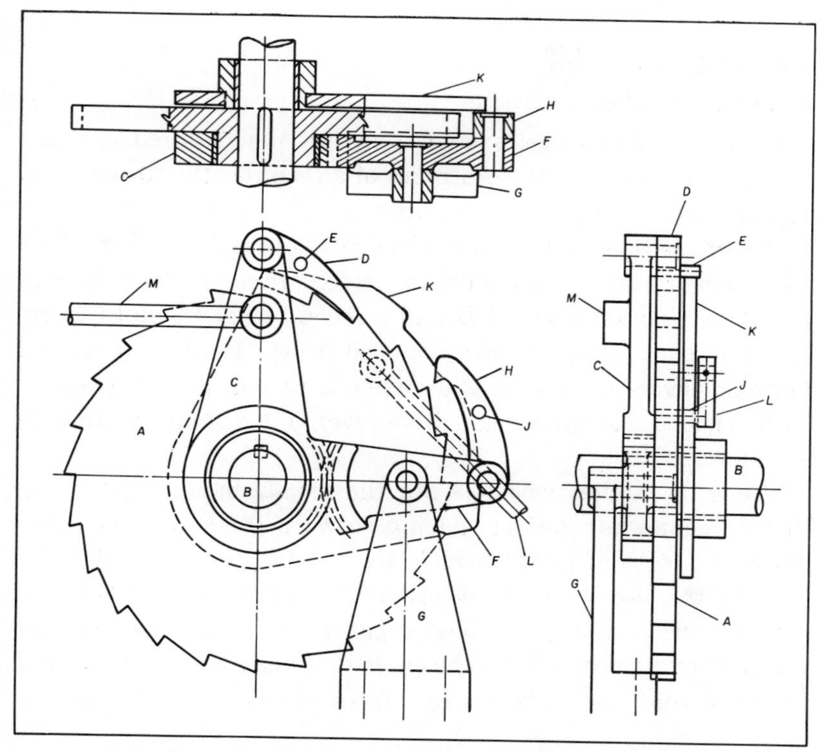

FIG. 8. Ratchet mechanism has a mask that allows automatic changes in ratchet stroke or complete temporary stoppage.

for each reciprocation of arm *C*, as a simple ratchet movement. In such movement pin *J* on hook pawl *H* rests on the lobe of mask *K* so that pawl *H* is held out of engagement with the ratchet.

When conditions arise that require increased movement of shaft *B*, the control mechanism, through connecting-rod *L*, moves mask *K* clockwise so as to permit pawl *H* to function. The mask holds the new setting until changed.

Conversely, if stoppage of shaft *B* is called for, the mask moves counterclockwise until pawl *D* is lifted out of engagement. This occurs when pin *E* rises on the lobe of mask *K*. Size of the lobe on mask *K* for pin *J* permits sufficient counterclockwise movement to hold both pawls out of engagement with ratchet wheel *A*.

Ratchet Operates on Alternate Strokes

A dual ratchet-wheel system provides the required rotation of a shaft only on alternate strokes of a reciprocating drive lever. Figure 9 shows the mechanism at the end of a power stroke.

Driven shaft *A* and ratchet wheel *B* are keyed together. This wheel has a hub on each side; one side carries one lever *C*, and the other side, pilot wheel *D* and a second lever *C*. Both levers and the pilot wheel are free on the hubs. Pawl *E* is pinned between levers *C*, and is wide enough to engage the teeth of both wheels. Reciprocating drive lever *F* transmits motion to both levers *C*.

Teeth of ratchet wheel *B* are the usual shape, except that there is somewhat greater spacing between them. On the other hand, the teeth of pilot wheel *D* are a special shape, as shown.

With the levers in the position illustrated, at the end of a power stroke, the pawl has engaged one tooth of ratchet wheel *B* and rotated it to the limit of lever movement. It will be noted that the radial contact faces of the teeth of wheels *B*

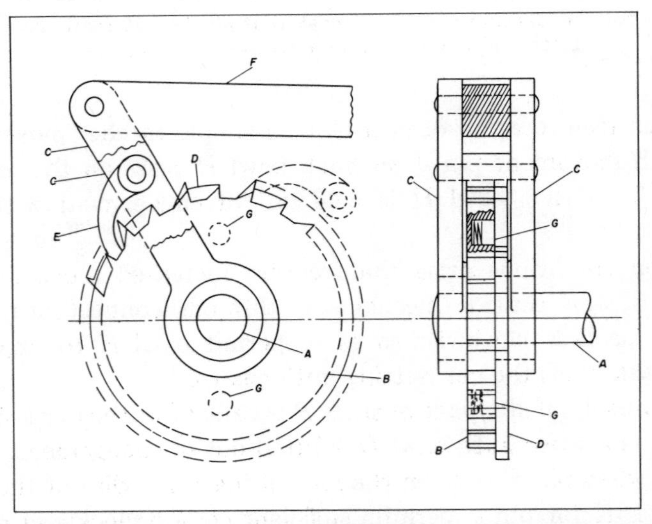

FIG. 9. The design of the teeth of pilot wheel *D* keeps pawl *E* out of engagement with ratchet wheel *B* on alternate oscillations of levers *C*.

and *D* coincide, and since the pawl is wide enough to engage both wheels, they have been rotated in unison.

The pawl is shown in broken line at the end of the subsequent return stroke. Here, it is in contact with one tooth of pilot wheel *D*, but is raised out of contact with the ratchet wheel *B*. On the next power stroke, the pilot wheel is rotated, but no motion is transmitted to the ratchet wheel, and therefore no motion to the driven shaft. At the end of this power stroke, the pilot wheel will come to rest so that the contact face of the tooth will coincide with the contact face of one of the teeth in the ratchet wheel.

Then, at the end of the next return stroke, the pawl will again be in position to fall into contact with one tooth on both wheels. In this way, the required shaft rotation on alternate reciprocations of the drive lever is obtained: on one power stroke, both wheels are rotated in unison, and the motion is transmitted to the driven shaft; but on the subsequent power stroke, the ratchet wheel is not rotated, since the pawl is held up.

Two spring-loaded plungers *G* are contained in the ratchet wheel, bearing against the adjacent face of the pilot wheel. By applying a light frictional resistance to the pilot wheel, they prevent any backward rotation due to the drag of the pawl on the return stroke.

The mechanism will operate regardless of the angular oscillation of levers *C*, with the limitation that the pawl always must move an uneven number of teeth, such as one, three, or five. If the pawl were to move an even number of teeth, such as two, four, or six, the ratchet wheel would be rotated on each oscillation, rather than on alternate oscillations as required. This, in itself, may be an advantage, in some instances, in that it is possible to vary the driven-shaft rotation from alternate to consecutive action merely by changing the range of oscillation of the levers, to increase or decrease the movement of the ratchet wheel. There must, of course, be an even number of teeth or contact faces on both the ratchet wheel and the pilot wheel of this mechanism.

Variable Intermittent Movement Derived from Gear Drive

An unusual mechanism which provides intermittent movement from a standard gear drive is shown in Fig. 10. The amount of movement of the driven member is variable within wide limits and is easily adjusted to any degree of arc.

In this mechanism, driving gear *A* revolves freely on stationary shaft *B* and is retained in position by collar *C*. Three concentric bores in gear *A* contain the elements of the intermittent movement device. A rectangular flange *D*, mounted integrally on shaft *B*, provides a mounting surface for slide *E*. Two screws *F*, passing through elongated holes in the slide, fasten it to the base of a deep slot machined across the flange face. These holes permit the degree of intermittent movement obtained from the mechanism to be varied.

A lever *G* pivots freely on shoulder-screw *H*. The screw is threaded into slide *E* at a point below the center of gear *A* as indicated by dimension Z. Ratchet wheel *J* is keyed to a shoulder on the end of output shaft *K*.

Motion is transmitted between gear *A* and ratchet wheel *J* by means of pawl *L*. The pawl is secured to gear *A* by shoulder-

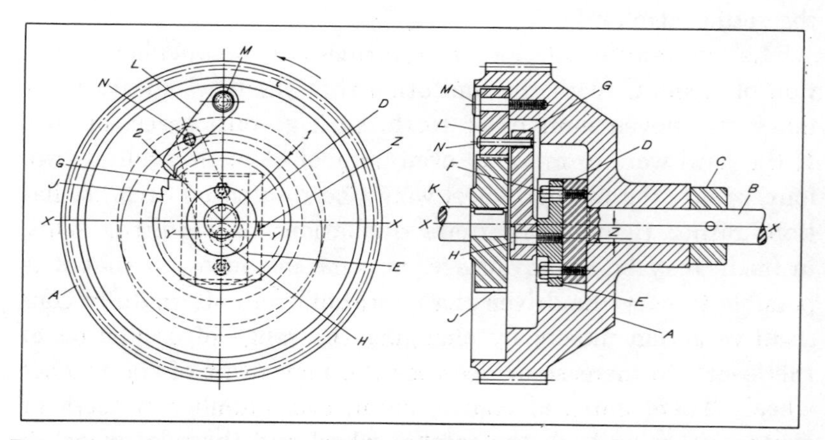

FIG. 10. Drive mechanism designed to convert constant rotary motion into variable intermittent rotary motion.

screw *M*. Driving pin *N* is pressed into the pawl and passes through a short elongated hole in lever *G*. The purpose of this connection is to control the swinging movement of the pawl.

As driving gear *A* rotates in the direction indicated by the arrow, ratchet wheel *J* is forced to revolve in the same direction due to the engagement of pawl *L*. Because driving pin *N* is connected with eccentrically mounted lever *G*, the pin moves in a circular path that is not concentric with shaft *B*. Thus, as gear *A* rotates, the pawl is successively drawn closer to, then farther away from, the teeth of the ratchet wheel. In this way, an intermittent motion is imparted to output shaft *K*.

With slide *E* located so as to provide an offset equal to distance Z, the ratchet wheel will rotate approximately 45 degrees during each revolution of gear *A*. The length of engagement between the pawl and the ratchet wheel is denoted by numbers 1 and 2 in the left-hand view. By adjusting the position of slide *E*, the distance traveled by the ratchet wheel can be varied.

Intermittent Motion Derived from Continuously Rotating Shaft

On a wire-forming machine, it was necessary to interrupt the feed of the wire at certain intervals in the cycle. To accomplish this, the shaft operating the feeding mechanism was cut at one point, and the mechanism illustrated was then installed.

As indicated in Fig. 11, shaft *A*, the driving member, transmits its motion to shaft *B*, which operates the feeding mechanism. Keyed to shaft *A* and rotating with it is a disc *C*. The disc carries a pawl *D* that is normally held in contact with a ratchet *E* by a spring *F*. Ratchet *E* is attached to shaft *B*. There are eight teeth spaced around the ratchet. A ring *G*, which is mounted on a stationary part of the machine, is slotted at eight equally spaced points to receive studs carrying rollers *H*, which contact the tail of the pawl.

For purposes of explanation, the rollers *H* are numbered 1 to 5. As shaft *A* and disc *C* rotate in the direction indicated by the arrow, the pawl engages one of the teeth of the ratchet, causing shaft *B* to rotate in unison until the tail of the pawl contacts

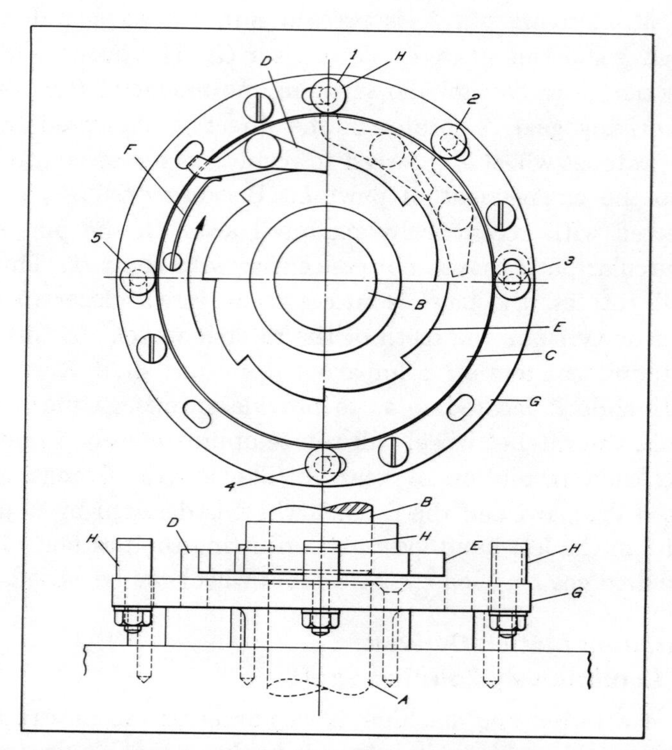

FIG. 11. The pawl D is disengaged from ratchet E by contact with one of the rollers H.

roller No. 1. The pawl at this and subsequent locations is represented in broken line. In contacting the roller, the pawl is released from the ratchet and the motion of shaft B is interrupted until the pawl again engages a ratchet tooth.

Assuming for the present that roller No. 2 has been removed, the movement of the ratchet again begins when the pawl engages ratchet tooth adjacent to roller No. 3. Thus far, shaft A has rotated 90 degrees, but shaft B has rotated only 45 degrees, the other 45 degrees having been lost by the pawl passing over one tooth of the ratchet. Continued movement of the disc causes the ratchet to again be rotated until the tail of the pawl contacts roller No. 3, when movement is once more interrupted. With rollers Nos. 1, 3, 4, and 5 positioned as shown, there are four movements of shaft *B* and four rest periods of 45 degrees each in every rotation of shaft *A*. On the machine involved, this was the particular intermittent motion required.

The design, moreover, lends itself to other variations. Assuming, for example, that roller No. 2 is placed as shown, the pawl is prevented from engaging the ratchet tooth. It will be noted that this roller has been moved to the upper end of its slot, the purpose being to operate the pawl before tooth engagement. If roller No. 3 were moved to the upper end of its slot, disengagement would continue until the tooth adjacent to roller No. 4 is reached, thus producing a rest period of 135 degrees. Likewise, if roller No. 4 were transferred from its present position to the slot immediately to the right, engagement would again be prevented, producing a rest period for shaft *B* of 180 degrees.

It is evident that various combinations of intermittent motions can be obtained, depending on the number of rollers used and their locations. This mechanism can be adapted to a wide variety of intermittent motions by increasing the number of teeth in the ratchet and providing a continuous slot in ring *G*, so that rollers are able to be placed in any position.

Cam and Ratchet Intermittent Mechanism

A mechanism that employs a cam, ratchet, and pawl for converting constant rotation into intermittent rotary motion is shown in Fig. 12. Drive-shaft *A* of this mechanism revolves continuously. Keyed to it is disc *B* which carries pin *C* which is the pivot for pawl *D*. On the driven shaft *E* there is keyed ratchet wheel *F*. The inner end of the pawl engages the ratchet teeth while the outer rides around a cam path cut into disc *G*.

As pivot-pin *C* is carried around disc *G* by the rotation of the drive-shaft in the direction indicated by the arrow at *M*, shaft *E* is turned by pawl *D* during the period when end *K* of the pawl is riding along the high section of the cam *P*. A spring, not shown, exerts sufficient pressure to hold the cam tooth in engagement with one of the teeth on the ratchet during this period of pawl movement.

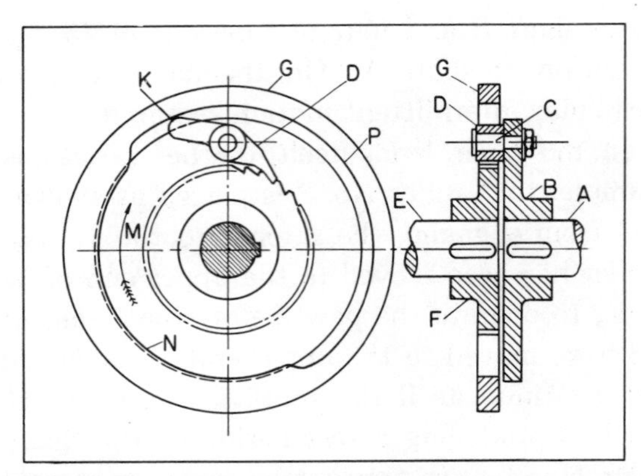

FIG. 12. Mechanism that employs a cam, ratchet, and pawl for converting constant rotation into intermittent rotary motion.

When end *K* of the pawl reaches the low portion of the cam indicated by letter *N*, the pawl is swung on its pivot and its tooth is disengaged from the ratchet. The driven shaft *E* then remains stationary until end *K* of the pawl again rides on the high section of the cam and the pawl tooth has once more been swung into engagement with a cam tooth.

Geneva Drive in which Gear Ratios Control Motion Time

Geneva drives used to a wide extent in automatic machinery generally consist of a driving roller at the end of a crank and a slotted member which is moved when the driving roller enters into a slot. The conventional Geneva drive has some disadvantages, one of them being that time for motion and dwell of the driven member is usually determined for a given number of slots or stations.

In Fig. 13 is shown a modified Geneva drive of which the time for motion is dependent upon gear ratios. In this drive, input shaft *A* rotates with uniform velocity and drives gear *B*, which in turn drives sun gear *C*. The latter is free to rotate on the shaft *D*. Shafts *A* and *D* are supported in the gear housing. Sun gear *C* drives planet gear *E*, and as long as the roller *H* is outside of the slotted member *F*, gear *E* rotates with uniform velocity because the planet block G is detented. The detent device is not shown. Roller H is just starting to enter a slot in Fig. 14, while Fig. 15 shows the mechanism some time after the roller has entered the slot.

At the moment that the roller enters the slot, the planet carrier becomes unlocked. The roller, however, is now in the slot, and because of the angular motion of link J, which is driven by planet gear E, the roller will penetrate deep into the slot and cause shaft K to rotate counterclockwise around shaft D.

In general, the size of an idler gear has no influence on gear ratio, but in this case the size of gear C is of importance because planet gear E rolls on gear C during motion.

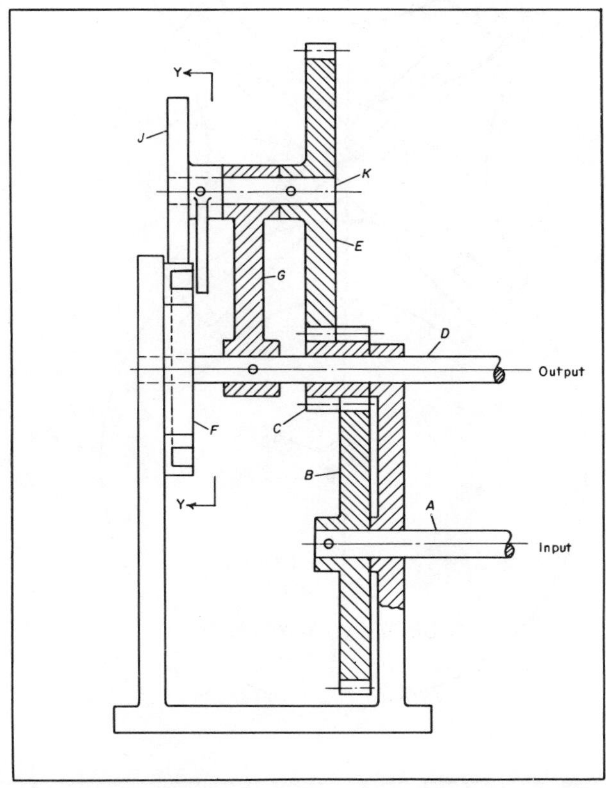

FIG. 13. Modified Geneva drive with motion time controlled by gear ratios.

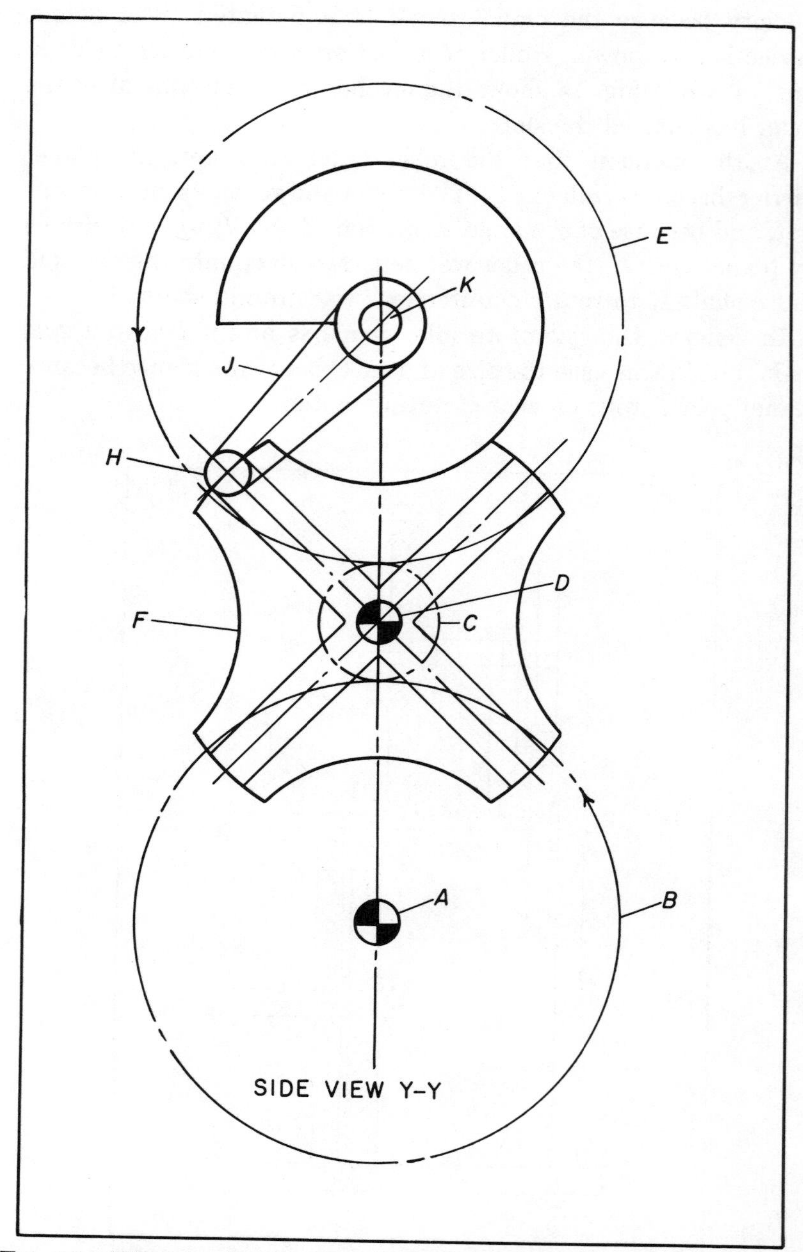

FIG. 14. Diagram showing roller H entering slot of planet block at beginning of Geneva motion.

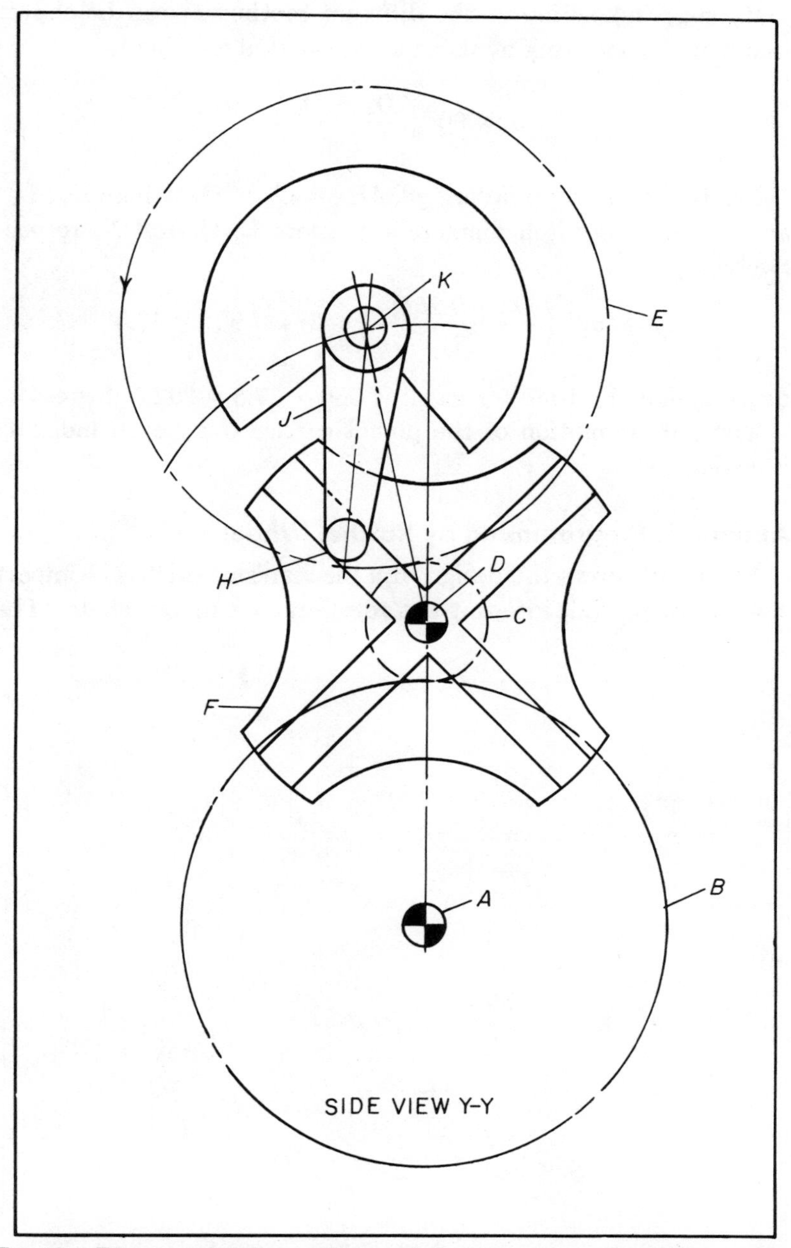

FIG. 15. Diagram showing positions of elements after roller *H* has moved along slot of planet block.

By superimposition of the different motions it can be shown that time for indexing by using a four-slotted member is

$$T = 90^\circ \left(\frac{D_4 - D_3}{D_2} \right)$$

For the mechanism described, $4D_3 = D_2 = D_4$, where D_2, D_3, and D_4 are the pitch diameters of gears B, C, and E, respectively.

$$T = 90^\circ \left(\frac{D_2 - 0.25D_2}{D_2} \right) = 3/4 \times 90^\circ = 67.5$$

degrees, and the time for dwell is $360 - 67.5 = 292.5$ degrees.

The output motion of the planet carrier G by each indexing is 90 degrees.

Automatic Programming by Ratchet Wheel

Figure 16 shows the design of a mechanism which will impart a variable, partial, intermittent rotation to a driven shaft. The

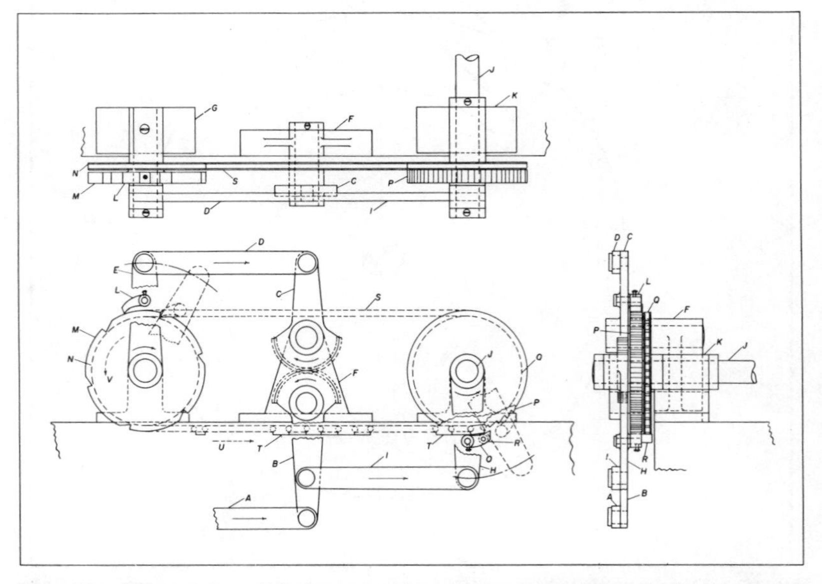

FIG. 16. Three views of the programming mechanism which gives variable, partial, intermittent motion to shaft J as driven by bar A and sequenced by chain S.

purpose is to produce a variable spacing of the strands of woven wire in a pattern.

Referring to Fig. 16, which shows three views of the mechanism, the reciprocating bar link *A* supplies the mechanism with motion which causes shaft *J* to deliver the required variable, partial, intermittent rotation. The movement of bar *A* is transmitted to lever *B*, which carries a gear sector at its upper end and is keyed to a shaft, which is free to rotate in bearing block *F*. The gear sector on lever *B*, in mesh with a mating gear sector on lever *C*, transmits linear motion to the lever arm of *C*. Lever *C* moves lever *E* through link *D*. Lever *E* swings freely on a shaft. The motion of lever *B* is also transmitted to lever *H* through link *I*. Lever *H* swings freely on the driven output shaft *J*, which is mounted for free rotation in bearing block *K*.

Lever *E* carries pawl *L* to engage in the notches on the periphery of ratchet disc *M*. The ratchet wheel is attached to sprocket wheel *N*, the pair being rotatable on a shaft. The number of notches on ratchet disc *M* is determined by the angular movement of lever *E*. Its angular movement, in turn, is governed by the amount of angular movement of lever *H* needed to produce the maximum partial revolution of shaft *J* that will give the required spacing of the strands of wire.

Lever *H* carries pawl *O*, which engages teeth of ratchet wheel *P*, keyed to output shaft *J*. On the side of the pawl is lifter pin *R*, which overhangs sprocket *Q*. This sprocket free-wheels on shaft *J*. Sprockets *N* and *Q* are linked by roller chain *S*. The length of this chain is governed by the number of movements of shaft *J* in one "repeat" of the required complete program for the wire-mesh pattern. It must be of such length that the number of links will be a multiple of the number of teeth of sprockets *N* and *Q* included in the angular movement of levers *E* and *H*. Chain *S* is equipped with special pawl-lifter links *T*, placed on opposite sides of the chain where necessary. In operation, the high links contact the pawl lifter, which causes pawl *O* to fail to contact ratchet wheel *P*. Pawls *L* and *O* are equipped with springs (not shown) that normally insure engagement with their ratchet teeth.

In the drawings, the mechanism is shown at the mid-point of its motion. Bar *A*, moving in the direction of the arrow, transmits motion to the various links and levers in the directions indicated by the arrows. At this point, there is no movement of either the sprockets or the ratchet wheels and, therefore, there is no movement of shaft *J* because the pawl is moving opposite to the direction required for engagement. Also, pawl *O* is held out of engagement with ratchet wheel *P* by the three lifter links shown.

Continued movement of bar *A* in the same direction causes the pawl *O* to end contact with the chain links *T*. The pawl then drops into one of the teeth of ratchet *P*. Rotation of shaft *J* starts, and continues until lever *H* reaches the end of its stroke. Levers *E* and *H* finally take the positions shown by the broken lines. At this point, pawl *L* has been brought into position to engage one of the notches in disc *M*.

On the return stroke of bar *A*, disc *M* and sprocket *N* rotate in the direction of the dotted arrow, causing chain *S* to move in direction *U* (dotted arrow) so that any links *T* attached to chain *S* will pass under pawl *O* moving in the opposite direction. In this manner, the movement of chain *S* is brought into position for the next working stroke of lever *H*. The number of chain links which pass under pawl *O* governs the dwell period of shaft *J*. Its rotation can be started only after pawl lifter *R* on pawl *O* has moved off the lifter links.

Adjustable Indexing Mechanism with 180-Degree Dwell

On certain types of printing presses it is often desired to incorporate a variable indexing motion in which a ratchet is moved during 180 degrees of rotation of the driving shaft and then dwells over the rest of the motion. A mechanism designed to accomplish this result is here shown.

In this mechanism gear *A* is half the size of the internal gear *B*; see Fig. 17. Gear *A* is carried in a circle around the internal gear by arm *C* which is attached to driving shaft *D*. This shaft rotates continuously.

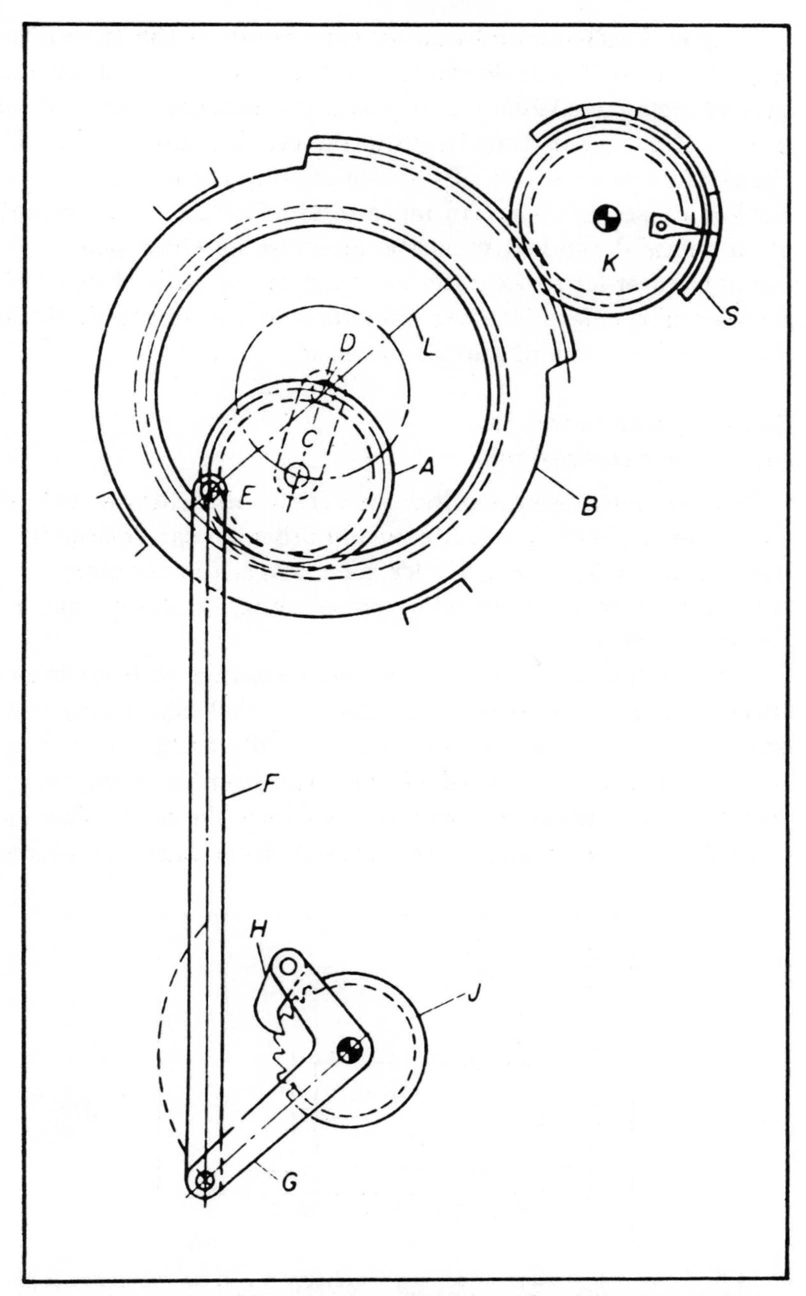

FIG. 17. An adjustable indexing mechanism with 180 degrees of dwell.

As gear *A* rolls around gear *B*, each point on the circumference of the small gear describes a straight line. Point *E* on the circumference describes a path along the straight line *L*. The motion of point *E* is transferred to the crank *G* through link *F*. Crank *G* carries pawl *H*. By the oscillating motion of lever *G*, pawl *H* causes ratchet *J* to move intermittently. The amount of motion is dependent upon the direction in which point *E* is moving. In order to vary the direction of the path of point *E*, gear *B* can be indexed limited amounts by means of gear *K*. Scale *S* indicates the amount gear *J* is indexed.

Counting Device for High Speed Operation

Counters have been required for various applications such as on computers, servo mechanisms, and other similar devices that function at high speeds. Under such usage, the counting unit must impart an absolute minimum of drag, or shock load, to the driving members.

A conventional, low-speed counter, operating with an intermittent motion, builds up momentary shock loads during normal cycling, as can be seen in the graphic illustration at X, Fig. 18. As the operating speed of this type counter increases, so does the acceleration and, as a consequence, the load. This increased load is dissipated in the form of either elastic or plastic

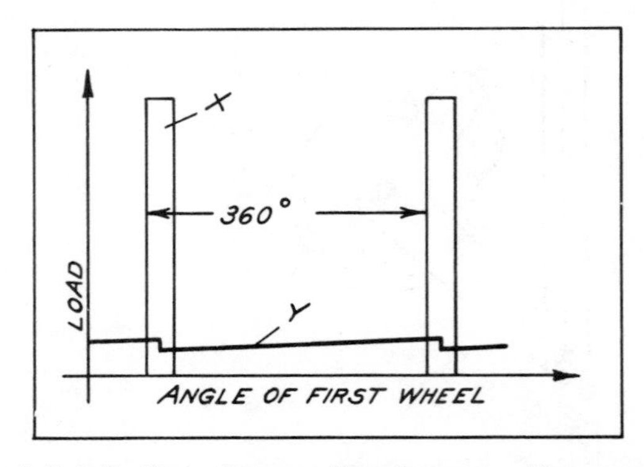

FIG. 18. Load distribution of conventional counter X as compared to the load distribution Y of the high-speed counter.

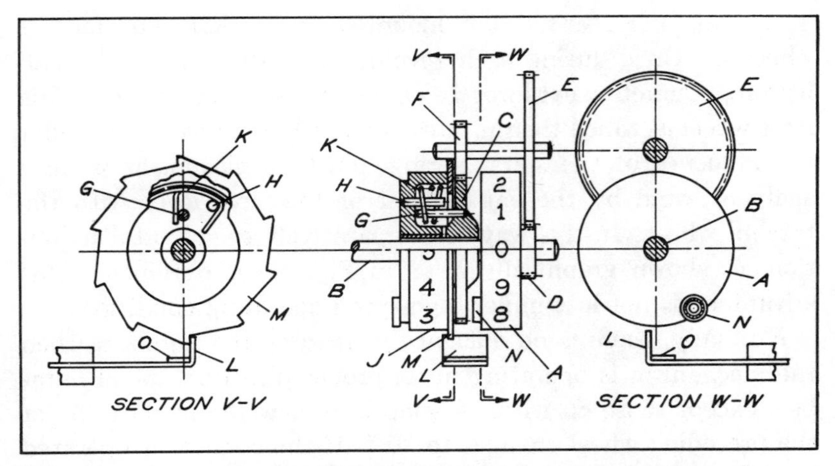

FIG. 19. Spring-loaded wheel J jumps forward one number for each complete revolution of wheel A.

deformation of the counter mechanism. Even if the preferable case of elastic deformation should result, the driving mechanism to which the counter is attached is liable to suffer from the shock of being momentarily halted or rapidly decelerated.

To overcome these problems associated with high-speed operation, ranging up to 12,000 R.P.M. and with a load of approximately 1 ounce-inch, the counter mechanism described was developed. Wheel A, which can be seen in the diagrammatic representation of the counter, Fig. 19, is keyed to input shaft B, thereby making one complete revolution for each made by the shaft. Gear C, which rotates independently of the input shaft, is driven at one-tenth the speed of wheel A by the action of reduction gears D, E, and F.

Pin G, projecting from the hub of gear C, and pin H, projecting from the bottom of a recess in wheel J, engage the bent ends of spiral spring K. As gear C rotates, wheel J tends to rotate in unison through the spiral spring linkage. However, movement of wheel J is prevented by pawl lever L which engages a tooth of ratchet wheel M. The ratchet wheel is attached to the counter wheel. While in this position, energy necessary to turn wheel J is being built up in a potential form in spring K.

A small ball bearing *N* is mounted on the left-hand face of wheel *A*. Once during each revolution of this wheel, the ball bearing contacts pawl lever *L*, depressing it momentarily. This frees wheel *J*, which then rotates one-tenth of a revolution under the influence of the spiral spring, until its movement is once again arrested by the engagement of the pawl lever with the ratchet wheel. In this way, a comparatively even load distribution, as shown graphically at Y in Fig. 18, is obtained. This advantage is not lost under high-speed operating conditions.

Normally, there is no difficulty in reading the numbers when the mechanism is operating in its proper direction, as all numbers except those on wheel *A* snap into view the instant "9" of the preceding wheel changes to "0." If the counter is operated in the reverse direction, however, the numbers will no longer swing into position but will move continually. At higher speeds the numbers on wheel *A* are no longer readable. A small weight *O*, attached to the pawl lever, retards return of the pawl to the ratchet wheel, thereby smoothing out the movement of wheel *J*. It is then possible for the operator to satisfactorily interpolate the readings of this wheel.

Multiple-Revolution Ratchet Movement

Generally, a ratchet movement is limited to a partial revolution of the driven shaft, since the rotation of the oscillating lever which carries the pawl is necessarily restricted in order to avoid a dead-center effect. The device shown in Fig. 20, however, incorporates an epicyclic gear train to produce multiple revolutions of the driven shaft with only a conventional magnitude of oscillation of the driving lever.

Gear *A* and a ratchet wheel *B* are both mounted and keyed to the driven shaft *C*. An oscillating lever *D* pivots on shaft *C* and carries two gears *E* and *F*. These gears are keyed together to rotate as a unit on stud *G*. Gear *F* rotates in mesh with gear *A*, and gear *E* with an internal gear *H*, which is free to rotate on shaft *C*. A pawl *J* engages with the teeth of ratchet wheel *B*, preventing rotation of the shaft in a clockwise direction. Another pawl, member *K*, engages ratchet teeth on the periphery

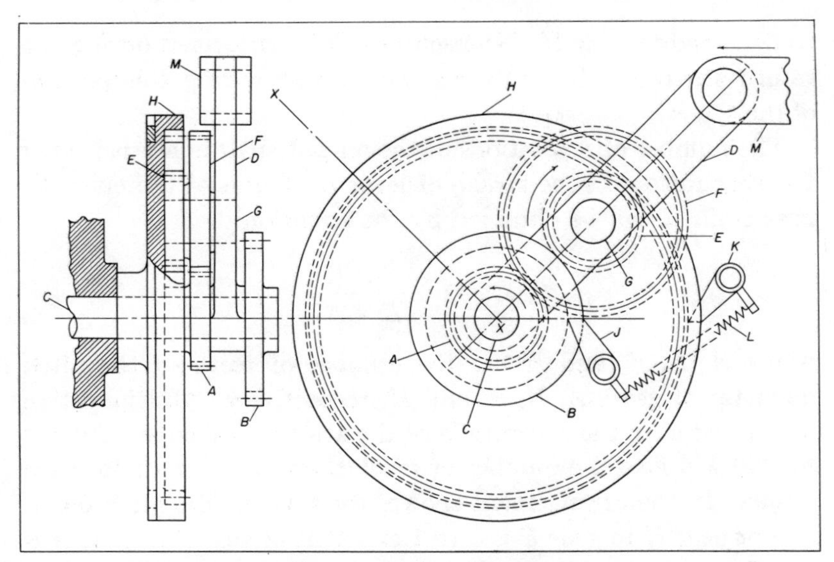

FIG. 20. Ratchet movement that can produce a number of revolutions of the output shaft with each working stroke.

of a ring secured to left side of gear *H*, preventing counterclockwise rotation. Both pawls are mounted on a stationary part of the machine and springs *L* hold them in contact with the ratchet teeth.

In the position illustrated, the mechanism is at the beginning of its cycle. Reciprocating rod *M*, which furnishes the operating power, moves lever *D* to the left until it occupies a position centered on line X-X. As the lever swings to the left, internal gear *H* is prevented from rotation in the same direction by pawl *K*, and gear *E*, being in mesh with the internal gear, is caused to rotate clockwise. Gear *F* rotates with gear *E* as a unit producing a counterclockwise rotation of both gear *A* and shaft *C*. On the return stroke of lever *D*, clockwise movement of the shaft is prevented by pawl *J* which engages a tooth in the ratchet wheel *B*. Since both ratchet wheel *B* and gear *A* are keyed to the shaft and held stationary, the motion of rod *D* is transmitted through gears *A*, *F*, and *E*, to produce a clockwise rotation of the internal gear. Thus, by locking gear *H*, motion is transmitted to the shaft, and by locking the shaft in the reverse direction, motion is transmitted to gear H. No useful work is performed during the return stroke as the shaft remains at rest during this portion of the cycle.

The number of revolutions of the output shaft is a function of the gear ratio and the stroke of lever D. Ratio of the epicyclic gear train R can be obtained by the equation:

$$R = 1 + \frac{F \times H}{E \times A}$$

where A, E, F, and H are the number of teeth or the pitch diameter of gears A, E, F, and H, respectively. Multiplication of this ratio by the magnitude of the stroke, in degrees, divided by 360 will give the number of revolutions of the shaft for each stroke. In the arrangement shown, the ratio of the pitch diameter or gear H to gear E is 4 to 1 and that of gear F to gear A is 2 to 1. The ratio R is, therefore, $1 + \frac{2 \times 4}{1 \times 1}$ or 9. Since the stroke is 90 degrees, the shaft will evolve $9 \times \frac{90}{360}$ or $2\frac{1}{4}$ times in a working stroke.

Two-Speed Double-Action Ratchet Mechanism

A ratchet mechanism had to be designed to move a conveyor belt intermittently so as to carry two parts of an assembly to a number of assembly stations. The two parts vary considerably in size, and so the conveyor belt had to be given a certain movement for the placement of one part and a greater movement for placing the larger part. The mechanism operates the conveyor belt in one direction during two oscillations of a lever and alternately imparts long and short movements to deliver the assembly parts.

In Fig. 21, shaft A, which operates the conveyor belt, carries gear B and ratchet wheel C, both of which are keyed to it. Ratchet wheel D is free on shaft A. Internal ring gear E is fastened to this ratchet wheel. Bracket F, attached to a stationary part of the machine, carries a short rod on which pinion G

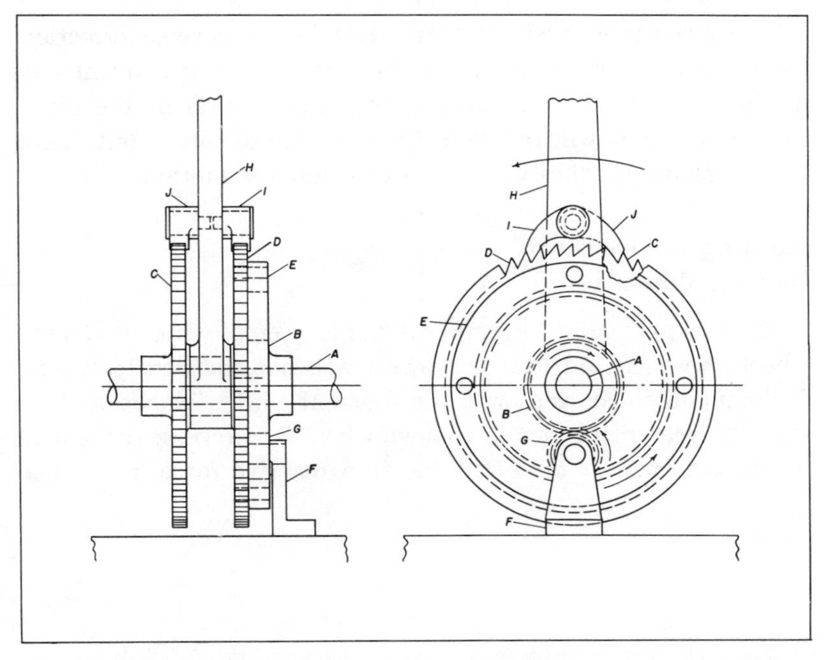

FIG. 21. Ratchet mechanism designed for imparting two rotations of different amounts in the same direction.

rotates freely, meshing with gear *B* and ring gear *E*. Ratchet wheel *D* has a hub on its inner face, on which lever *H* is free to oscillate. Attached to this lever is pawl *J*, which engages the teeth of ratchet wheel *C*, and pawl *I*, which engages the teeth of ratchet wheel *D*.

When lever *H* is moved in the direction indicated by the arrow in the right-hand view, the long stroke is made which produces the longer movement of the conveyor belt. Pawl *I* turns ratchet wheel *D* in the direction indicated, and the motion is transmitted to shaft *A* in the reverse direction through ring gear *E*, pinion *G*, and gear *B*. As the ratio of the tooth count between gear *B* and ring gear *E* is 2 to 1, the angular rotation of gear *B* is twice that of lever *H*. It will be noted that during this portion of the cycle, the rotation of ratchet wheel *C* is in the reverse direction to that of ratchet wheel *D* so that pawl *J* cannot engage.

When the movement of lever *H* is in the reverse direction, pawl *J* engages the teeth of ratchet wheel *C*, and transmits its motion directly to shaft *A*. During this portion of the cycle, ratchet wheel *D* will rotate in the opposite direction but, being free on shaft *A*, takes no part in transmitting motion.

Blocking Device for a Geneva Wheel

In one particular driving mechanism employing a Geneva wheel, it was found that the wheel was not sufficiently locked. It frequently occurred that the moment pin *A*, shown at Z in Fig. 22, cleared the slot in Geneva wheel *B*, a reverse movement would take place. This was due to a reactive force in the ma-

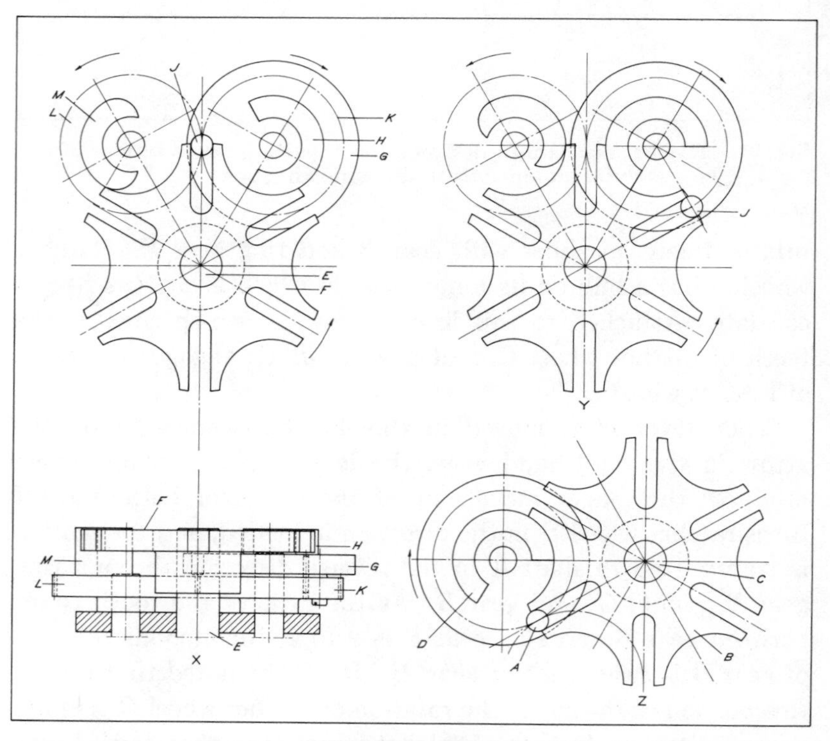

FIG. 22. Geared auxiliary blocking segment prevents reverse movement of Geneva wheel.

chine being driven from shaft *C*. The reverse movement was unchecked because blocking flange *D* is effective in one direction only.

The improved Geneva mechanism shown at X and Y, incorporating a reverse-motion stop, has been designed to eliminate this condition. Shaft *E*, which drives the machine, is fitted with a six-station Geneva wheel *F*. Driving wheel *G*, having a conventional blocking flange *H* and drive-pin *J*, is screwed and doweled to spur gear *K*. Meshing with this gear is a similar spur gear *L* on which is located a crescent-shaped reverse-motion stop *M*.

At X is shown the position of the components at the instant the Geneva wheel has been indexed one station. As drive-pin *J* leaves the slot, wheel *F* is blocked in the forward direction by a portion of flange *H*. The wheel is also blocked in the reverse direction by reverse-motion stop *M*.

The position of the components as the Geneva wheel is about to be indexed another station is shown at Y. Drive-pin *J* enters the appropriate slot in wheel *F* just as crescent-shaped stop *M* is disengaged from the wheel. Due to its shape, stop *M* disengages at a rate that will not impede the forward motion of the Geneva wheel, thus permitting smooth functioning of the mechanism.

Ratchet and Two Pawls Control Movement of Indexing Fixture

Ratchet-controlled positioning and an expanding, work-holding stub-arbor are two features of the unique indexing fixture shown in Fig. 23. From two to eighteen indexing positions can be obtained, depending on the number of notches in ratchet plate *A*.

In this fixture the work-piece is gripped internally on an expanding stub-arbor *B*, shown in the enlarged section X-X. The projecting portion of the arbor has three radial slots, giving it the action of a collet. Knob *C* is prevented from turning on shaft *D* by a full-dog set-screw. The dog point of the set-screw has a sliding fit in a keyway machined in the shaft. When the

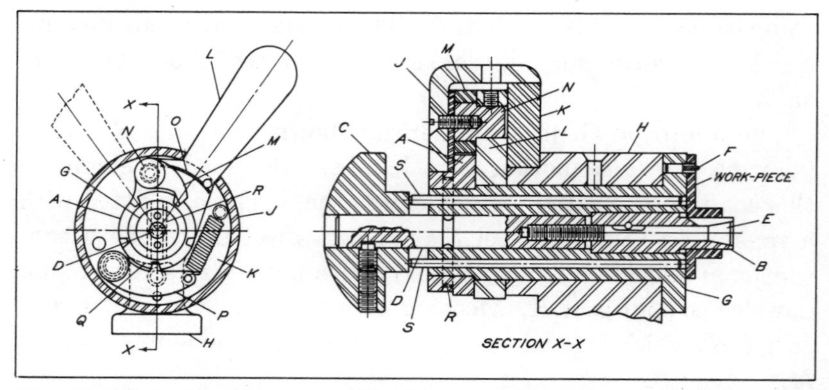

FIG. 23. Indexing fixture functions around the action of a notched-plate type ratchet wheel and two pawls — one driving and one locking.

knob is turned, the threaded end of tapered plug *E* is drawn into the shaft, causing the arbor to expand. Shaft *D* is restrained from sliding by dowel-pins *R*.

The work-piece illustrated is located through a hole in its flange by means of a diamond type locating pin *F*. This pin is pressed into the flanged face of rotating housing *G* — the entire subassembly being contained within fixture base *H*. The complete indexing mechanism is located between moving cover *J* and stationary plate *K*, and functions in the following manner.

Indexing of the work-piece is effected by movement of lever *L*. Pawl *M* rides on shoulder-stud *N* which, in turn, is locked to the enclosed part of lever *L* by a cone-point set-screw (Section X-X). Cover *J* also is locked to this shoulder-stud by means of a flat-head machine screw as shown.

When the lever is moved to the left, pawl *M* is disengaged from its notch in ratchet plate *A* and slides over to the next notch. Flat spring *O* is brazed to the pawl at one end and backed up by a pin at the other end to maintain downward pressure on the pawl at all times.

During this initial thrust of lever *L*, the ratchet plate is prevented from rotating by a tooth on the lower, spring-loaded pawl *P*. However, as the lever moves to the left, a cam surface *Q* at the lower end of the lever gradually disengages pawl *P*. Complete disengagement is timed to occur when pawl *M* drops into the next notch in the ratchet plate.

At this point, returning the lever to its original position will cause the ratchet plate to rotate clockwise a distance equal to the space between two adjacent notches. This indexing movement is imparted to housing *G* by two long dowel-pins *R* that connect the housing to the ratchet plate. As the lever moves to the right, the receding slope of cam surface *Q* permits spring-loaded pawl *P* to re-enter a notch in the ratchet plate, thus securing the new position of the work-piece.

After machining operations on the piece have been completed, it is released by first backing off knob *C* to relieve the expansive forces on stub-arbor *B*. Then, by striking the knob, ejector-pins *S* will move to the right and drive the work-piece off the arbor. Altering the number of index positions handled by this fixture would necessitate the replacement of ratchet plate *A* for one with the appropriate number of notches, and lever *L* for one with a modified cam surface *Q* that will effect engagement and disengagement of the lower pawl at the proper moment.

Half Revolution Geneva Mechanism

Three slots are the minimum number that can be used in the conventional type of Geneva mechanism. Or in terms of motion, 120 degrees is the greatest possible angle of rotation of the driven member for each revolution of the driver. It is for this reason that many designers resort to other mechanisms when 180 degrees of intermittent rotation is required of the driven member. For some applications, however, it is possible to use a modified form of the conventional Geneva mechanism, of the type shown in Fig. 24, so that the necessary half-revolution is obtained.

In Fig. 24, driver *A* is a circular disc to which pin *c* is bolted. For better performance a roller on a sleeve or an anti-friction bearing, instead of the pin, which is shown for simplicity, should be used. A segment *d* on driver *A* serves to lock wheel *B* in position during the idle period of the cycle.

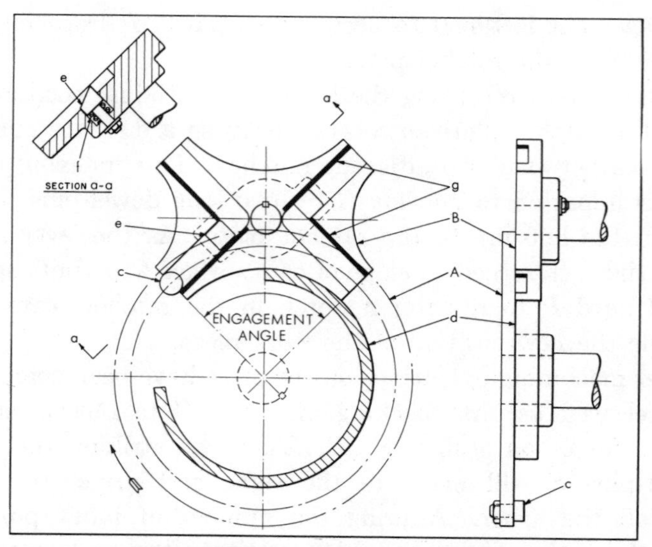

FIG. 24. **Diagram of half-revolution Geneva mechanism showing driving pin about to enter slot in driven member.**

Wheel *B* has two V-shaped slots. When the wheel is in the idle position, the center line of one leg of each slot is tangential to the circular path of the roller, as shown. Mounted on wheel *B* are two spring-loaded dogs *e*, one in each of the V-slots. These dogs have beveled tops, and, under pressure of driving pin *c*, are forced into a recess in the V-groove so that the pin may pass over them.

When driver *A* rotates, pin *c* enters the slot which lies in its path, and begins to turn wheel *B*. Also, at this point, the segment *d* passes the center line, leaving the wheel free to rotate as the pin continues to enter the slot. Approaching the center of the V-slot, the pin passes over dog *e*, pressing it down as it passes. In Fig. 25 is shown the position of the mechanism when pin *c* has reached the bottom of the V-slot. In this position the pin has passed over the dog and the dog has been returned to its initial extended position by the force exerted through spring *f*, shown in section a-a, Fig. 24.

When the pin is rotated further, it presses against the vertical side of the dog, which now forms an extension of the side of the

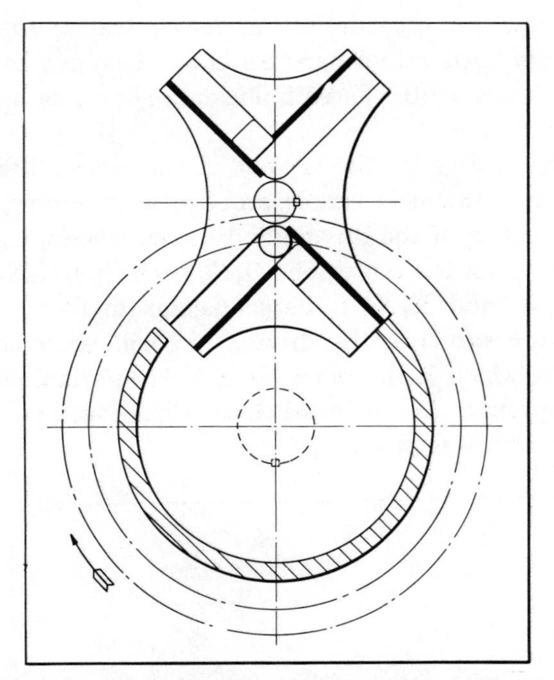

FIG. 25. When the driven wheel has been rotated 90 degrees, as shown, the driving pin has passed over the dog and is at the bottom of the V-slot.

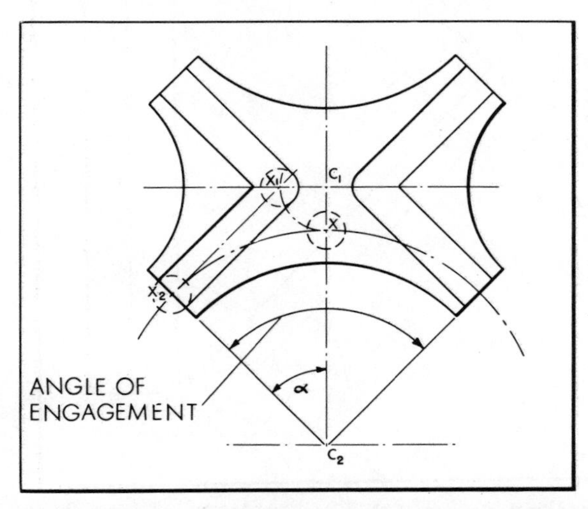

FIG. 26. The correct proportions for the Geneva mechanism involve simple geometric considerations.

V-slot. In practice, the sides of the slot exposed to the pressure of the pin are lined with a hard material which can be replaced when it becomes worn. These linings are shown as heavy lines marked *g* in Fig. 24.

As shown in Fig. 26, the lay-out of the mechanism may be considered in terms of a simple problem in geometry, namely: Given the centers of the driving and driven wheels, C_1 and C_2, find a point X on the center line C_1C_2 such that $C_1X_1 = C_1X$, $C_2X_2 = C_2X$, and X_1X_2 is perpendicular to C_2X_2. If these conditions are satisfied, the driving pin will enter the V-slot tangentially, which is the most favorable condition, and it will be at the bottom of the V-slot when the driven member has been turned through 90 degrees.

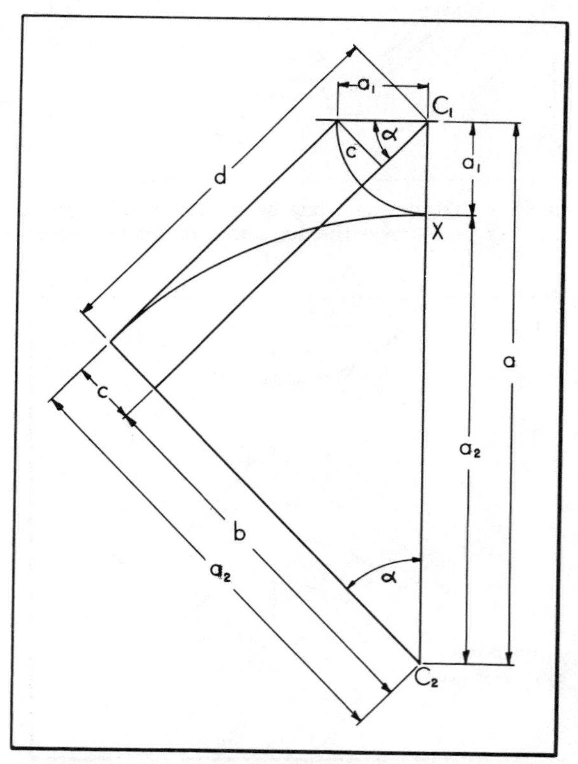

FIG. 27. Simplified lay-out of the mechanism illustrated in Fig. 24, which is employed to determine its proportions.

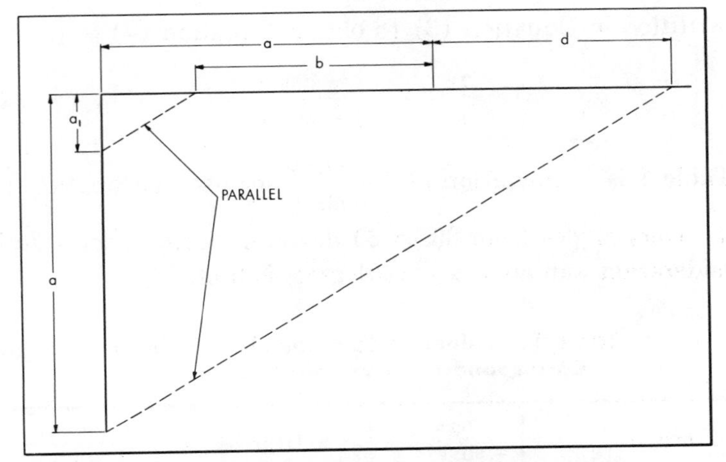

FIG. 28. Graphical method of determining length a_1 shown in Fig. 27.

In Fig. 27, the basic geometric form shown in Fig. 26 has been redrawn with two auxiliary lines added. From the illustration,

$$\frac{d}{a} = \frac{c}{a_1} \tag{1}$$

$$c = a_2 - b = a - a_1 - b \tag{2}$$

Substituting in Equation (1) the value of c from Equation (2), and simplifying,

$$\frac{d}{a} = \frac{a - a_1 - b}{a_1}$$

$$a_1 d = a^2 - aa_1 - ab$$

$$a_1 d + aa_1 = a^2 - ab$$

$$a_1 (a + d) = a(a - b)$$

$$\frac{a_1}{a} = \frac{a - b}{a + d} \tag{3}$$

This last equation suggests the graphical method shown in Fig. 28 as a means of finding length a_1 in Fig. 27. Instead of this graphical solution, however, an analytical method may be used: Since $b = a \cos \alpha$ and $d = a \sin \alpha$, these values for b and d may be substituted in Equation (3) to obtain Equation (4):

$$\frac{a_1}{a} = \frac{1 - \cos \alpha}{1 + \sin \alpha} \qquad (4)$$

Table 1 is a tabulation of $\frac{1 - \cos \alpha}{1 + \sin \alpha}$ for use in Equation (4), and covers angles from 30 to 60 degrees. From general design considerations, an angle $\alpha = 45$ degrees is best.

Table 1. Values of $(1 - \cos \alpha)/(1 + \sin \alpha)$ Corresponding to Various Values of α

α, Degrees	$\frac{1 - \cos \alpha}{1 + \sin \alpha}$	α, Degrees	$\frac{1 - \cos \alpha}{1 + \sin \alpha}$
30	0.08931	46	0.17759
31	0.09427	47	0.18367
32	0.09932	48	0.18981
33	0.10445	49	0.19601
34	0.10964	50	0.20227
35	0.11493	51	0.20858
36	0.12028	52	0.21495
37	0.12571	53	0.22138
38	0.13121	54	0.22786
39	0.13677	55	0.23441
40	0.14242	56	0.24101
41	0.14812	57	0.24766
42	0.15389	58	0.25436
43	0.15972	59	0.26113
44	0.16561	60	0.26795
45	0.17157	..	......

The angular displacement, angular velocity, and angular acceleration of modified Geneva mechanisms having 60-, 90-, and 120-degree engagement angles are shown in Figs. 29, 30, and 31. The velocity and acceleration curves for each of these mechanisms are based on the driving member having a uniform angular velocity of 1 radian per second (9.55 rpm). The velocity curves shown were obtained by graphical differentiation of the displacement curves, and the acceleration curves by graphical differentiation of the velocity curves, since the equations involved in an analytical solution were not easy to handle.

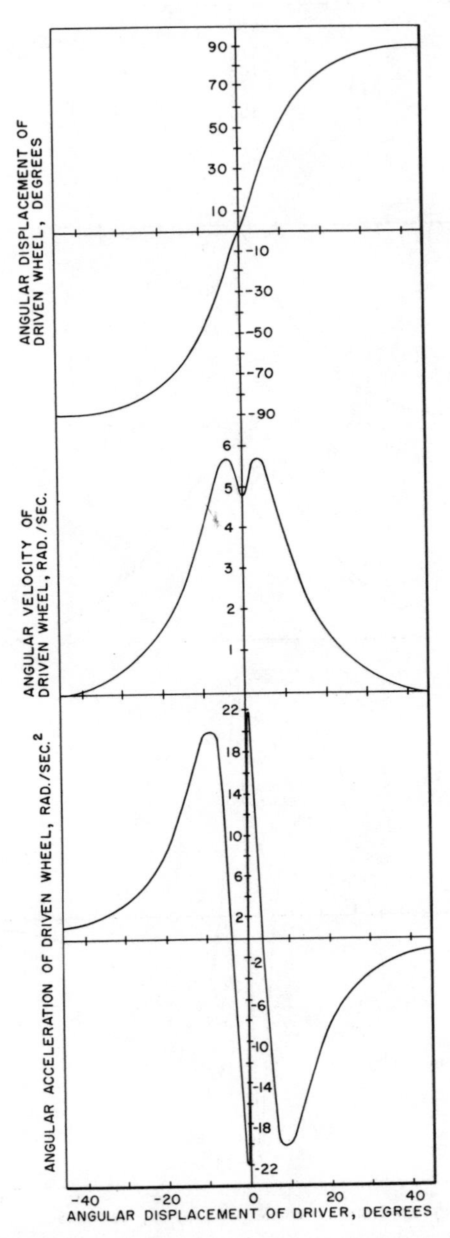

FIG. 29. Displacement, velocity, and acceleration diagrams for a half-revolution Geneva mechanism having an engagement angle of 60 degrees.

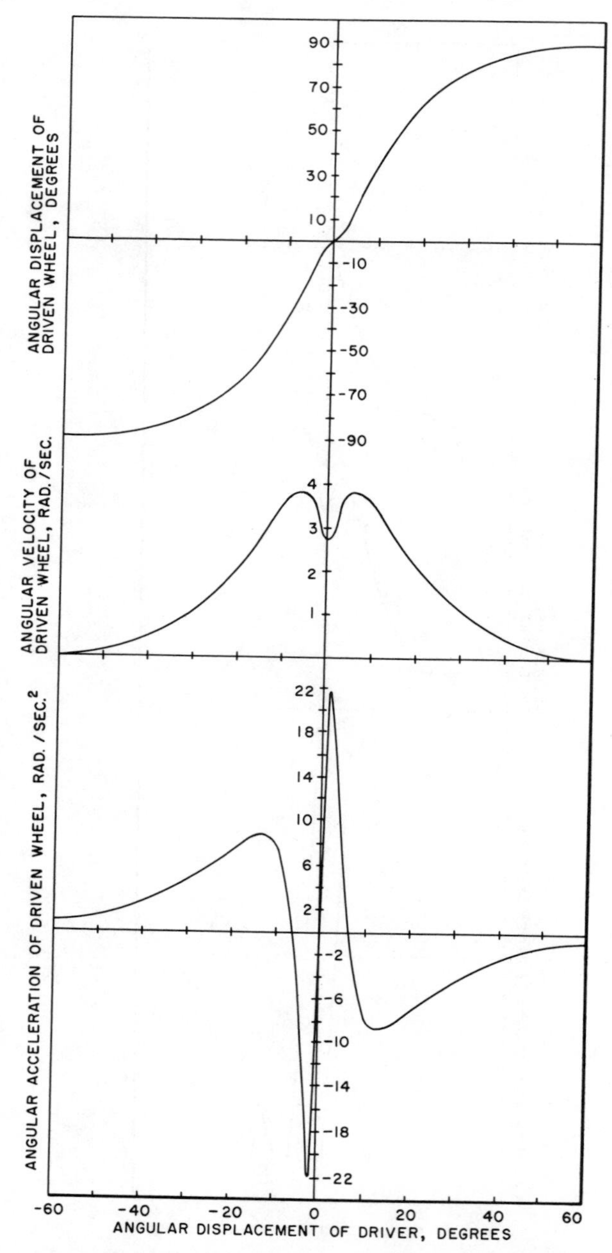

FIG. 30. Displacement, velocity, and acceleration diagrams for a half-revolution Geneva mechanism having a 90-degree engagement angle.

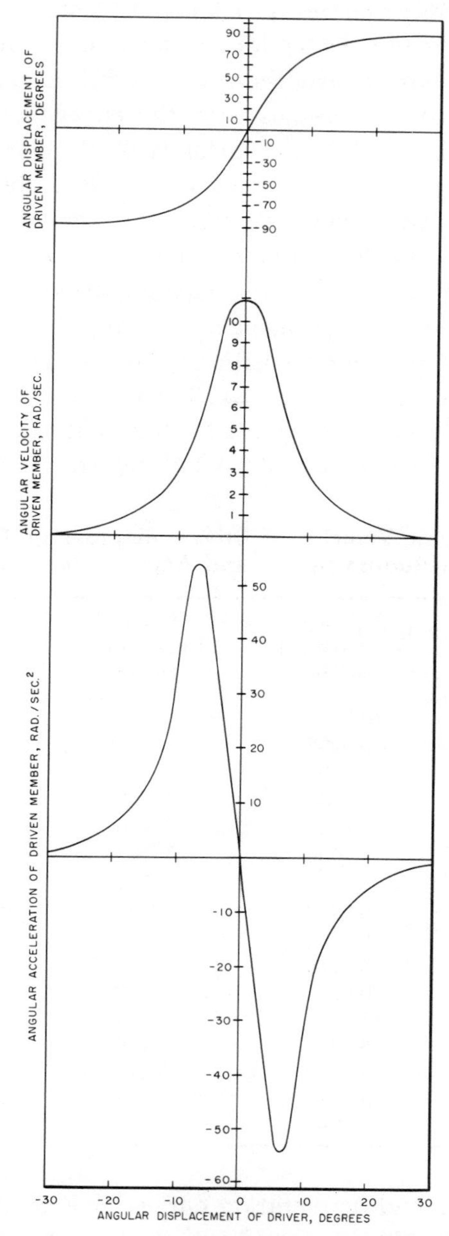

FIG. 31. Displacement, velocity, and acceleration diagrams for a half-revolution Geneva mechanism having an engagement angle of 120 degrees.

The displacement curves have a slight "bend" near the center position which is responsible for two maximums and a minimum in the velocity curves near the center position. The "hump" in the velocity curves decreases with the engagement angle, and disappears entirely when the latter is 60 degrees. In practice, this change in velocity occurring over a very short portion of the cycle will cause some roughness in the operation of the mechanism around the center position.

The curves also show that the maximum velocity of the driven member, for angles of engagement greater than 60 degrees, is not achieved at the center position. The angular velocity in the center position equals a_2/a_1 which value is $[(1+\sin\alpha)/(1-\cos\alpha)]-1$. Table 2 gives values of these velocities for a number of engagement angles from 60 to 120 degrees.

Table 2. Angular Velocity of Driven Member in Center Position Based on 1 Radian per Second Angular Velocity of Driver

Angle of Engagement, Degrees	Velocity, Radians per Second	Angle of Engagement, Degrees	Velocity, Radians per Second
60	10.9695	92	4.6309
62	9.6078	94	4.4445
64	9.0684	96	4.2684
66	8.5739	98	4.1017
68	8.1207	100	3.9438
70	7.7009	102	3.7943
72	7.3139	104	3.6522
74	6.9548	106	3.5171
76	6.6213	108	3.3886
78	6.3115	110	3.2660
80	6.0214	112	3.1492
82	5.7512	114	3.0377
84	5.4981	116	2.9314
86	5.2609	118	2.8295
88	5.0382	120	2.7320
90	4.8285	...	....

The velocity and acceleration curves in Figs. 29, 30, and 31 may be used to obtain the velocity and acceleration values for

N revolutions per minute of the driving member by multiplying the ordinates on these curves by $\left(\frac{\pi N}{30}\right)$ and $\left(\frac{\pi N}{30}\right)^2$, respectively.

Ninety Indexes per Minute

Simple in design, the high-speed indexing mechanism illustrated in Fig. 32 is designed for rotary type transfer machine applica-

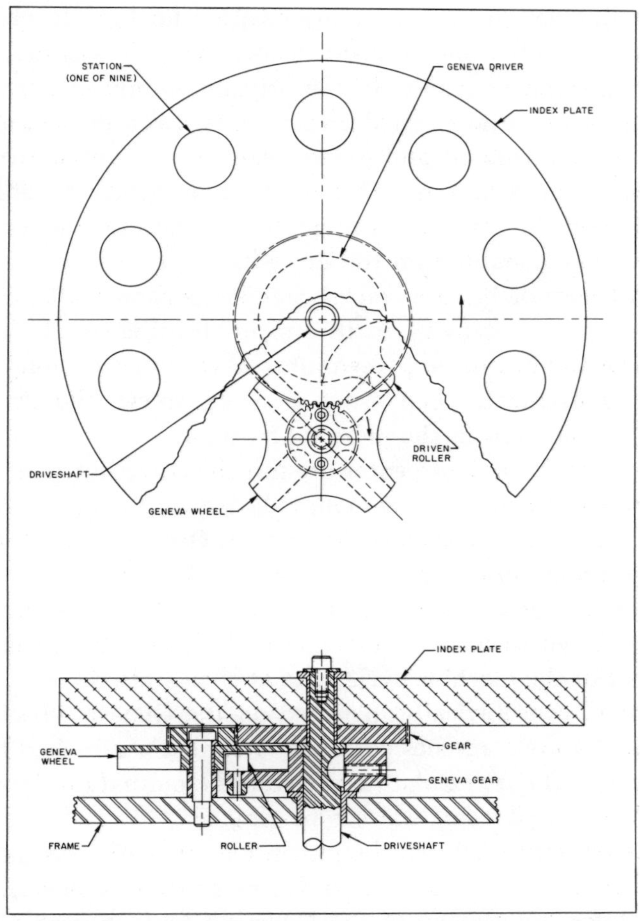

FIG. 32. Driven roller indexes Geneva wheel which rotates the index plate at a slower rate through reducing gears.

tions in which the index plate and the driveshaft are concentric. This arrangement permits tool slides at any or all stations to be actuated from cams mounted on the driveshaft. Each revolution of the driveshaft indexes or cycles the machine to the next station. Operational speeds up to ninety or more indexings per minute can be achieved.

Indexing is in a period equal to one-quarter of the total cycle time for each station regardless of the number of stations.

While this design can be readily adapted for light machining, stamping, assembly and inspection operations, it was first used on an inspection machine. In the original equipment, the part to be inspected is hopper fed and automatically loaded, gaged, rotated, gaged, rotated and gaged again in six consecutive stations. Ejection occurs at either of the four remaining stations depending on the results of gaging. All operations are actuated by three cams mounted on the driveshaft.

A front view of the original mechanism is shown with part of the index plate cut away in the upper drawing on the facing page. The Geneva drive is seen after having just completed an indexing movement. The lower drawing shows the drive at exactly its midpoint in the indexing motion.

All machine functions center around the driveshaft, on which any number of cams can be mounted. Keyed to this shaft, the Geneva driving member moves the driven Geneva wheel through a 90-deg. arc in 90 deg. of its own travel. The concentric diameter of the driver and the concave cutouts in the driven member mesh as shown to locate the Geneva wheel radially during the remainder of the machine cycle time at each station.

A pinion, concentrically mounted on the Geneva wheel with screws and dowels, meshes with a gear similarly attached to the index plate. The index plate and gear are mounted on the driveshaft by means of a single sleeve bearing.

In the original machine, the pinion has 28 teeth and the gear has 70 teeth. As a result, the index plate rotates 36 deg. with every 90-deg. movement of the Geneva wheel. Suitable gears could, of course, be selected to produce the angular rotation necessary for the other members of index plate stations.

Sleeve bearings and the Geneva drive provided sufficient radial accuracy for the purpose of the original machine. Needle roller bearings and an auxiliary shot bolt operated from a cam on the driveshaft could yield greater positioning accuracy.

CHAPTER 4

Overload, Tripping, and Stop Mechanisms

Mechanisms which automatically operate to stop an operation when overload occurs, to trip and start a new sequence or operation when a certain position or part of a cycle is reached, or to bring an operation to a halt at the end of a given cycle or when a given amount of motion has occurred, are described in this chapter. Other mechanisms performing similar functions are described in Volumes I, II and III of "Ingenious Mechanisms for Designers and Inventors."

Shock Absorber for a Rotating Shaft

Shock loads are isolated from the driving gears by the mechanism shown in Fig. 1. Driving gear *C* rotates gear *B* which has a slide fit over shaft *A*. Stud *D* which is rotatably attached to *B* transmits motion to collars *H* which in turn transmit motion to shaft *A* through compression springs *G′* and *G* and bracket *E*. Springs *G′* and *G* are located over rods *F′* and *F* which can slide in retaining holes in bracket *E*. Collars *H* retain the spring.

Should shaft *A* receive a shock, bracket *E* will be caused to rotate with respect to gear *B* and one or the other spring will be compressed. As the springs transmit the motion, the force of the shock will be limited.

An Overload Slipping Ball-Clutch

The diagrams in Fig. 2 show the effective design, construction and operation features of an adjustable slipping type of ball-clutch, which was successfully incorporated in the original drive

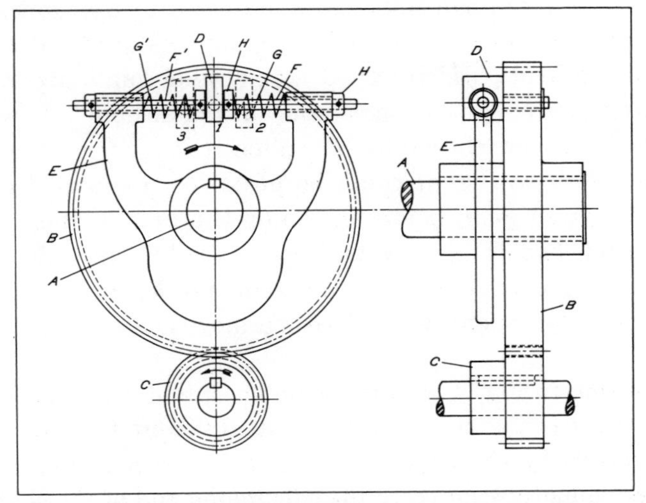

FIG. 1. The mechanism is able to isolate shock loads from the gearing without interfering with the over-all timing of the shaft.

transmission of a machine to safeguard the driven elements against overloading. The clutch was required to replace an existing positive chain-sprocket type clutch for transmitting the drive between two shafts mounted in axial alignment with each other.

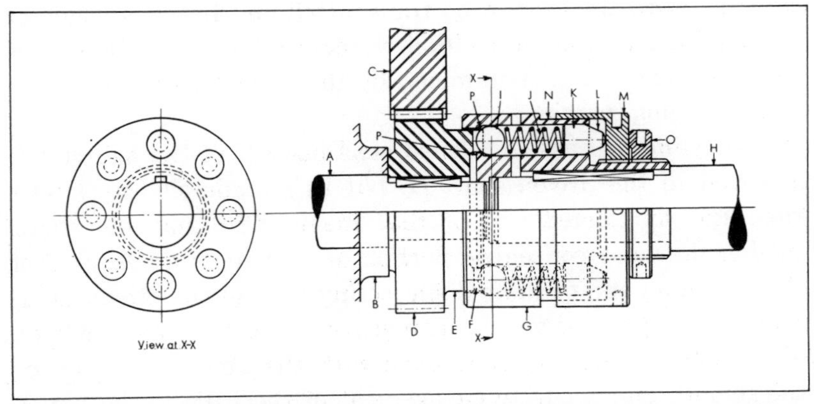

FIG. 2. Contained balls *I* are pressed by spring *J* into notches in a flange of gear *D*. Tension is adjusted by rotating nut *M* which shifts part *K*.

The new clutch had to be easily adjusted to transmit a range of different torques and also for setting to slip at various predetermined loads, which varied within wide limits according to the particular operation carried out on the machine. It was also essential for the clutch to be an entirely self-contained and compact unit so that it could be preset on a special fixture to slip at a given load, before installing the unit in the machine. The clutch had to operate with equal facility and efficiency in either direction of rotation, at different speeds, and to accommodate itself for slight axial floating movements of the driven shaft.

Referring to Fig. 2, *A* is the main shaft and is mounted in the horizontal bearing *B*. It is driven at different forward and reverse speeds by the train gear *C* meshing with pinion *D* keyed and permanently secured to the shouldered end of the shaft. The right-hand side of the pinion has an integral flange *E* which is slightly hollowed out on its end face to leave a narrow annular band at the periphery. A series of radial vee shape serrations, *F*, of identical size and shape and at equal pitch spacings apart, is milled across the annular band as shown.

In the illustrated example, the sides of the serrations are inclined 37½ degrees relative to the centre axis of the pinion that is, with an included angle of 75 degrees. This important dimension can, of course, be varied within certain limits in accordance with the load to be transmitted and the magnitude of the overload at which the clutch is required to slip. The size of the serrations is also determined by the diameter of the driving balls engaging therein.

The cylindrical case-hardened steel body *G* of the ball clutch is keyed to the driven shaft *H*, but it is made slightly longer than the shouldered end of that member so that the short, smaller diameter concentric portion of its bore is a slip fit over the adjoining end of shaft *A* projecting beyond the pinion *D*, as shown. The purpose of this arrangement is to maintain the body of the clutch perfectly concentric with the pinion for ensuring the smooth and accurate engagement of the balls in the pinion serrations. The left-hand end of the body is recessed a small depth to admit the serrated portion of flange *E*, the outside diameter of which has a tight clearance fit in the recess. This overlapping part of the body serves to enclose the serrations and prevents the ingress of dirt and cuttings.

With this particular example, eight hardened and ground steel balls, *I*, are employed, each being fitted closely in a drilled and reamed hole passing axially through the body. The holes are spaced exactly 45 degrees apart around the same pitch circle, the diameter of which is equal to the pitch diameter of the serrations in the annular band of the flange *E*, thus the balls are disposed radially so as to engage centrally in the width of the serrations, as shown in the half-section diagram. The number and diameter of balls may be varied to suit loading requirements. A stiff coil spring *J* is interposed in each hole behind the ball and backing against the hardened steel plunger *K* fitted in the opposite end of the hole. All the plungers are the same overall length, and the conical portion *L* normally projects about ⅜ inch beyond the body. The end face of each plunger is slightly domed and well polished after hardening.

The right-hand end of the body is reduced in diameter and threaded to receive the hardened steel sleeve *M* of the same outside diameter as the front end of the body. The sleeve is deeply bored at one side to be a close fit over the reduced portion *N* ground on the outside of the body. By fitting the sleeve over the body at that point its correct and accurate location relative to the body is not determined by the fit in the threads. The eight plungers bear simultaneously against the inner left-hand face of the sleeve; thus as that member is adjusted longitudinally, all the springs *J* will be compressed or expanded the same amount. A smaller threaded ring *O* is also screwed on the body behind the sleeve for locking the latter member in any desired setting.

One of the main disadvantages of ordinary type ball-clutches is that when the driving and driven elements are separated, the balls can move completely out of the body of the clutch, and for that reason they generally cannot be preset on the bench or in a fixture, since there is no easy means available for holding the spring-loaded balls in the correct operating position. With the design of clutch described, this limitation is eliminated in a simple yet effective way.

Before machining the shallow recess in the left-hand side of the body, the eight axial holes for receiving the balls and spring were drilled, starting in from the right-hand end of the body and extending to a carefully predetermined depth, namely, to within about 3/8 inch of breaking through the opposite end. The recess was then bored to a controlled depth to ensure breaking into the eight holes a certain amount so as to leave the small lips *P* as shown in the half-sectioned view. When the balls correctly mesh in the serrations for driving purposes, the lips *P* are approximately 1/32 inch clear of the balls, thus allowing them to make full contact. As the body is moved axially away from the pinion, a very small amount, each ball is pressed against the lips and thus cannot move farther out of the body, which member can then be removed without fear of displacing or losing the balls and springs.

To set the removed body for slipping at a different overload, it is simply mounted on a keyed plug fastened in a fixture and sleeve *M* is adjusted in the appropriate direction, during which the balls will remain pressed against the lips *P*; thus the setting operation is simple, rapid and reliable.

To meet another application, this design of ball-clutch was slightly modified for transmitting the drive between two shafts in conditions where the driven shaft was required to operate for certain periods in a slightly axial off-set relationship to the driving shaft. The amount of eccentricity of the shafts varied from zero to 0.050 inch. This requirement prevented the use of any type of rigid clutch or one employing a chain and sprockets.

The modification consisted of shortening the overall length of the clutch body to equal that of the shouldered end of the driven shaft on which it was to be fitted. That eliminated the short, smaller diameter bearing portion at the left-hand end of the body for fitting over the end of the driving shaft. The shallow recess in that end of the body was also machined suitably larger than the diameter of flange E to allow for the above-mentioned degree of off-setting.

Safety Devices Protect Slides Against Overloads

In the design of a machine, it may be important to protect a slide and its elements against excessive overloads. Where the slide is driven from an oscillating shaft, either of the two safety devices illustrated in Figs. 3 and 4 proves highly satisfactory. For clarity, the slides themselves have been omitted from each illustration.

In Fig. 3, the movement of the slide, connected to the right end of link A, is transmitted from an oscillating drive-shaft B through a lever C. Lever C swings in an arc, an equal distance to each side of vertical. The lever and link are joined by a pin

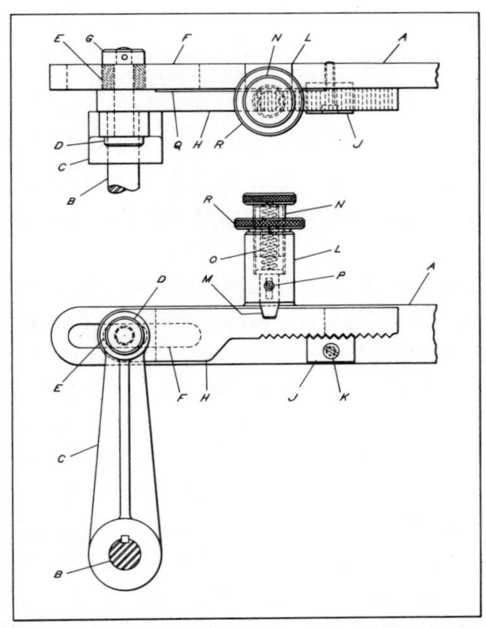

FIG. 3. The original path of transmission through this safety device is resumed automatically when the overload has been removed.

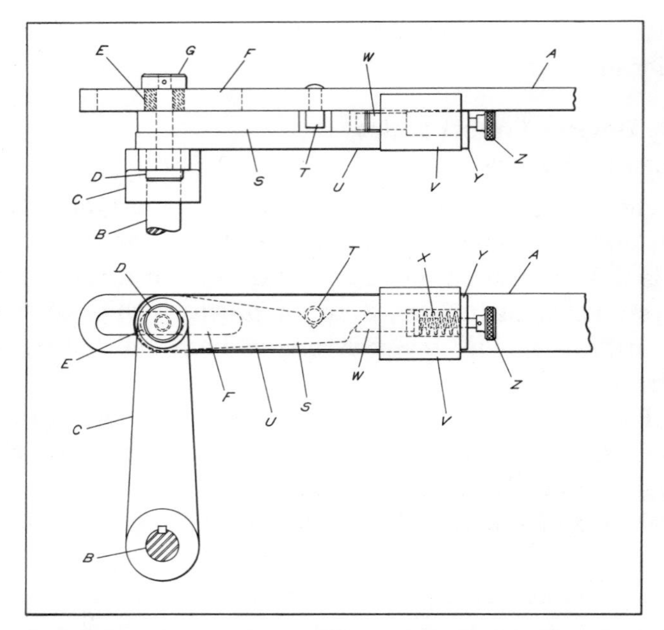

FIG. 4. This safety device can be used where a stud projecting above the links would be impractical.

D which passes through close-fitting holes in the end of the lever and in a bushing *E* located in a slot *F* in the link. A collar *G* is doweled to the protruding end of the pin, keeping the bushing in position.

The pin also passes through a hole in the left-hand end of a second link *H*. This link is shorter and somewhat narrower than link *A*. Also, its right half is further reduced in width, and has a fine-pitch tooth rack machined out. The teeth have a 45-degree side angle and engage mating teeth cut along the upper edge of a rectangular block *J*. This block is thicker than link *H* so that it can be registered in a shallow depression in the face of link *A* to which it is secured by a set-screw *K*.

Welded to the top edge of link *A* is a hollow stud *L*. A detent *M* in the bottom of the stud has two flat surfaces fitting a tapered slot cut across the top edge of link *H*. The top of the stud is tapped to receive an adjusting screw *N*. A blind hole in this screw contains a spring *O* which forces the detent to remain in the slot, thus keeping the tooth rack and block in engagement. Within the stud, a key *P* prevents the detent from rotating.

In the normal operation of the slide, there is no independent travel of the bushing *E* in the slot *F*. Instead, the path of transmission from lever *C* to the slide is through link *H* and block *J* to link *A*. Should the slide become overloaded, further movement of lever *C* and link *H* in either direction will raise the detent out of the channel, and the tooth rack will disengage block *J*.

Lever *C* and link *H* are then free to continue their movements, with the bushing *E* now traveling back and forth in the slot *F*. (The length of the slot is made slightly more than the stroke of the slide.) Since no movement is transmitted to link *A*, the slide is protected.

Once the overload is removed, the original path of transmission is resumed automatically. Adjusting screw *N* is pre-set to have spring *O* impart just enough pressure to permit the detent to be raised at a specified amount of overload. A jam nut *R* maintains the setting of adjusting screw *N*.

The safety device shown in Fig. 4 can be used where, because of space limitations, a stud cannot be located over link *A*. Here, link *A*, drive-shaft *B*, lever *C*, pin *D*, bushing *E*, slot *F*, and collar *G* are the same as their counterparts in Fig. 3, and similarly identified.

The left end of link *S* (corresponding to link *H*, Fig. 3) fits over pin *D*. The upper edge of link *S* tapers to the right, and also forms a 90-degree vee around a hardened pin *T* pressed into the face of link *A*. A third link *U*, also fitting over pin *D*, is considerably longer than link *S*, and has an integral block *V* at its right-hand end. The back of this block straddles the upper and lower edges of link *A*.

A plunger *W*, contained in a hole in the block, has a 45-degree point, matching a 45-degree slope on the end of link *S*. A spring *X*, which is retained by a cover plate *Y*, causes the plunger to exert a pressure contact against link *S* so that the sides of the vee normally bear on pin *T*.

In operation, the path of transmission from lever *C* to the slide is through link *S* and pin *T* to link *A*. Should the slide become overloaded, the resistance of *A*, acting through *I* pushes *S* down and temporarily pushes *W*.

Lever *C* and link *S* are then free to continue their movements, with the bushing *E* now traveling back and forth in the slot *F*. No movement is transmitted to link *A*. But unlike the first device, the original path of transmission here is not resumed automatically once the overload is removed. Instead, the knurled knob *Z* on the end of the plunger shaft must be pulled out while lever *S* is raised to position. The rear of the knob is milled flat in order to clear link *A* and keep the plunger point in alignment with the slope on the end of link *S*.

Safety Overload Mechanism Permits Adjustable Dwell on Reciprocating Drive

Instant, safe, and automatic disengagement of the drive for a reciprocating machine slide when subjected to excessive load is obtained by means of the mechanism illustrated. A useful feature of this safety overload mechanism is that the period of dwell at each end of the reciprocating stroke can be varied by a simple adjustment. Also, the drive is instantaneously and automatically re-engaged when the overload has been removed. The mechanism is smooth and quiet in operation.

Driving lever *A* (see Fig. 5), which is fastened at its lower end to an oscillating shaft (not shown), swings through a constant arc. The upper end of the lever is bored to be a free swiveling fit on the cylindrical boss *B* of bracket *C*. Collar *D* is pinned to the boss to retain the lever without binding. The rectangular body of bracket *C* is slotted to hold rectangular sliding member *F*. Slide *F* is retained in the slot by a plate *G*, which is secured to the body by four screws.

Connecting rod *H* can slide through a hole in *F* for a distance determined by the positions of lock-nuts *J*. The length of dwell varies with slide distance. The end of the connecting-rod is attached to the reciprocating slide of the machine (not shown) by shaft *K*.

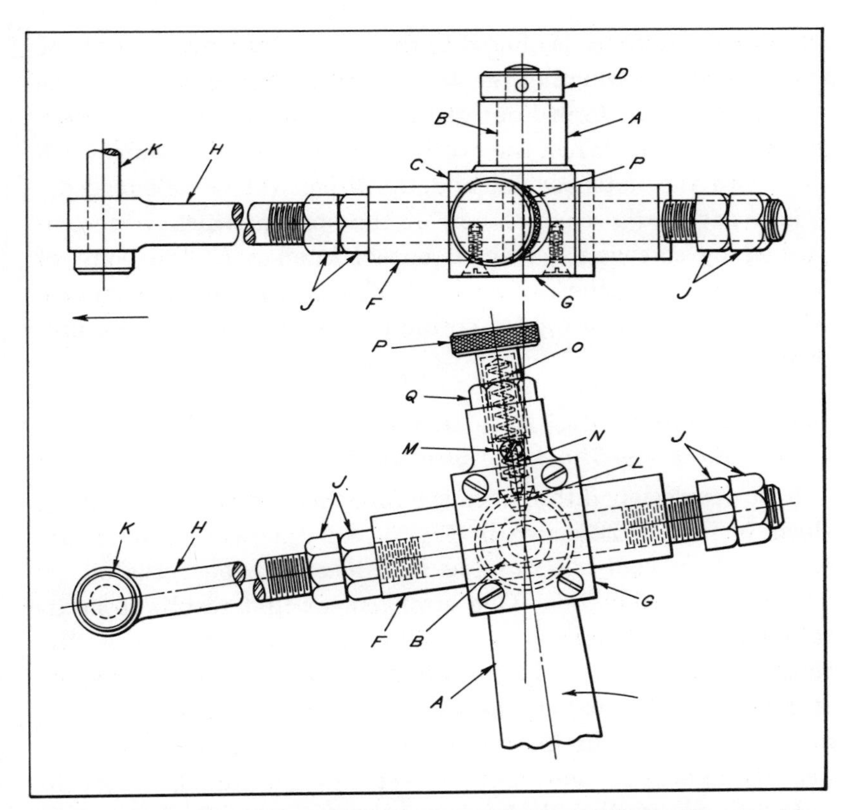

FIG. 5. The drive from oscillating lever A to a reciprocating machine slide which is attached to shaft K is disengaged by a spring-loaded plunger L when overloads are applied to the slide.

Slide F is connected to bracket C by a spring-loaded plunger L, the tapered nose of which fits into a V-notch machined across the slide. The plunger is a sliding fit within a boss on top of the bracket. It is prevented from rotating by a dog-point set-screw M which enters a shallow keyway N cut along the slide of the plunger. Spring O is seated in a blind hole in the plunger, and retained by a knurled-head adjusting screw P, which engages a threaded hole in the bracket boss. Screw P is held in any desired setting by lock-nut Q.

In operation, slide F and connecting-rod H move with driving lever A, thus reciprocating the machine slide. However, when additional resistance is offered to the horizontal movement of the machine slide — whether on its forward or return stroke — the plunger *L* will be forced out of the notch in slide *F*, thus disengaging the drive. When the overload is removed, the plunger will snap into the notch again, and the drive will be re-engaged.

By varying the compression of the spring, which is accomplished by screwing *P* into or out of bracket *C*, the point of loading at which the drive will be disengaged can be changed. Also, a heavier or a lighter spring can be used to suit requirements.

Device Reduces Initial Acceleration of Flying Shear — *Shockless Startup of Inertia*

A device designed to reduce the force necessary to set a flying shear in motion is shown in Fig. 6. In practice, flying shears employed on cold-roll forming machines may be actuated by any of several methods. One arrangement that results in accurate cutting-to-length of the roll-formed sections allows the shearing mechanism to be started and pulled by the rolled strip. In many cases, however, the strip does not have sufficient stiffness to overcome the inertia of the shear without buckling. By permitting acceleration to occur over a longer period, the mechanism shown in the illustration decreases the initial force required to bring the shear up to the speed of the strip and, thereby, reduces the tendency of the strip to buckle.

The mechanism is arranged as follows: A light flag, or lever, mounted on a track at the outgoing end of the strip runout table, is connected to the flying shear *A* by a cable to a hinged lever *B*, see Fig. 6. This lever, in turn, is attached to a second lever *C*. As the roll-formed section starts the flag moving, a roller *D*, mounted on lever *C*, pushes against a bar *E* attached to the machine base. This setup will gradually accelerate the shear carriage, which is mounted on rollers or on machine ways. The shearing mechanism is pneumatically or hydraulically operated to sever the rolled strip.

When lever *C* reaches the vertical position, it comes into contact with a stop and the roller leaves bar *E*. The shear carriage

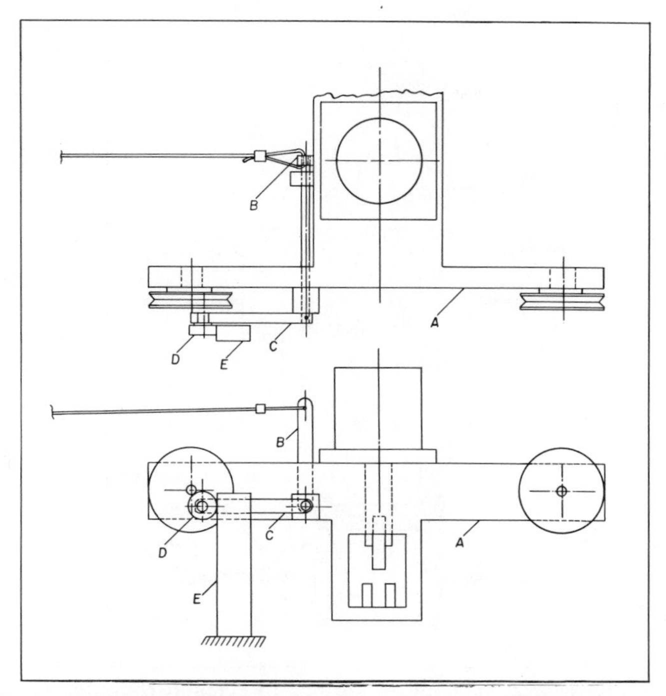

FIG. 6. Device that reduces the initial force necessary to accelerate a flying shear. When lever *B* reaches a horizontal position the shear *A* and the strip are moving at the same speed.

has then been accelerated to the same speed as the moving strip and is being pulled by it. At this point, the shearing mechanism is immediately actuated by a micro switch and the flag is triggered, releasing it from the end of the strip. A spring returns the shear to the starting position and the cycle is repeated. Lengths of the lever arms may be varied to suit the acceleration required.

Springs Cushion Shock Loads in Gear Drive

Shock loading of a gear train in either direction can be greatly reduced with the arrangement shown in Fig. 7. The mechanism features a drive that operates under increasing spring

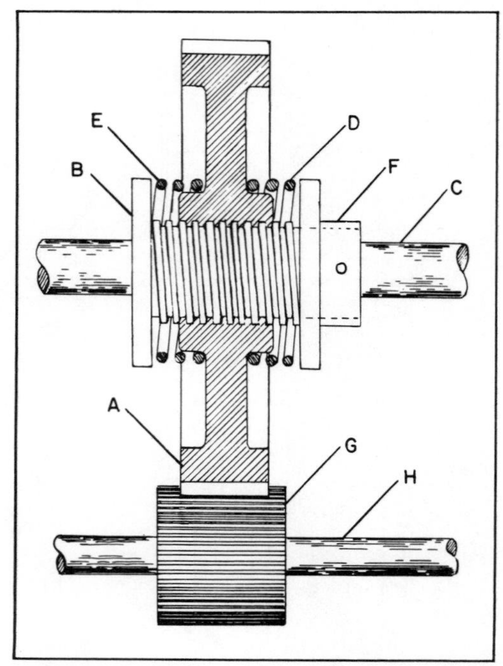

FIG. 7. Gear drive that employs springs to cushion the effects of shock loading in either direction.

pressure as greater angular displacement or slippage occurs between the driving and the driven shafts.

A square thread is machined into the bore of gear *A* to fit a threaded sleeve *B*, which is press-fitted on shaft *C*. A short compression spring *D*, gear *A*, and a second similar spring *E* are mounted on the sleeve in that order. This assembly is held together by a stop-collar *F*, which is pinned in place on the sleeve, the pin passing through the collar, the sleeve, and the shaft.

The pinion *G*, which is mounted on shaft *H*, should be made wide enough to insure full tooth contact with gear *A* as it moves to either the right or the left. To prevent excessive lateral movement of gear *A*, the springs should be under compression when it is centered on the pinion. The springs should be heavy enough to prevent the shock load from causing the gear hub to jam against either the collar or the flange of the sleeve. This condition would render the shock-absorbing feature of the mechanism inoperative.

In normal operation, when the machine is started under load, the gear will move laterally and compress one spring until the initial loading is overcome. As the machine picks up speed, gear *A* will move back toward the center of the pinion, the distance depending on the running load applied by the machine. Intermittent shock loading of the machine will cause gear *A* to move back and forth on the threaded sleeve. Loading of the drive in the opposite direction will cause the gear to compress the other opposing spring with similar results. If the shock loading is in one direction, the mechanism may be modified to operate with only one spring.

This arrangement is being successfully employed on a drive for a tumbling barrel, the springs having been selected by trial and error. A 0.250-pitch thread and a 10-pitch gear train are used. The driving gear *G* and the driven gear *A* are 3 and 9 inches in diameter, respectively, each gear being mounted on a ⅞-inch-diameter shaft. A ¾-hp, 60-rpm, geared head motor drives the machine.

Torque-Controlled Drive Release for Tapping

A chuck that automatically disengages the drive when a pre-set torque is applied to the tap is shown in Fig. 8. When properly adjusted, this device can effectively reduce tool breakage.

Spindle *A* has equally spaced slots for axial location of three keys, the outer edges of which engage the inclined bases of keyways in sleeve *C*. The flanged lower end of this sleeve is fitted with three bushings with tapered bores for part of their lengths. For driving purposes, these bushings are held in engagement with steel balls *D* by the action of the compression spring *E*. The balls are housed in pockets in the base of the large-diameter bore at the upper end of body *F*.

Spring *E* can be pre-set to release at the maximum torque that can be applied to a tap by adjusting the internally threaded cover *G*, which threads onto the body *F*. When this setting

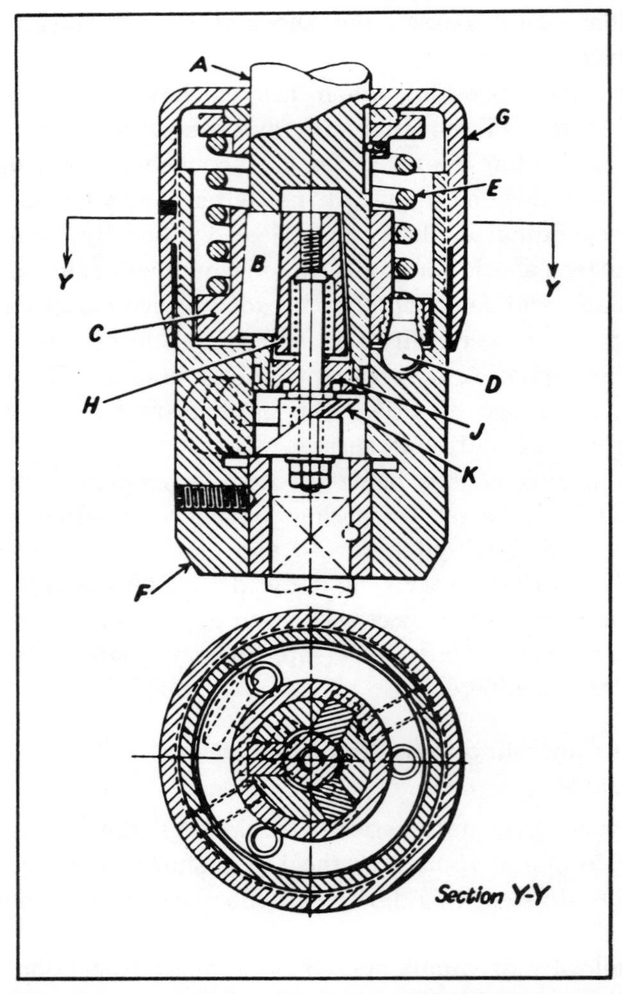

FIG. 8. Excess torque transmission is prevented by sleeve C rising to clear the driving balls D. Re-engagement is prevented by the outward movement of keys B.

has been made, the cover is secured to the body by means of a set-screw.

The inner edges of keys B make contact with cone H, which is housed in the tapered bore of spindle A. When the pre-set torque is applied to the tap, continued rotation of the spindle causes the sleeve C to be moved upward against the action of spring *E*. When the bushings rise high enough to clear balls *D*, the drive is disengaged. At the same time, cone *H* is caused to move upward by the action of a second compression spring, the lower end of which bears against the threaded plug *J* in the spindle. As a result, the keys are moved radially outward by the wedging action between their inner edges and the cone. In this way contact is maintained between the keys and the bases of the keyways in sleeve *C*. The cone and the keys have a taper of 1 in 20 and are self-locking. The result is that any downward movement of the sleeve and, consequently, engagement of the drive are prevented.

When the spindle has been stopped, the sleeve can be released to bring the bushings and the balls *D* into engagement again by depressing a plunger which passes through a cross-hole in the body *F*. This action causes wedge *K*, which is attached to the plunger, to be moved at right angles to the spindle axis and into contact with a mating wedge. The latter is carried on the lower end of a pin attached to member *H* and is therefore caused to move downward to give the releasing action.

The tap is mounted in a bushing in the lower end of body *F*. A cross-pin, which passes through a hole formed partly in the square-shaped end of the shank and partly in the bore of the bushing, holds the tap in place.

Mounting Provides Double Action for Compression Spring

A compression spring can be mounted so that either a push or a pull will put the spring under compression. In Fig. 9, spring *A* is contained between washers *B* and *C*, and is secured to the end of shaft *D* by fillister-head screw *E*.

While shaft *D* remains stationary, force is applied to shaft *F*, which is pinned to one end of step-bored thimble *G*. The other end of the thimble has an internal thread engaging a slotted externally threaded bushing *H* which has a slide fit over shaft *D*.

When shaft *F* is pulled to the right, thrust is transmitted through the set-screw and washer *B* against the left end of the spring. Or, when shaft *F* is pushed to the left, thrust is trans-

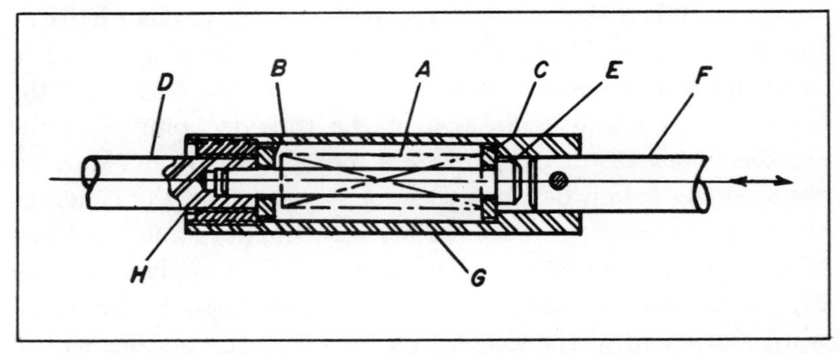

FIG. 9. The mounting permits spring *A* to be compressed when shaft *F* is moved in either direction.

mitted through the step in the thimble bore and washer *C* against the right end of the spring. Thus, the spring is compressed either way shaft *F* moves.

Linkage for Combined or Independent Lineal Travel

A critical element and an auxiliary element of a mechanical system can be linked to a common actuator in such a way that either the two operate together, or the critical element operates alone, if the auxiliary element is jammed.

Heart of the linkage device is cylinder *A*, held in a fixed position in bracket *B* (see Fig. 10). Within the cylinder are two tubular slides — inner slide *C* and outer slide *D*. Cable *E*, entering the cylinder from the left, is joined to plunger *F*. Around the

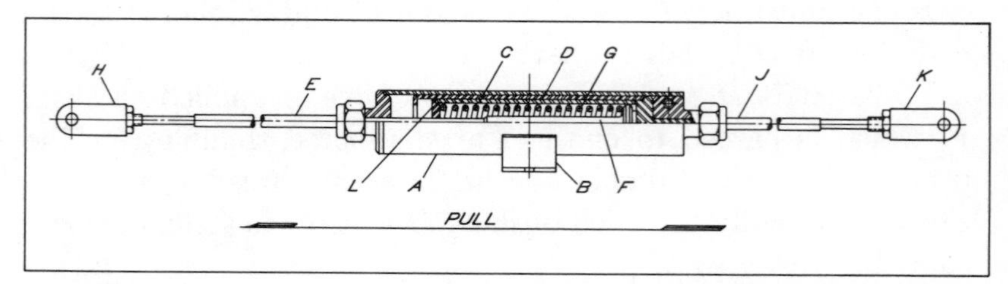

FIG. 10. During the second 1/4 inch of travel, slide *D* normally moves with slide *C*. If the auxiliary element is jammed, neither slide moves, and spring *G* is compressed.

plunger is coil spring *G*. The other end of this cable is connected to the critical element (not shown) through clevis *H*.

Another cable *J*, entering the cylinder from the right, is joined directly to slide *D*. The other end of this cable is connected to the auxiliary element (not shown) through clevis *K*.

The device functions as follows: When the critical element is operated, it pulls the plunger to the left. The spring, being heavy, resists compression under normal load, and restricted by end plate *L*, causes the inner slide to move as a unit with the plunger for ¼ inch of free travel. At this point, the end plate contacts the bottom of the outer slide.

Then, for a second ¼ inch of travel, the plunger, inner slide, and outer slide move as a unit. Since cable *J* is joined to the outer slide, the auxiliary element operates with the critical element during the second ¼ inch of travel as is desired.

On the other hand, assume that the auxiliary element is jammed. Then, in the second ¼ inch of travel, the plunger will travel independently of the inner slide, which has now been immobilized by the stalled outer slide. During this movement, the spring is compressed by the pull on the plunger through the operation of the critical element.

A practical application of this device is found in fighter aircraft. The critical element of the system is a seat ejector, and the auxiliary element, a headrest latch release. In this instance, the spring around the plunger has a 100-pound pre-load. Spring rate is 100 pounds per inch. The additional 25-pound load, when created by the freezing or jamming of the headrest latch release, is reduced to about 1.4 pounds at the input because of a 1 to 18 ratio of the pulling force.

Pressure Governor for Handwheel of Lathe Tailstock

The danger of exerting excessive axial pressure when adjusting a lathe tailstock for a "between centers" operation can be averted with the governor illustrated. It can also be used to advantage where a small drill or reamer must be supported in a relatively large tailstock — increasing the sensitivity of the feed and thus reducing tool breakage. The device is of simple construction and retains all the standard parts of the tailstock.

In Fig. 11, which shows the right-hand end of the tail-stock, the spindle *A* has a sliding fit with the casting *B*. The end of the spindle is threaded and engages the feed-screw *C* which rotates in the cap *D*. A key *E* holds the spindle from turning as it is advanced or retracted. The handwheel *F* is removed from its normal position near the end of the feed-screw and is replaced by a cast-steel disc *G*, feathered to the feed-screw by an existing Woodruff key *H*. Shaft *J*, threaded to the right-hand end of screw *C*, retains disc *G*.

Sleeve *K* clears disc *G* and tight slide fits shaft *J* and is retained on *J* by pinned collar *L*. Handwheel *F* is rigidly attached to sleeve *K*.

At a point on its periphery, the disc *G* has a 90-degree V-notch *O*. A bossed section *P* of the sleeve contains a detent plunger *Q*. This plunger is kept in position in the V-notch by a spring *R*. During normal operation, handwheel *F*, sleeve *K*, disc *G* and feed-screw *C* turn as a unit.

Should the advancing spindle meet with excessive pressure, plunger *Q* will rise out of the V-notch. Thus, if a revolving center is supported by the tailstock spindle, the handwheel ad-

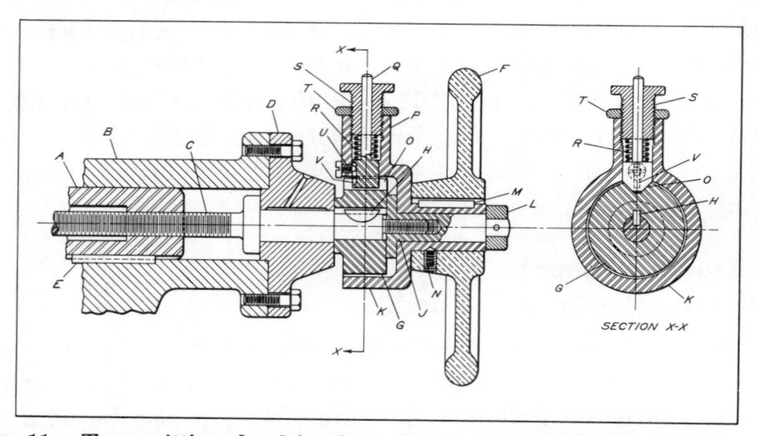

FIG. 11. Transmitting the drive from the handwheel *F* to the feed-screw *C* through a detent plunger *Q* limits the axial pressure that can be exerted.

vances the spindle until the center engages the conical opening in the end of the work. Further advance is resisted by the center, and the detent plunger automatically rides out of the V-notch.

The load that the detent plunger can carry before it will disengage is adjusted by the vertical setting of a bushing *S* threaded to the inside of the bossed section. A lock-nut *T* serves to maintain the setting. For a revolving center, the mechanism is set at a point that is well below the safe loading on the balls and races, but will permit proper support for the work. This point is readily established by testing the revolving center for free rotation under load.

When the detent plunger is disengaged, its alignment with the V-notch is maintained by a set-screw *U* engaging slot *V* in the plunger. Should it be desired to operate the handwheel without the regulating action of the governor, the bushing *S* can be lowered to fully compress the spring *R*.

Press Clutch Automatically Disengaged after Required Number of Strokes

An indexing die was designed for piercing a number of equally spaced holes in drawn sheet-metal parts or shells. The shells were rotated a partial revolution with each stroke of the press, and the operator disengaged the press clutch when the required number of holes had been pierced. The human element soon became apparent by the number of parts with one or more holes missing. It was then decided to control the number of strokes per shell automatically: the mechanism shown in the accompanying drawing was designed for this purpose.

In the end views seen at the bottom in Fig. 12 and in Fig. 13, the crankshaft bearing and other details have been omitted for clarity. The press used for this operation has a sliding key clutch. Flywheel *A*, rotating in the direction indicated by the arrow, transmits motion to the crankshaft *B* through the sliding key *C*. The spring-loaded key is engaged and disengaged by the action of a wedge-ended lever *D*, which is operated by a foot-pedal through the clevis-rod *E*.

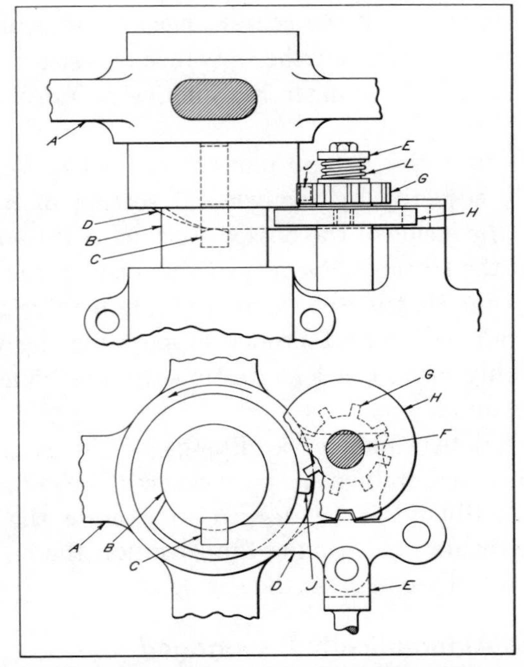

FIG. 12. Mechanism employed on a punch press to disengage the clutch automatically after eight strokes of the ram.

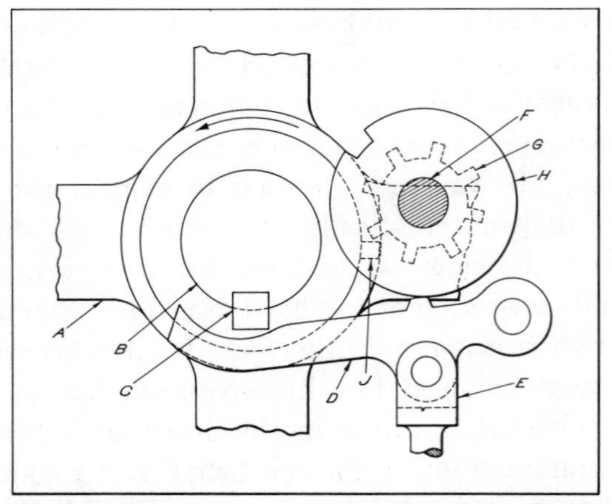

FIG. 13. Spring-loaded sliding key *C* shown in its engaged position, with the flywheel *A* rotating crankshaft *B*. When the ram has completed eight strokes, the projection on top of lever *D* will enter the groove in disc *H*, thus disengaging the clutch.

In Fig. 12, the sliding key is shown disengaged from the flywheel, and the press ram is stationary, permitting loading. A toothed disc *G* and an attached grooved disc *H*, are free to rotate on a stud *F*, which is attached to the side of the press frame. Spring *L* applies frictional resistance to the rotation of discs *G* and *H*. Pin *J*, inserted in the crankshaft flange, contacts one tooth of disc *G* with each revolution of the crankshaft. A projection on the upper edge of lever *D* engages the groove in disc *H*.

When the work-piece has been placed in the die, the foot-pedal is depressed and lever *D* is withdrawn from the groove in sliding key *C*, as seen in Fig. 13. The key then engages the rotating flywheel and locks the crankshaft to it. As soon as the crankshaft starts to rotate, pin *J* contacts one tooth on disc *G*, causing it and disc *H* to rotate a partial revolution. The projection on lever *D* is now in contact with the periphery of disc *H*, thus preventing the wedge end of lever *D* from entering the groove in key *C*. Continued rotation of crankshaft *B* causes disc *G* to be rotated, one tooth per revolution, until the projection on lever *D* again enters the groove in disc *H*. In this position, lever *D* is permitted to return to the disengaging position.

The number of strokes per cycle is governed by the number of teeth on disc *G* and the number of grooves in disc *H*. It is necessary, however, that the number of teeth on disc *G* be a multiple of the number of grooves in disc *H*.

Safety Attachment Designed for a Reciprocating Movement

On a machine for producing a wire product, short lengths of wire are drawn into the machine from a magazine, moved in one direction for one operation, and then in the opposite direction for the next operation. Occasionally, a defective wire fails to release properly from the magazine, resulting in breakage. An attachment designed to eliminate jamming in cases of misfeed is illustrated in the accompanying drawings.

A plan view and front elevation of the attachment during normal operation are shown in Fig. 14. Rod *B* imparts a re-

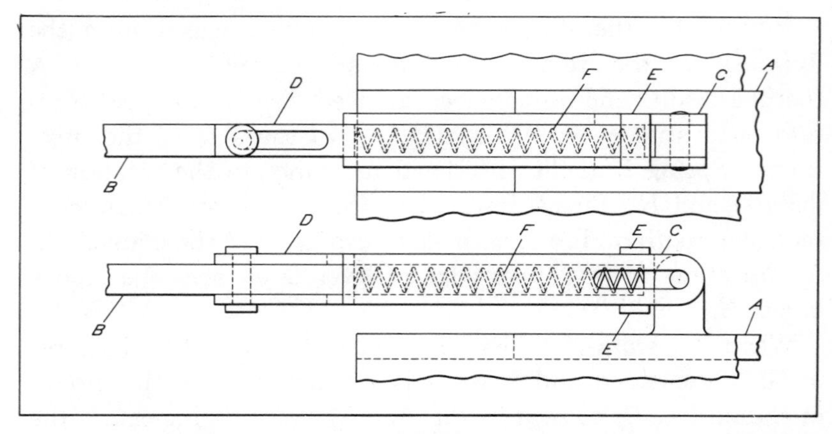

FIG. 14. Plan view and front elevation of attachment for eliminating jamming of the machine in the event of a misfeed.

ciprocating motion to slide *A*, which carries the feeding mechanism, not shown. Part *C* is a U-shaped piece, with its right-hand end slotted and supported on a pin mounted in lugs on slide *A*. Part *D* is another U-shaped piece, also slotted at its left-hand end and supported on a stud in bar *B*. Welded to part *D* are two straps *E*, which support the closed end of part *D* on part *C*. Part *D* is free to slide within part *C*, and both parts *C* and *D* are free to slide on their supporting studs.

A spring *F*, nested in the hollow box section formed by the assembly of parts *C* and *D*, is under compression at all times, so that the closed end of part *D* is held in contact with the lugs on slide *A*. Also, the closed end of part *C* is held in contact with the right-hand end of bar *B*. In the position illustrated, which is normal operation, movement of bar *B* is transmitted to slide *A* through spring *F*. The tension of this spring, determined by trial, must be sufficient to transmit the required movement and still permit further compression without resulting in the breakage of parts.

The operation of the mechanism under abnormal conditions is illustrated in Fig. 15. In the view at the top, slide *A* has been prevented from moving to the right because of a defective wire failing to release from the feeding mechanism. As bar *B* continues its movement to the right, movement of part *D* is prevented by its contact with the lugs on slide *A*, and the stud in bar *B* slides in the slots in part *D*. Also, since the end of bar *B* is in contact with the open end of part *C*, movement of the bar causes this part to move with it, against the compression of the spring. Thus, part *C* slides over the pin mounted in the lugs on slide *A*. As bar *B* again moves to the left, the assembly returns to its normal position, as shown in Fig. 14.

If slide *A* is prevented from moving to the left with bar *B*, a condition such as the one illustrated at the bottom in Fig. 15 is produced. Bar *B*, in moving to the left, draws part *D* with it, thus compressing the spring against the closed ends of parts *C* and *D*. When the abnormal condition has been corrected, the parts again assume the positions shown in Fig. 14.

In a subsequent application of this arrangement having space limitations, it was impossible to provide an attachment long enough to accommodate a compression spring of the required length and strength. Consequently, the design was altered, as shown in Fig. 16, to permit the use of externally supported compression springs. In general, the design has not been changed except by the addition of side extensions on part *D*, and

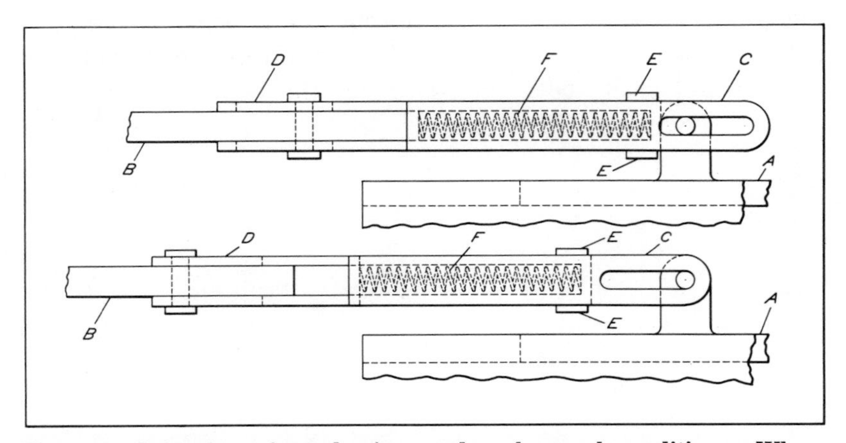

FIG. 15. Operation of mechanism under abnormal conditions. When slide *A* is prevented from moving to right, part *D* contacts lugs on slide *A*, as seen in Fig. 14. When slide *A* is prevented from moving to left, spring *F* is compressed, as illustrated in Fig. 15.

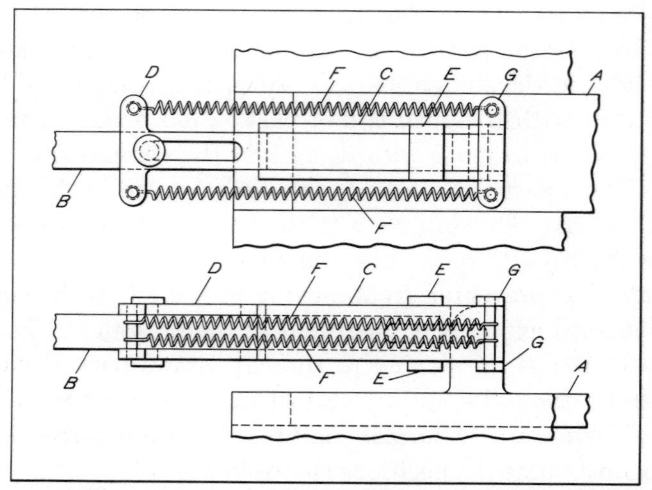

FIG. 16. Modified design of mechanism shown in Fig. 14, necessitated by space limitations that prevented use of long, strong spring.

members *G* to part *C*, for attaching springs *F*. Operation is the same as in the original design.

Although both of these designs will perform equally well, the choice must be governed by space limitations, as the latter design will require greater width, as indicated by the plan view in Fig. 16. However, the fact that the springs are supported externally in this design may prove a definite advantage in that the tension may easily be adjusted to suit the requirements.

CHAPTER 5

Locking, Clamping, and Locating Devices

Means of positively locking a mechanism, clamping a workpiece or part, and locating work in the proper position for some operation to be performed on it, or locating a carriage or table in the correct loading position, are described in this chapter. In some cases, the locking or clamping operation is performed automatically while in others hand operation is required. Similar devices are described in Volumes I, II and III of "Ingenious Mechanisms for Designers and Inventors."

Intermittent Drive with Reverse-Locking Feature

An arrangement that prevents reversal of an intermittent drive during the dwell period is shown in Fig. 1. Compact and quiet in operation, the device was designed for use as a high-speed indexing mechanism in shoe processing machinery.

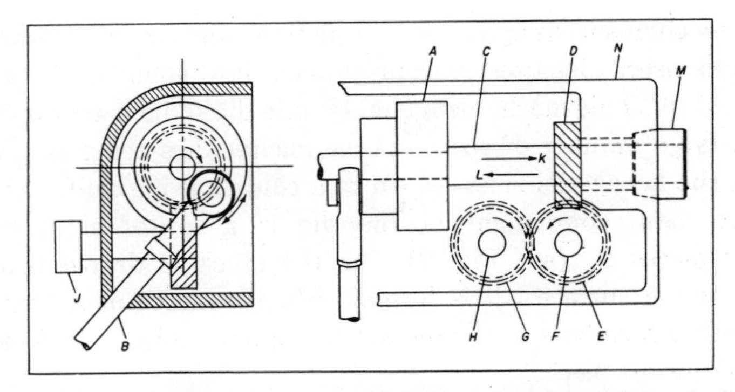

FIG. 1. Cone brake prevents reversal of this intermittent drive during the dwell period.

A roller type indexing clutch *A* is driven with a reciprocating motion by connecting-rod *B*. This causes shaft *C* to rotate clockwise intermittently. Right-hand helical gear *D* is fastened to shaft *C* and engages with a second right-hand helical gear *E* attached to shaft *F*. Left-hand helical gear *G* mounted on shaft *H* also meshes with gear *E*. Two feed rollers *J* are fastened to the ends of shafts *F* and *H*, which rotate intermittently in opposite directions.

When clutch *A* is driving gear *D*, a thrust is produced laterally in shaft *C* in the direction *K*. During the dwell portion of the indexing cycle, any attempt to make gears *G* and *E* the driving gears will produce a lateral thrust and displacement of shaft *C* in the opposite direction *L*.

A cone brake *M* attached to shaft *C* takes advantage of this reversal of thrust to lock the shaft and the rollers during the dwell period. Thrust in the reverse direction *L* causes the cone to be displaced slightly and become tightly held in a mating conical bore in frame *N*. An increase in the reverse thrust only increases the holding power of the cone brake. During the following index cycle, the thrust produced on shaft *C* by helical gear *D* is again in direction *K*, and cone *M* is released from the conical bore. Lateral movement of shaft *C* is held to the minimum displacement necessary to free the cone.

Mechanism for Adjusting Size of "Iris" Drawing Dies

A mechanism designed for adjusting the size of hexagonal carbide-insert dies used in cold-drawing hexagonal stock is here illustrated. This mechanism enables one die to be used for drawing a large number of sizes. Three master dies cover the range of all the hexagonal sizes drawn in a cold-drawing mill.

The main component of the die is a set of six carbide-insert pieces *A*, (see Fig. 2). As the stock is drawn through these insert surfaces, it is formed into the shape of a hexagon. The other members of the die act to support, adjust, or lock the carbide-insert pieces.

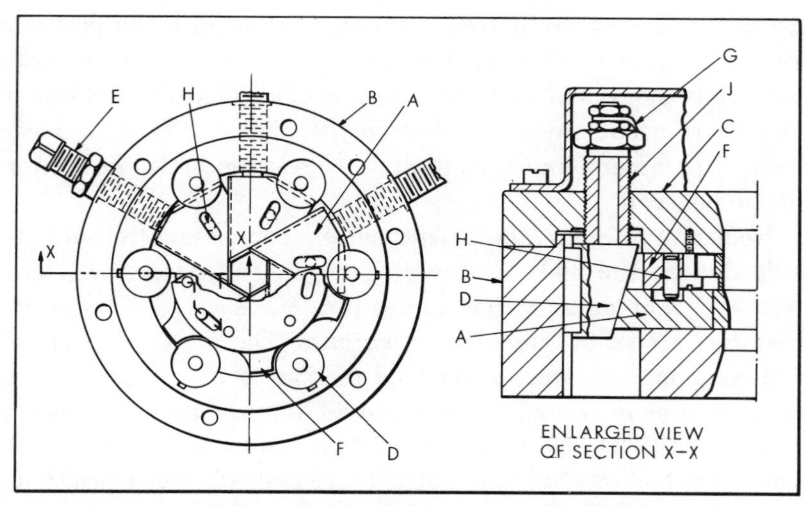

FIG. 2. Plan view and enlarged cross-section of the die showing the arrangement of the various components.

To understand the principle on which the adjustment of the carbide-insert pieces is based, reference should be made to Fig. 3. Here the relative movement of two of the six pieces is indicated by the arrows. Initially, the pieces are in the position indicated by the solid lines, forming two sides of the solid-line

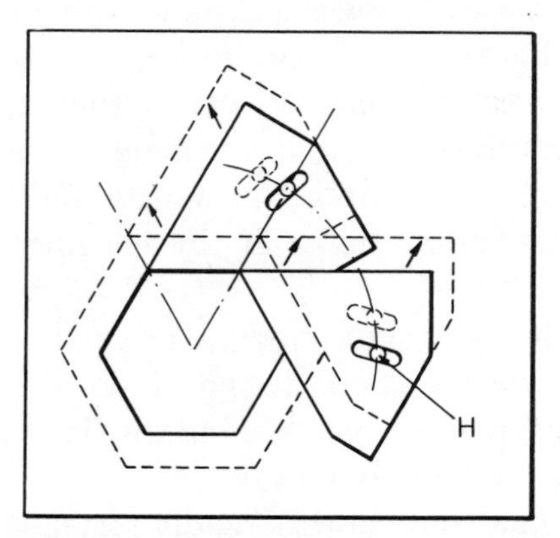

FIG. 3. Carbide-insert pieces are moved in a straight line from the solid- to the dotted-line positions as indicated by the small arrows. Pins acting in elongated slots effect this motion.

hexagon. If they are moved without rotation into the position shown by the dotted lines, they form two sides of a larger hexagon. The movement of these pieces is controlled by the action of pins *H* inside elongated slots in the pieces. The pins move along the circumference of a circle whose center is the center of the hexagon.

Referring to Fig. 2, the other members comprising this adjustable die are the body *B*, cover *C*, keyed bar wedges *D*, lock-screws *E*, and a disc *F* that carries pins *H*. Each of these members has a specific function to perform. Die body *B* holds all the component parts, so that the motion of the carbide-insert pieces can be restricted within required limits. Cover *C* supports disc *F*, which is free to rotate a small distance. Cover *C* also holds the hollow stud and nut arrangement *J*, which positions the wedges *D* laterally.

The purpose of the wedges is to provide a means whereby the carbide-insert pieces can be located properly. Adjustment is necessary when the drawing surfaces of the carbide-insert pieces wear. The lock-screws *E*, provide a locking function.

Once the die has been set, which is usually done in the maximum open position to minimize any error in shape, it is a rather simple operation to adjust it to the required size. Lock-screws *E*, together with their lock-nuts, are loosened. Three alternate keyed bar wedges *D*, the tops of which have indexing pointers *G*, are next loosened to unlock the six carbide-insert pieces. The purpose of the indexing pointers is to insure the proper locating of these three bar wedges for true hexagonal positioning of the insert pieces in the ensuing locking operation.

The other three alternate wedges are left undisturbed, in order that the true hexagonal shape of the die will be maintained in sliding the insert pieces along the flat surfaces of these wedges from one position to another.

Two knurled pins (not shown) rigidly fastened to disc *F* and extending through elongated slots in cover *C* are used to impart rotation to disc *F*. The pins *H* that are engaged in the slots of the six carbide-insert pieces slide these pieces simultaneously into the new position. When in the required position, the three alternate wedges *D* are properly located by the use of their indexing pointers, and the six lock-screws *E* and their lock-nuts are tightened, thereby locking the die.

Three-Axis Adjusting Mechanism

For applications where a single supporting member must be adjustable in all directions, use can be made of the device shown in Fig. 4. This device consists primarily of a special eyebolt that can be swiveled around a spherical surface and locked in any desired position. The eyebolt can also be adjusted lengthwise.

Bracket *A* serves as an attaching component that is fastened to the object which requires an adjustment feature. Swiveling of the eyebolt *E* is accomplished after backing off nut *B*, thus relaxing washers *C* and allowing spherical washer *D* to slide around the spherical end of housing *F*. The pivot point is about the center of the spherical bushing *G*. Washer *D* carries belt *F*.

Lengthwise adjustment is accomplished by rotating spanner nut *H* to advance or retract housing *F*. Pin *J* rides in a slot of housing *F* and restrains rotation of the housing.

As shown, this mechanism permits adjustment to within any point in the space of a 2-inch cube.

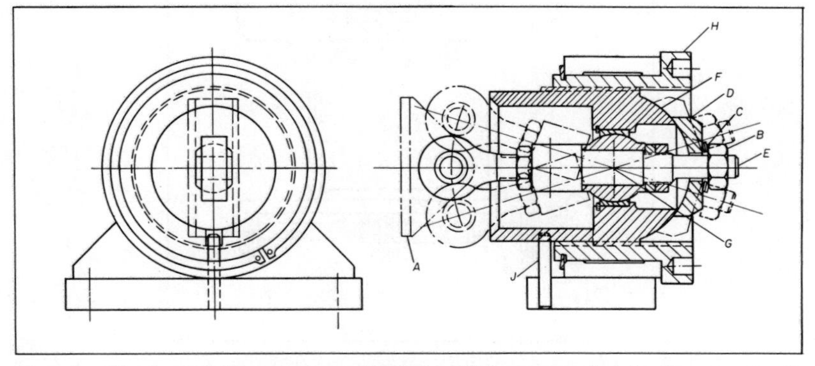

FIG. 4. Device which can be adjusted within any point in the space of a 2-inch cube.

Adjustable Disc-Stacking Magazine

For a special machine designed for an operation on leather discs, a fixture had to be provided to hold discs from 1 to 6 inches in diameter. Figure 5 shows an adjustable "nest" developed to meet this requirement.

An arm from a molding press moves into the position indicated by line *P* to pick up discs successively and then transfer them, one at a time, to a mold cavity located to the right of the stacking device. Regardless of size, the centers of the discs must always be in the same place.

In adjusting this magazine to suit a change in work diameter it is necessary to swivel arms *A*, *B*, and *C* inward or outward. Movement of the arms is accomplished by rotating threaded

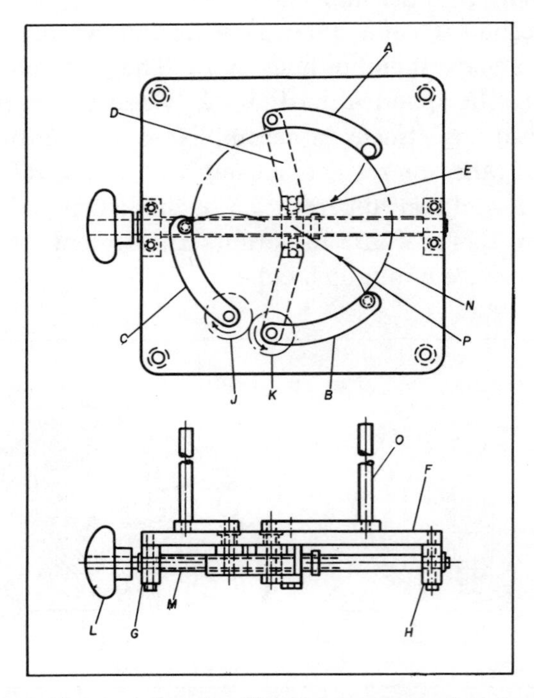

FIG. 5. Adjustable magazine for discs can accommodate work sizes from 1 to 6 inches in diameter.

shaft *M* to move block *E* laterally. Arms *D* are pivoted on this block. Their outer ends are attached to arms *A* and *B*. Gear *K* on the outer end of Arm *B* meshes with gear *J*. This gear is mounted on the pivot shaft of arm *C*.

Movement of arms *D* and gears *J* and *K* causes arms *A*, *B*, and *C* to move toward or away from the center of the stacking unit to suit discs of various diameters. Vertical rods *O*, mounted on arms *A* and *B*, allow for stacking discs on top of each other to convenient heights.

Tape Reel Has Quick Action and Constant Gripping Pressure

Tapes used in programming machine tools and in other operations are wound on spools, which, in turn, are positioned on a reel. Tapes and spools are made in different widths, but the spools necessarily have the same inside diameter.

To grip the spool, most reel designs involve the tightening of a cap-nut on a thread which is part of the reel spindle. The nut presses against a rubber cylinder which expands in the bore of the spool, securing it to the reel. Disadvantages of such action are that engagement and disengagement time is relatively long; the firmness of the grip depends on how much the cap-nut is tightened, which might vary from operator to operator; and some of the components of the reel have to be changed whenever the spool width is changed. An added shortcoming is that the clutching action can only be performed manually, and cannot be made automatic.

On the hand, the reel design proposed in Fig. 6 is quick acting, assures a constant gripping pressure, accommodates spools of different width without adaptation, and can be readily converted to operation by a solenoid.

The device is driven by a motor through a timing belt (not shown) running around pulley *A* feathered to shaft *B*. Hub *C* is pressed on the shaft. Flange *E* bolted to the frame is lined with a bearing which supports the shaft. This bearing extends through the bore of the hub. The portion of the shaft which

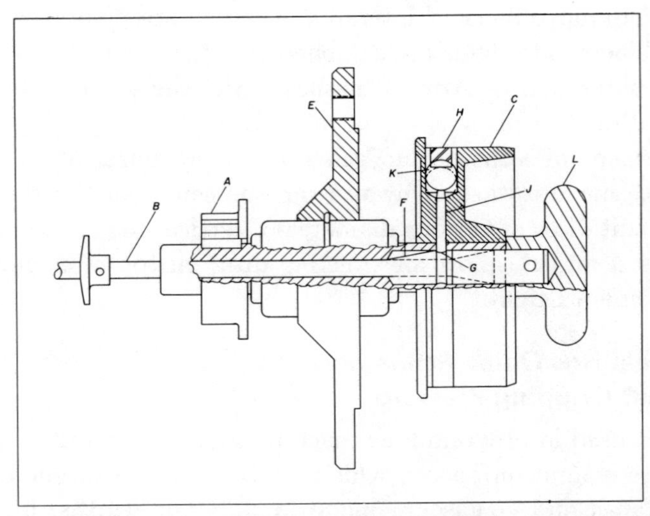

FIG. 6. With ball *K* up (solid line), part of rubber ring *H* is displaced; with ball down (broken line), all of rubber ring is contained in neck of hub *C*.

runs through the hub has a tapered surface *F* and an annular groove *G*.

Rubber ring *H* has a snug fit in a neck in the hub. Pressure is exerted around the hub by the ring, since the inside diameter of the ring is smaller than the diameter of the neck. There are six counterboard holes equally spaced radially in the neck, each containing a dowel-pin *J* and steel ball *K*. The rubber ring keeps the balls and pins in pressure contact with shaft *B*.

Reel spools have a slip fit over hub *C*. To position a spool on the reel, knob *L*, pinned to the shaft, is pulled to the right, and the rubber ring forces the balls and dowel-pins down radially on taper *F*. Since the rubber ring is now completely contained within its neck, the spool is able to slip over it.

To grip the spool, the knob is thrust to the left, the dowel-pins being forced out by the taper, settling in groove *G* in the shaft. (This is the position illustrated in Fig. 6.) Simultaneously, the balls move out radially, partly displacing the rubber ring in the neck. The displaced rubber fills the clearance between the outside of the hub and the inside of the spool, and is sufficient to exert a firm grip on the spool. This grip remains constant from spool to spool, since it is outside the control of the operator.

Total radial movement of the dowel-pins is calculated so that the volume of the penetration of the balls in the rubber equals the clearance area. Relatively large tolerances are permissible, because of the compressibility of the rubber, which takes up the variations.

It is possible to connect the left end of shaft *B* to a two-directional solenoid for automatic operation. The solenoid has to be actuated only when a spool is being positioned or removed. A micro switch can be used to sense the axial position of the shaft, and the machine can be wired in series with the micro switch so that it will not start if the spool is not firmly gripped.

Quick-Acting Clamp with Wide Work Capacity

An unusual gripping and releasing mechanism which enables instant adjustment of a clamping jaw to suit widely different sizes of work is a feature of the special vise-like assembly fixture illustrated in Fig. 7. This clamp was primarily designed

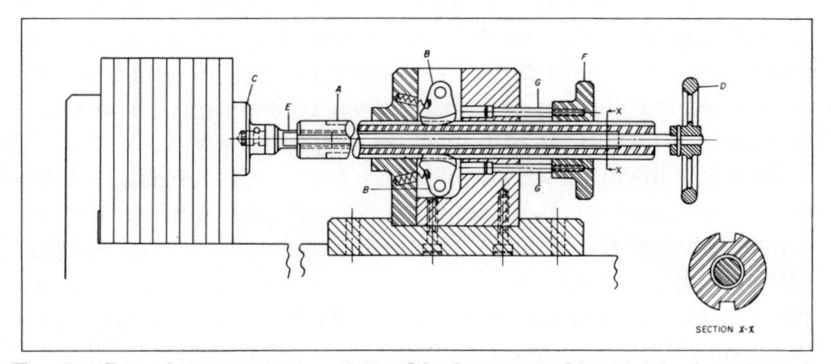

FIG. 7. Ingenious arrangement on this fixture enables quick clamping of work despite large variations in work thickness.

for holding packs of thin sheets that vary considerably in width. Packs vary in over-all width from approximately 5 to 16 inches.

The different pack sizes are made up by the operator in the assembly fixture, according to requirements. To minimize clamp setting time it was desirable to provide a clamp which, although hand-operated, would be capable of being adjusted rapidly. It had to be suitable for imparting a sufficiently powerful grip on packs of all sizes.

The principal working member of this device is a hollow ram *A* which can be readily slid to the right or left except when retarded by the action of pawls *B*. In loading the fixture, the ram is pushed by hand until clamp *C* bears against the pack of work sheets. Then handwheel *D* is revolved to apply pressure through rod *E*. This rod extends completely through hollow ram *A*. At the left-hand end, there is a threaded enlarged diameter on the rod which engages a thread in the ram.

Consequently, after clamp *C* has been positioned against the work, it is positively tightened by revolving handwheel *D*. This action causes an adjustment of the threaded portion of rod *E* in the threaded left-hand end of the ram, and thus exerts pressure on clamp *C* and on the work. This can happen because pawls *B* prevent any right-hand movement of the ram until they are released.

Release of the pawls is effected by striking a sharp blow against ring *F*. This causes ejector pins *G* to strike against the pawls with sufficient force to overcome the pressure of the springs which tend to force the pawls to the right. The pawls operate in grooves cut in ram *A*, as seen in section X-X.

In a modification the two pawls *B* are substantially of the same shape and size as those in Fig. 7, in respect to their contacting peripheries (see Fig. 8). However, each pawl has an integral tail at right angles to the contacting portion. In each case, the pawl tail passes through a clearance slot machined in the right-hand side of the fixture body.

Fastened to the top of the body over both slots is a steel plate which bridges the slot. Screwed into this plate is a fine-pitch

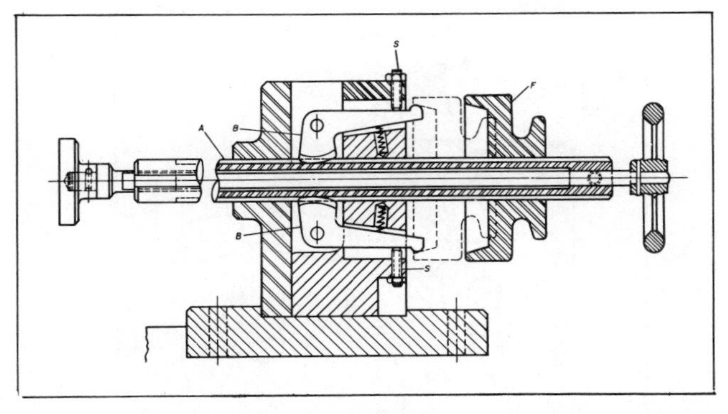

FIG. 8. Safety overload on this modified design guards against work distortion and damage.

headless set-screw, the lower end of which bears against the side of the pawl tail. A lock-nut secures the screw in any desired height setting. The purpose of the two screws *S* is to restrict the amount of swiveling movement of each pawl in the direction of its frictional contact with the ram grooves. A light compression spring bears against the inner side of each pawl tail so as to maintain the contacting peripheries of the pawls in a light frictional engagement with V-grooves of the ram.

Instead of two ejector pins, the ring mechanism *F* has a conical bore which is large enough to pass over the rounded tips of the pawl tails, as shown by dotted lines. When the ring is moved swiftly toward the body, the sides of the conical hole strike the pawl tails and cause the pawls to release.

Cam-Jaw Chucks for Twisting Rod

Two chucks having cam jaws furnish a powerful grip which twists steel rod used in reinforced concrete. One of the chucks, Fig. 9, anchors one end of the rod. The other, Fig. 10, rotates the opposite end of the rod to produce the required twist.

The anchoring chuck has a bracket *A* mounted on the frame of the machine. Rod *B* to be twisted is located against a pad *C* transversely movable within the bracket by means of a toggle arrangement with lever *D*. At two other points on its periphery the rod is under the pressure of knurled cam jaws *E* and *F*.

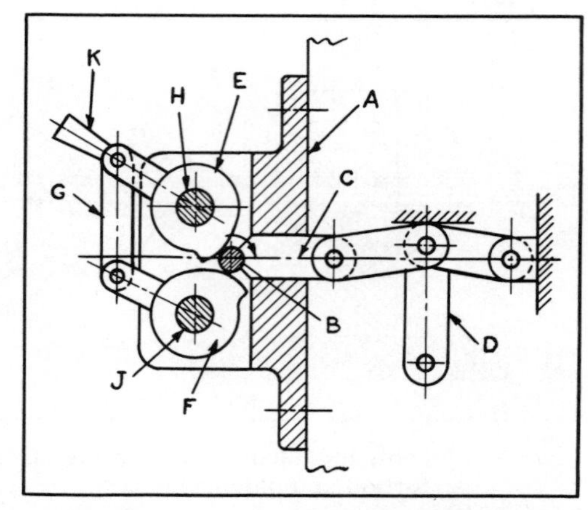

FIG. 9. This anchoring chuck prevents one end of the steel rod from rotating.

A link *G* joins the jaws so that they can pivot in unison around their respective shafts *H* and *J*. Extending from the link is an operating arm *K*.

When slipping a rod into position in the machine, lever *D* and arm *K* occupy the positions illustrated. The arm is then moved down, and the jaws pivot counterclockwise, to lock

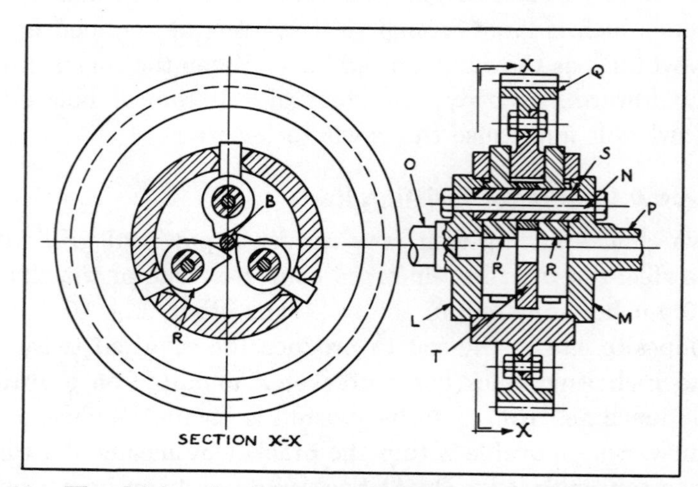

FIG. 10. The driving chuck rotates the opposite end of the rod to produce the required twist.

the end of the rod against the pad. As the machine starts to twist the opposite end of the rod, the cams tend to tighten their grip. To release the rod, lever *D* is lowered. This action causes locating pad *C* to retract.

The driving chuck, Fig. 10, has two cover plates *L* and *M*, tied together by bolts *N*. The outside of each plate is turned down to form integral bearing diameters *O* and *P*. Diameter *P* is bored to receive the end of rod *B*. Located between the cover plates is a large gear *Q*. This gear has two hubs; each is drilled radially and beveled at three points to receive the lobes of cam jaws *R*. The jaws pivot on bushings *S*, which are mounted over the bolts *N*. Also, the two rows of jaws are separated by a disc *T*.

When the machine is running, a pinion, which is engaged to the gear, rotates it in the direction indicated by the arrow. The cam jaws immediately pivot in and start twisting the rod, the opposite end of which is fixed in the anchoring chuck. To release the twisted rod, the gear is reversed momentarily. Like the cam jaws in the anchoring chuck, those in the driving chuck have knurled bearing surfaces to provide a better gripping action.

Variable Stroke and Quick-Action Lock for Reciprocating Slides

The device of Fig. 11 was designed to cause two slides to be moved by hand and then be securely locked in a predetermined position.

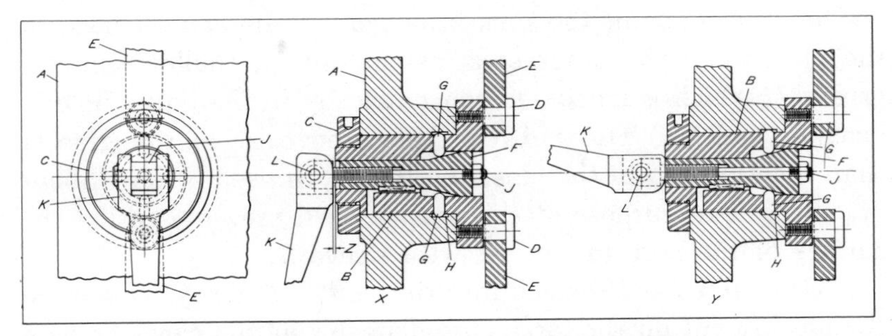

FIG. 11. Manually operated mechanism provides stroke adjustment and rapid locking for two opposed reciprocating slides.

The cross-section at X shows the construction of this manual drive mechanism and the positions occupied by working members when in the unlocked position. A section of the vertical machine wall *A*, on which an integral boss is located, is bored to receive flanged sleeve *B*. The portion of the sleeve extending beyond the machine wall is reduced in diameter and threaded for circular lock-nut *C*, which holds the sleeve in place and may be adjusted by means of a spanner wrench.

Two identical links, *E*, see Fig. 11, can pivot about shoulder studs *D*. The opposite ends of links *E* are attached to the slides.

Sliding within a hole bored through sleeve *B* is cylindrical plug *F* which is keyed in place to prevent independent rotation. The plug has a conical head which rides within a counterbore in the flanged end of the sleeve. The largest diameter of this head is ground with parallel sides for a short distance to provide a close sliding fit within the counterbore. This lends additional support to the head and helps to maintain its accurate alignment during locking movements. The surface of the conical head, which is formed at an angle of 10 to 12 degrees from the axis, should be hardened and polished smooth.

Four equally spaced holes are drilled radially through the side walls of sleeve *B* into the counterbore. These holes are the same diameter and are located in the same plane. Sliding freely within each hole is a pin *G*, both ends of which are rounded. All four pins must be accurately machined to the same over-all length and hardened.

The inner ends of the pins bear against the conical head of the plug while the outer ends extend into a shallow annular groove *H* which is machined concentrically in the bored hole in machine wall *A*. The width of this groove is slightly greater than the diameter of the pins, and the depth need be only about 1/16 inch. The purpose of this groove is to prevent scoring the surface of the bore in contact with sleeve *B*.

A clevis-pin *J* is threaded into the center of plug *F*, fine-pitch threads are cut on the larger pin diameter at the clevis end. A standard hexagon nut is screwed on the opposite (right-hand) end of the clevis-pin to lock it in position. This arrangement permits both radial and endwise adjustment of the clevis-pin.

Operating lever *K* has a forked end for fitting over the end of pin *J*. The width of the lever end is almost the same as the smaller diameter of sleeve *B*. A cam-like curvature is formed on the upper right-hand corner of the lever as shown at X in Fig. 11. The cam radius increases gradually as the curve approaches the top surface of the fork. The thickness of the forked end of the lever, together with the location of pivot-pin *L*, are carefully determined so that a clearance *Z* of about 0.025 inch will be provided.

To lock sleeve *B* within its bearing hole, lever *K* is simply pivoted upward as shown at Y. This action causes plug *F* to be drawn to the left. The conical head of *F* contacts pins *G* forcing them outward and into annular groove *H*. Sleeve *B* is thus locked within the bearing hole in wall *A*, and further movement cannot be transmitted to the two links *E*.

To operate the machine-slides, lever *K* is depressed to the vertical position, thereby releasing the locking pressure. The lever, clevis, plug, and sleeve members can then be rotated in unison to impart the desired motion to the machine-slides.

Finger Holds Down Paper Stack on Printing Press

On printing presses, some type of suction device generally feeds the paper by lifting the leading edge of the top sheet in the stack and drawing the sheet forward into grippers. To avoid the tendency of the top sheet to pull the next sheet with it — as sometimes happens because of static electricity in the paper — a mechanical hold-down finger can be added to the press. Through a cam and double bellcrank construction, the finger operates in time with the suction device, and separates the top sheet.

Figure 12 shows three positions of the mechanism. In view X, the finger *A* holds down the stack of paper after the leading

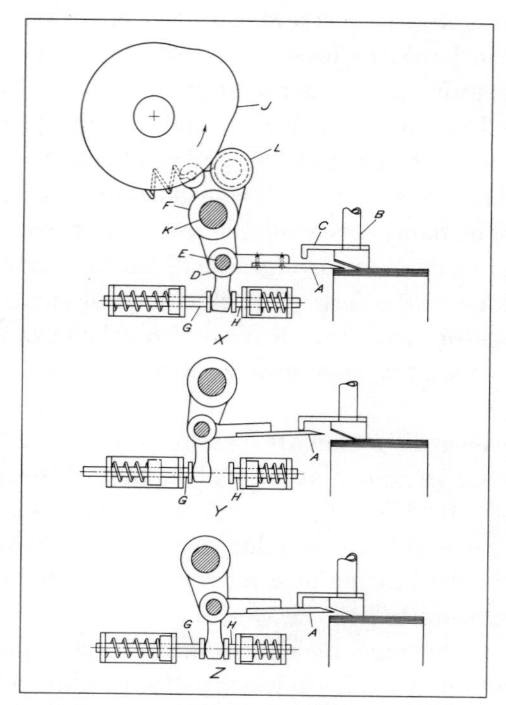

FIG. 12. Hold-down finger A is synchronized with suction cup B through cam J and bellcranks D and F.

edge of the top sheet has been lifted by suction cup B. The bottom of the cup is cut at an angle, as shown, so as to raise the sheet edge sharply. Guide C, fastened to the cup, moves up and down with it.

The finger is attached to one arm of small bellcrank D, which pivots on shaft E in the lower arm of large bellcrank F. Two spring-loaded plungers G and H control the position of the lower arm of the small bellcrank. Cam J, revolving continuously in time with the movement of the suction cup, causes the large bellcrank in turn to pivot on shaft K under the direction of follower L.

When the suction cup comes down on the stack, the finger is pulled back, and under the pressure of plunger G, it is forced upward against the guide, as in view Y. At this point, the lobe of the cam bears on the follower, and the suction cup lifts the edge of the top sheet.

Then, as the lobe leaves the follower, plunger *G* loses control of the lower arm of the small bellcrank and plunger *H* takes over, as in view Z. The guide, meanwhile, forces the finger down, directing it into the lip formed by the raised edge of the top sheet. When the finger is fully in place, as in view X, the suction cup assembly raises the sheet into the grippers which feed it into the press.

Cycling is continuous, with the cam revolving once for each sheet fed. A feature of the mechanism is that the finger is located by the position of the feed cup, so is not affected by variations in the height of the paper stack.

Toggle-Action Drill Jig That Clamps Work at Four Points

In drilling hold-down bolt holes through the steam cylinder heads of duplex piston pumps, it was found that the location of the holes was often inaccurate. The original jigs employed for drilling such holes were simply flat plates of the same shape as the cast heads to be drilled. These bushing plates were equipped with vertical pads around their peripheries to form nests for the castings. However, due to variations in the size of the castings, many of the work-pieces fit loosely in the jigs, resulting in inaccurate location of the drilled holes. To overcome this difficulty, the drill jig seen in Fig. 13 was designed to accurately clamp the work at four points by means of a single toggle action.

The two clamping arms *A* are slidably mounted on bushing plate *B* by means of studs *C*. The central portion of these studs pass through large holes in the arms to permit their free movement. Pins *D* loosely fit in the centrally located projections of the clamping arms, and their lower, enlarged diameter ends are provided with flats to fit slots milled in the bushing plate. This permits the arms to pivot about these pins and to slide along the slots when operating handle *E* is rotated.

Cam *F*, which is rotated by handle *E* about stud *G*, is connected to clamping arms *A* by links *H*. These links can pivot

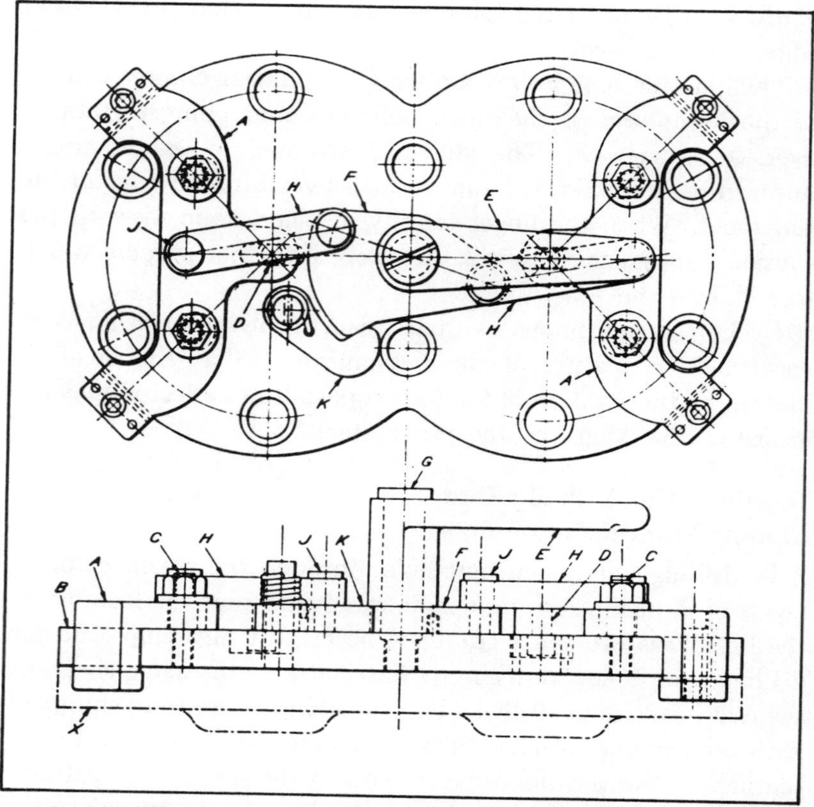

FIG. 13. Work-piece X, which is a cast steam cylinder head for a duplex piston pump, is rigidly clamped at four points in this toggle action drill jig.

about the loose-fitting studs J. A spring-loaded latch K holds the cam, levers, and arms in the work-clamping position shown (the loading position).

As the cam is rotated counter-clockwise, latch K will be rotated clockwise and links H will become aligned with each other. Clamping arms A are moved apart so that the jig can be placed over work-piece X. The cam is then turned clockwise to the position shown, and arms A are pulled together firmly to clamp the work for drilling.

Cam-Operated Stock Clamp for Piercing and Blanking Dies

Difficulty is often encountered in operating piercing and blanking dies if some means is not provided to keep the stock rigidly pressed against the back gages of the die. This is especially true if a high degree of accuracy is desired. The stock will usually weave when being pushed through the gages, or it will jump when struck by the punches. These conditions are particularly aggravating when handling heavy stock.

To overcome such trouble, a mechanism was designed which automatically presses the stock against the back gage without requiring any effort on the part of the operator. This device also has the advantage of reducing the number of scrapped parts.

Shown in Fig. 14 are two hardened and ground slides *A* mounted in ways at the front of the die-block *B*. These slides are spaced as far apart as possible. The lever *C* swivels in a clevis bracket *D*, which is screwed and doweled to the top shoe. A spring *E* in the top shoe keeps the lever up against a Z-shaped retainer *F* when the die set is in the open position.

As the press ram descends, the lower end of the lever *C* strikes the angular surface on the slide *A*, forcing the slide up against the stock. Of course, the device must be so designed that the slide is pressed firmly against the stock just before the punches enter the stock. On the up stroke, the slide *A* is retracted, allowing the operator to move the strip easily through the gages.

Swing Stop for Automatic Lathe

A unique mechanism for operating a swing stop on an automatic lathe is shown in Fig. 15. This mechanism permits the stop to be swung to a position in front of the headstock spindle to stop the axial feed of the bar stock at the beginning of the automatic cycle. Thereafter, the stop is held clear of the work, and, at a predetermined point in the cutting cycle, a mechanism

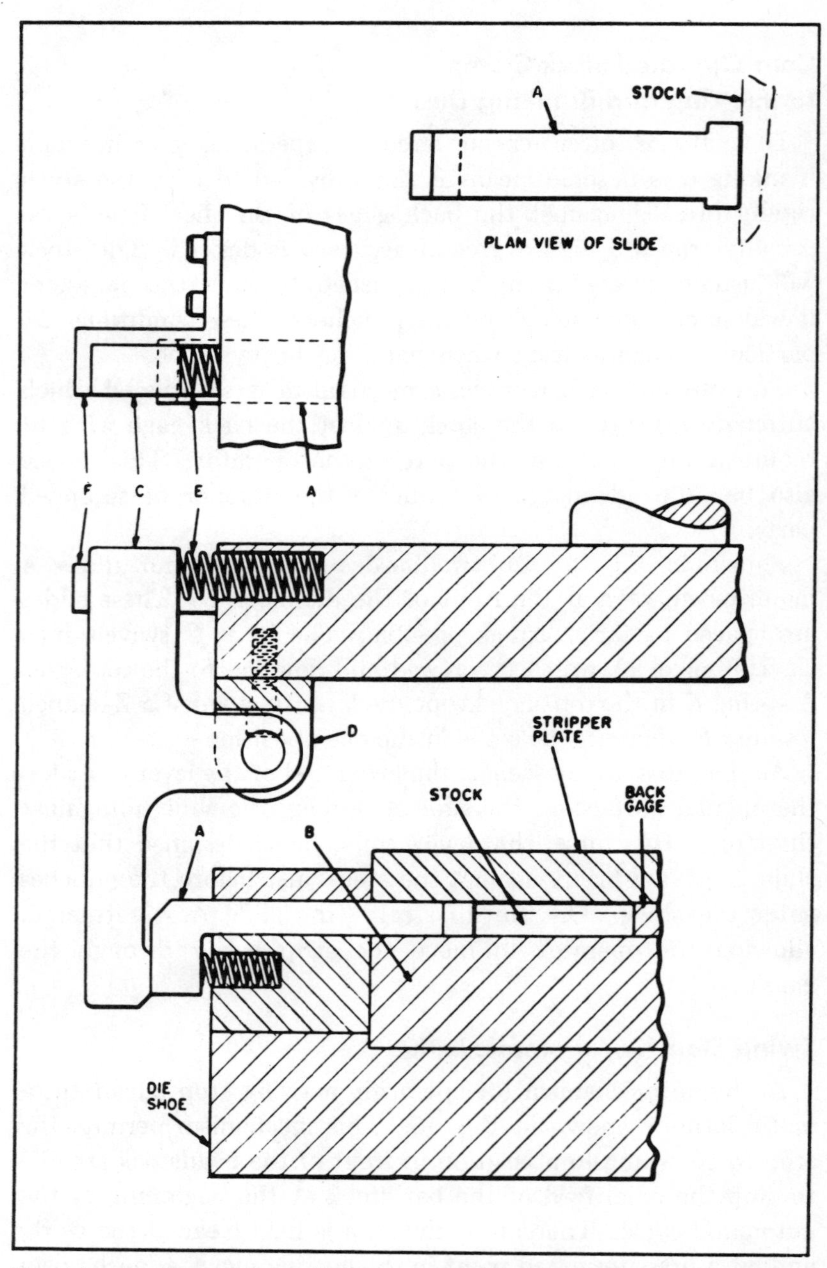

FIG. 14. As the top shoe of a die set descends lever *C* strikes *A*, pushing *A* to the right and clamping the stock.

(not shown) is operated to impart a second feed movement to the stock.

The swing stop *A* is secured to one end of a shaft which can swivel in a bearing in the headstock. The opposite end of this shaft carries an arm that is connected by link *B* to the follower arm *C*, the latter being pivoted at its left-hand end on a pin fitted to the frame. Downward movement of the stop *A* is imparted by a compression spring *D* enclosing a pin, the upper end of which makes contact with the arm *C*. The lower end of this pin is attached to the frame. Upon completion of the initial feed of the bar stock, stop *A* is swung upward by cam *E*, which is engaged by a roller on arm *C*.

The shaft on which cam *E* is mounted operates the mechanism that feeds the stock through the collet, and is driven intermittently from the back-shaft of the lathe, through a one-

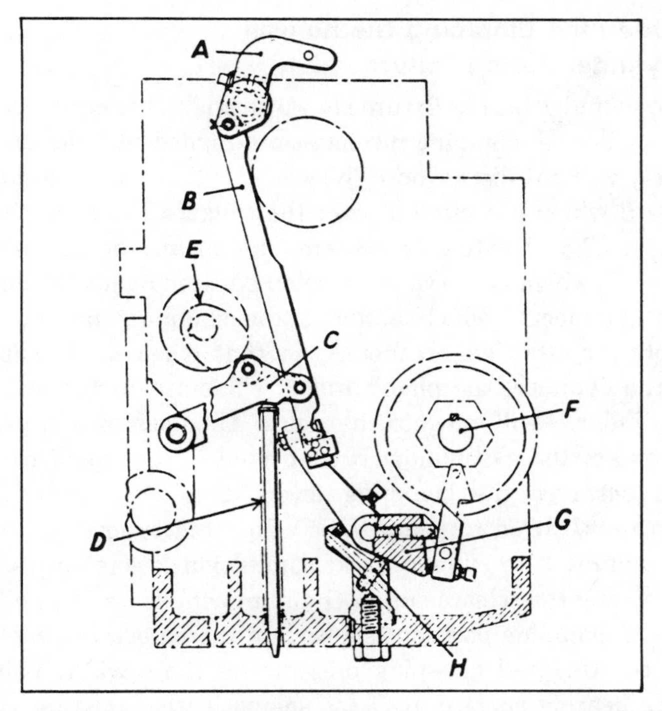

FIG. 15. Mechanism for operating swing stop *A* on an automatic lathe to control axial feed of bar stock.

revolution-and-stop clutch (not shown). With this arrangement, the cam *E* is rotated through two revolutions while the continuously driven front camshaft *F*, which controls the turret and cross-line motions, makes one complete revolution during each cutting cycle.

When the stock is to be fed a second time, a trip-dog, attached to a disc on the camshaft *F*, engages a spring-loaded pawl fitted to the right-hand end of the bellcrank lever *G*. The lever, which is carried on a forked bracket secured to the frame, swivels in a clockwise direction against the action of the spring-loaded plunger *H*. As a result, a stepped pin fitted to the left-hand end of lever *G* is swung to a position above an angle bracket attached to the lower end of link *B*, so that movement of the link and of stop *A* is prevented. Simultaneously, the shaft carrying cam *E* is rotated, causing the stock to be fed.

Air-Operated Clamping Mechanism for Cylinder Boring Fixture

The cylinder boring fixture shown in Fig. 16 is equipped with an air-operated clamping mechanism designed to hold the work securely without distortion. The work (cylinder *C*) is located on saddle *B* which is a close fit over the tongues *E′* on brackets *E*. Cylinder *C* is located or centered by means of the pivoting bracket *K*, which is shown in the open position by dotted lines *K′*.

Air cylinder *G* which actuates the clamping mechanism is slidably mounted on bracket *F*, so that when air is admitted into the cylinder, the piston rod *J* will move to the left while the cylinder *G* will move to the right. The open or non-pressure head end of the air cylinder is connected to the equalizer bar *N* which actuates the clamping levers *D′*. The outer end of piston rod *J* fitted to piston *H* is connected to lever *D*.

Air admitted to cylinder *G* at connection *P* acts on piston *H*, compressing the release spring *O* and moving levers *D′* and lever *D* to the clamping positions shown. In these positions the levers exert the required clamping pressure on the work at points *S*. A long bearing surface at *S* (as shown in the separate view in the lower right-hand corner of the illustration) distributes

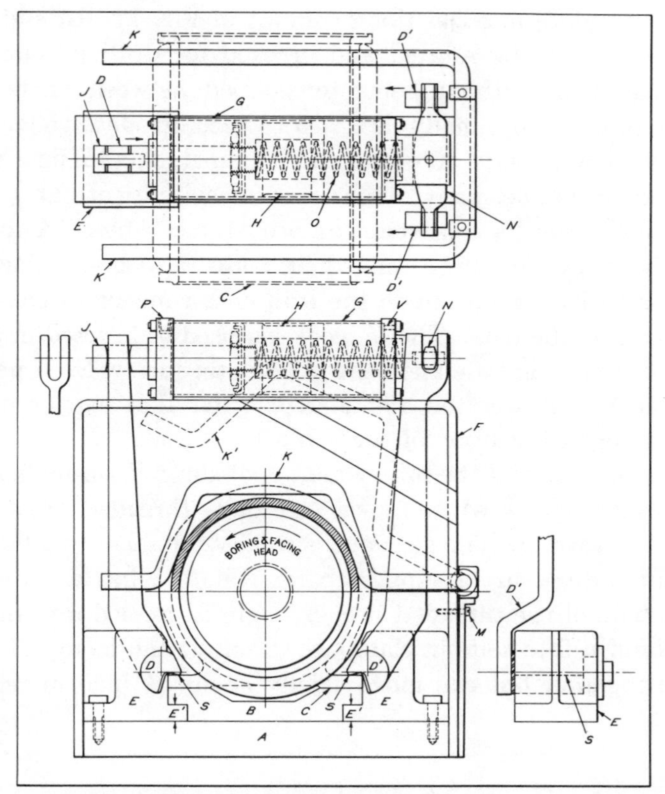

FIG. 16. Cylinder boring fixture equipped with air-operated equalizing clamping levers.

the load over a larger area. The vent *V* in the non-pressure end of cylinder *G* permits only atmospheric pressure to act on the right-hand side of piston *H*. When air pressure is released from the closed end of the cylinder, spring *O* causes the upper end of lever *D'* to move to the right and the upper end of lever *D* to move to the left so that the clamping pressure is released at points *S*.

Clamping and Indexing Mechanism for Drill Jigs

A multiple-purpose drill jig that incorporates an arrangement to automatically clamp a work-piece simply by lowering a hinged jig plate into position is shown in Fig. 17. In addition, components can be rotated and indexed for drilling a number of radial holes without being unclamped between operations. Although it was originally designed to accommodate collars and pinions in a variety of widths and diameters, the jig can be adapted to handle many other types of cylindrical parts.

Basically, the jig consists of an adjustable V-block *A* to support the work-piece *B*, a hinged jig plate *C* with a replaceable bushing to locate and guide the drill and a means of clamping and rotating the part. The V-block is raised or lowered in guide block *D* by turning the knob end of adjusting screw *E*. A pointer on scale *F* indicates the diameter of collar that can be drilled at each vertical position of the V-block.

Gear *G* transmits the rotary motion of shaft *H* through a gear train to gear nut *J* which moves externally threaded sleeve *K* in an axial direction. The gear nut is axially retained by a bushing and the outer-bearing support plate. Shaft *L* has a sliding fit in the bore of the threaded sleeve. A pin is pressed into shaft *L* and is fitted into slots in clamp *M*. Shaft *L* and clamp *M* must rotate together but can move axially about 1/8 inch in relation

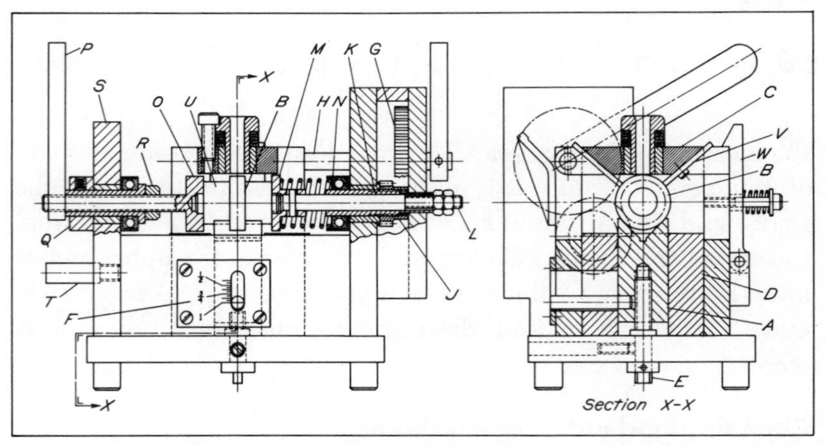

FIG. 17. Drill jig with mechanism for clamping and indexing a variety of cylindrical parts. Clamping action is concurrent with lowering of hinged jig plate.

to each other. A coil spring pushes clamp *M* toward the work-piece and is allowed by thrust bearing *N* to rotate freely with the clamp and the shaft. Forward motion of shaft *L* is restrained either by the work-piece through clamp *M* and its retaining pin or by threaded sleeve *K* through lock-nuts. The latter is the case when clamp *M* is not in contact with the work-piece.

Clamp *O*, lever *P*, and an internally threaded sleeve with its retaining collar *Q* and lock-nut *R* may be rotated as a unit in a fixed axial position. A thrust bearing allows free movement of the parts under any heavy clamping force applied to the work-piece by clamp *M*. Index holes are provided in bearing support plate *S* for insertion of threaded stop *T*. This stop is used to accurately position lever *P* and, therefore, the work-piece for the drilling of radial holes.

Axial position of clamp *O* can be adjusted to suit work of various widths by advancing or retracting this member within the internally threaded sleeve which is held in place by lock-nut *R*. In similar fashion the lock-nuts on the end of shaft *L* provide a means of adjusting the axial position of clamp *M*.

This jig is simple to use. A work-piece is placed on the V-block which is then set to the proper height by means of the adjusting screw. The rotary motion of lowering the jig plate into position is converted to the horizontal translation of threaded sleeve *K*. Clamp *M* (and its retaining shaft) are forced by the compressed spring to duplicate this movement and clamp the work-piece properly. Extension pieces *U* are screwed into the faces of the clamps to increase the range of the jig. Clamp *O* is pre-set axially to locate the work-piece properly. The lock-nuts on shaft *L* should be adjusted so that the pin is not in contact with either end of the slot in clamp *M* when the work-piece is securely gripped. This will prevent friction between the threads in parts *J* and *K* when the work is rotated.

After drilling the first hole in the work-piece, lever *P* is indexed to the stop for the second hole. Additional radial holes can be produced by moving the threaded stop to the next indexing position and repeating this operation. Clamping pressure is maintained on the work-piece since thrust bearing N allows the spring to rotate freely with clamp M. Adjustable work stops V prevent parts from being pulled up when the drill is retracted, and spring-actuated latch W holds the jig plate down in position for drilling.

CHAPTER 6

Reversing Mechanisms of Special Design

Described in this chapter are various arrangements for obtaining reversal of motion. Other reversing mechanisms are described in Chapter 6 of Volumes I and III and Chapter 7 of Volume II of "Ingenious Mechanisms for Designers and Inventors."

Sensitive Feed Arrangement for Coil-Winding Machine

Machines employed for winding fine wires into coils require sensitive feed arrangements that permit extremely quick reversals. For example, in winding wire 0.001 inch in diameter, with the bobbin rotating at 6000 rpm, reversing should be accomplished in a distance of 0.001 inch in a hundredth of a second. Although such quick reversals are usually obtained by fine screw feeds or by friction wheels, the arrangement shown in Fig. 1 is a unique solution to the problem.

A steel band C (see Fig. 1) is driven at constant speed by means of pulleys A and B. The band passes through two electromagnets, M_1 and M_2, which are mounted on wire guide carriage D of the coil-winding machine. When the carriage comes to the right-hand stop E, which is a micro switch, magnet M_1 is de-energized and magnet M_2 energized. The carriage is thus attracted to the lower portion of the band and moved to the left. When the carriage contacts the left-hand stop F (another micro switch), magnet M_2 is de-energized and M_1 energized, thus moving the carriage to the right.

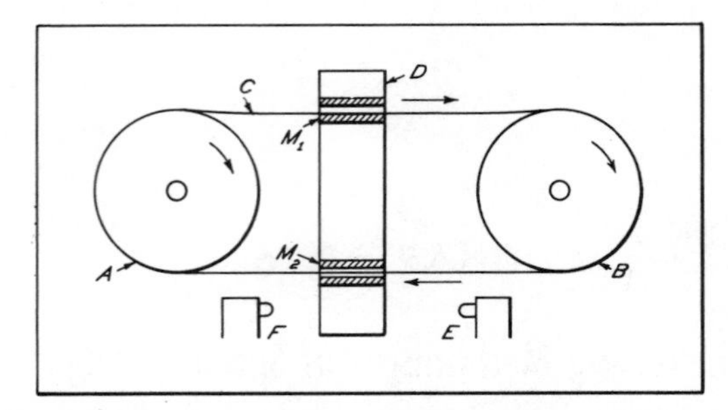

FIG. 1. Wire guide carriage *D* of a coil-winding machine is reciprocated between micro switches *E* and *F* by a steel band *C* which is alternately attracted to two electromagnets M_1 and M_2.

Speed of the steel band can be varied to obtain different rates of feed, and the location of the micro switches can be altered to change the length of carriage stroke.

Excessive-Torque Reversing Mechanism

A rotating drum type hopper, used to feed small molded parts into a chute, was subject to occasional jamming. The simplest way to free the jam was to reverse the direction of hopper rotation. To do this automatically, the illustrated mechanism was designed.

Two bevel gears *A* and *B* (see Fig. 2) are free to turn on drive shaft *C*. Smaller bevel gear *D*, which is keyed to the hopper shaft, is in constant mesh with the two larger bevel gears. Keyed to the drive shaft is a central driving member *E*. Two stepped hubs *F* and *G*, each free to turn on the drive shaft, carry pins *H* and *J*.

Spiral springs *K* and *L* connect the two hubs to their respective bevel gears by means of pin *M* in the case of spring *K*, and a similar pin, not shown, in the case of spring *L*. The primary function of the spiral springs is to absorb any shock load that might occur in the event of the hopper jamming. Driving dog *N* is secured to member *E* by means of a shoulder screw on which it is free to pivot. The dog is held against one of the two stop-pins *O*, by toggle spring *P*.

During normal operation, drive shaft C rotates in the direction of the arrow. With driving dog N in its left-hand position as shown, motion is transmitted from member E to stepped hub F, and from there, through spiral spring K to pin M. This causes bevel gear A to become the driving gear with respect to driven bevel gear D on the hopper shaft. Under these conditions, bevel gear B merely idles.

Any jamming that occurs during operation of the unit causes an increase in the torque necessary to drive the hopper. As a result, spiral spring K is placed under load. When sufficient pressure is built up between pin H and driving dog N to overcome the initial tension in toggle spring P, the dog will pivot about its mounting screw and come to rest against the right-hand stop-pin O. This releases the driving load from the components on the left.

Continued rotation of drive shaft C in its normal direction causes the right-hand side of the driving dog to engage pin J in stepped hub G. Through spiral spring L, and a pin similar to M, the driving force is now transmitted to bevel gear B. This, of course, causes the hopper to rotate in the opposite direction.

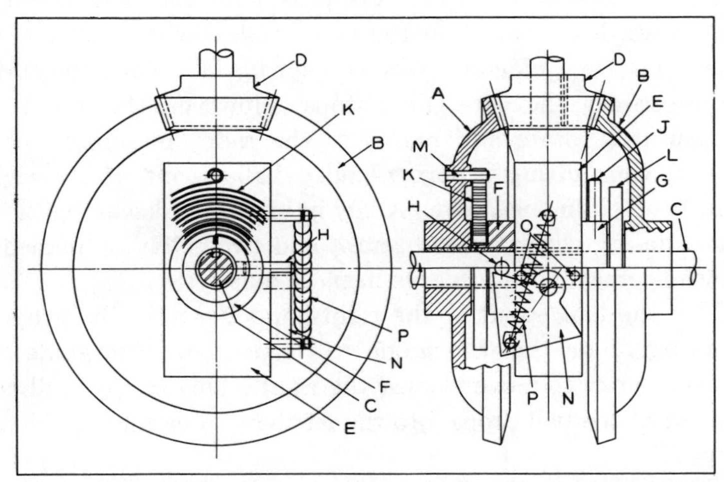

FIG. 2. Direction of hopper rotation, driven by gear D, is reversed by the pivoting action of a driving dog N whenever excessive torque is built up.

Reciprocating Traversing Device with an Adjustable Stroke

A mechanism designed for leading wire onto a spool in uniform layers is shown in Fig. 3. The arrangement incorporates a simple means of producing a smooth, reciprocating motion to the wire guide. In addition, the length of stroke of the guide is easily adjusted to accommodate spools of various widths.

The wire guide *A* is free to slide on a guide rod *B* and is traversed by a lead-screw *C*. Brackets *D* and *E* serve as bearings to support both the lead-screw and the guide rod. A bevel gear *F* is secured on shaft *G*, which is connected to the drive for the wire spool. Two additional bevel gears *H* and *J* are in mesh with gear *F* and rotate in opposite directions on the guide rod. Gears *H* and *J* each have a saw-tooth clutch plate attached to one face. A driving clutch member *K* having teeth on each face is pinned to the lead-screw between the gears. The direction in which the lead-screw is driven depends on the position of the driving clutch member *K*.

In the illustration, part *K* is shown in position to rotate the lead-screw in the direction that will cause the wire guide to move toward a collar *L* on the guide rod. Before reaching the collar, the wire guide compresses a spring *M*. When the spring is compressed, the wire guide stops. However, the lead-screw continues to rotate and moves to the right, pulling sleeve *N* with it, thus lifting a spring-loaded ball *O* out of the right-hand V-notch in the sleeve. A key in bracket *E* keeps the sleeve from rotating with the lead-screw, and two collars *P* pinned to the lead-screw hold the sleeve in place axially.

Once the ball is out of the right-hand V-notch, the pressure of spring *M* on the wire guide will cause both the guide and the lead-screw to move toward the right. This motion will continue until the ball drops into the left-hand V-notch provided in sleeve *N*.

Clutch member *K* will then be engaged with the clutch plate attached to gear *J*, and the lead-screw will rotate in the opposite direction. This will cause the wire guide to move toward collar

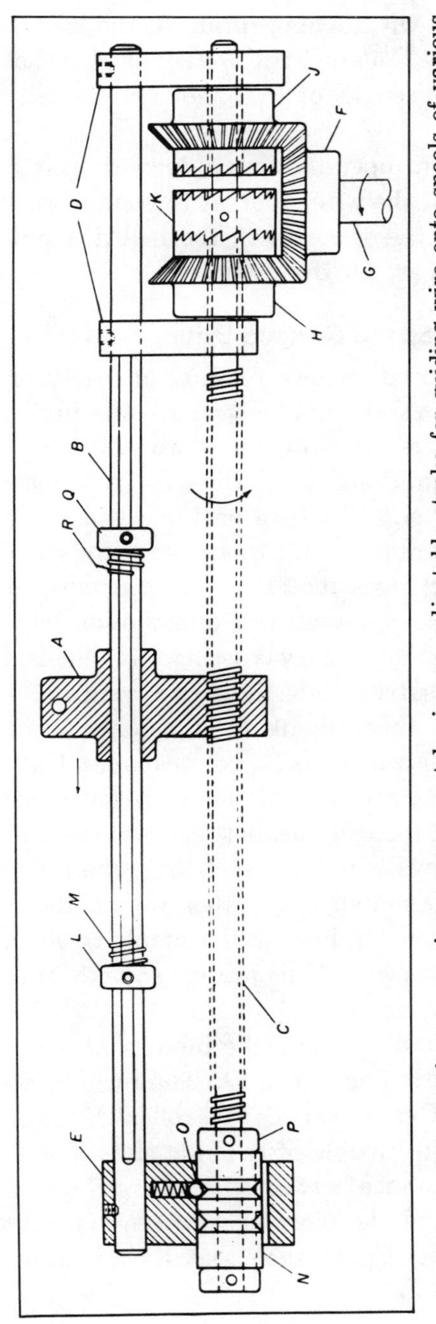

FIG. 3. A reciprocating, traversing arrangement having an adjustable stroke for guiding wire onto spools of various widths.

Q and spring *R*. On reaching spring *R*, the lead-screw reversing cycle is repeated. Collars *L* and *Q* may be placed at any distance apart within the length of the guide rod to suit various spool widths.

The mechanism operates smoothly with just a slight pause before reversal of the wire guide at the end of each stroke. The rapid motion of the guide when the ball lifts out of the notch compensates in part for the pause.

Reversing Two-Speed Geneva Drive

A proposed aerial camera required a mechanism to drive a prism in a certain series of movements. Specifically, these were: (1) turn 60 degrees counterclockwise, (2) stop momentarily, (3) rotate an additional 60 degrees in the same direction, (4) pause for an instant, (5) turn back clockwise 120 degrees, (6) stop again momentarily, and then repeat the cycle. The device was to be driven by a motor having constant, uniform speed; and the transition between rest and motion had to be shock-free. The compound Geneva mechanism illustrated in Fig. 4 was designed to satisfy all of these requirements.

The driving member, crank *A*, is fastened to the drive shaft. This crank carries two rollers *B* and *C* capable of entering slot *D* in a Geneva wheel *E*, which serves as the prism carrier. The length of the crank and the distance between its center of rotation and that of the Geneva wheel are such that the rollers engage and disengage slot *D* radially. In other words, the angle between slot *D* and the center line of the crank is 90 degrees at the moment of engagement. This insures smooth, shock-free operation.

Attached to crank *A* (and the input shaft) is a spur gear *F* which meshes with a spur gear *G*. The ratio between these two gears is 2 to 1. Gear *G* carries two rollers *H* and *J*, located 120 degrees apart and capable of entering slots *K* and *L* in Geneva wheel *E*. The distance between the rollers *H* and *J* and the center of gear *G*, and the distance between the center of gear *G* and Geneva wheel *E*, are such that the rollers engage slots *K*

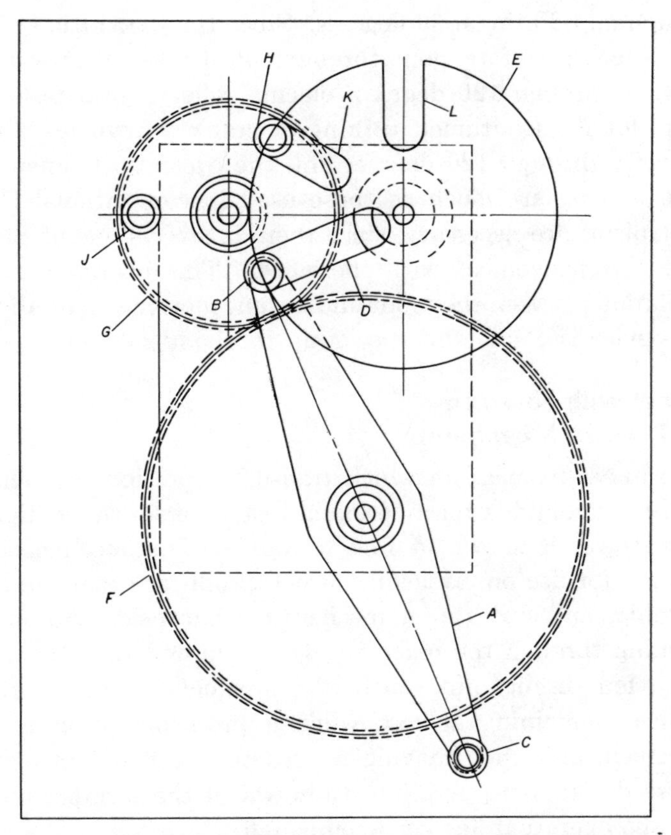

FIG. 4. Plan view showing details of Geneva mechanism. Members are in position to start cycle of movements required for a prism in an aerial camera.

and *L* radially. These slots are positioned 60 degrees apart on member *E*.

In operation, the drive starts its cycle in the position shown in Fig. 4 and crank *A* is rotated counterclockwise. This causes roller *B* to leave slot *D* and gear *F* to drive gear *G* clockwise. Roller *H*, mounted on gear *G*, simultaneously engages slot *K* and turns wheel *E* through 60 degrees. At this point, roller *H* moves out of slot *K*, wheel *E* stops momentarily, and roller *J* enters slot *L*. Once engaged with slot *L*, roller *J* turns wheel *E* through an additional 60 degrees. Since the gear ratio is 2 to 1, gear *G* has, therefore, gone through 240 degrees of rotation and crank *A* through 120 degrees, placing roller *C* in a position to enter slot *D*. Continued turning of crank *A* revolves wheel *E* clockwise through 120 degrees and the cycle starts anew.

No locking arcs such as those used in conventional Geneva mechanisms are necessary, since there is always one of the four rollers in engagement with the wheel. This insures a positive correlation between the input and output movements at all times. The device is capable of speeds up to about 250 rpm.

Ratchet with Forward and Reverse Movements

A ratchet mechanism which transmits rotation to a shaft in one direction and, after a period of rest, reverses the motion to a lesser degree is shown in Figs. 5 and 6. This mechanism was designed for use on a machine which produces ornamental wire screening, and operates a mechanism that feeds the sheet of screening through the machine. It is required that the screening be fed through intermittently, advancing a predetermined distance, remaining at rest while a press operation is being performed, and then moving a predetermined distance in the reverse direction to permit withdrawal of the forming punches after the operation has been completed.

Front views of the mechanism at different positions of the cycle are shown in Figs. 5 and 6. In this mechanism, shaft *A* carries the ratchet wheel *B* which is keyed to it. Due to the reversing action of the mechanism, the conventional saw-toothed ratchet wheel cannot be used; therefore, the teeth on the ratchet wheel are a series of 90-degree notches. With this design of tooth, it is necessary for the thrust of the pawl to be approximately perpendicular to the contact surfaces, otherwise the pawl will jump out of the notches.

In operation, lever *C* swings freely on the hub of the ratchet wheel, receiving an oscillating motion from a cam-operated connecting-rod. Lever *C* carries pawl *D*, which is heavier on one side in order to be unbalanced. An extending arm of pawl

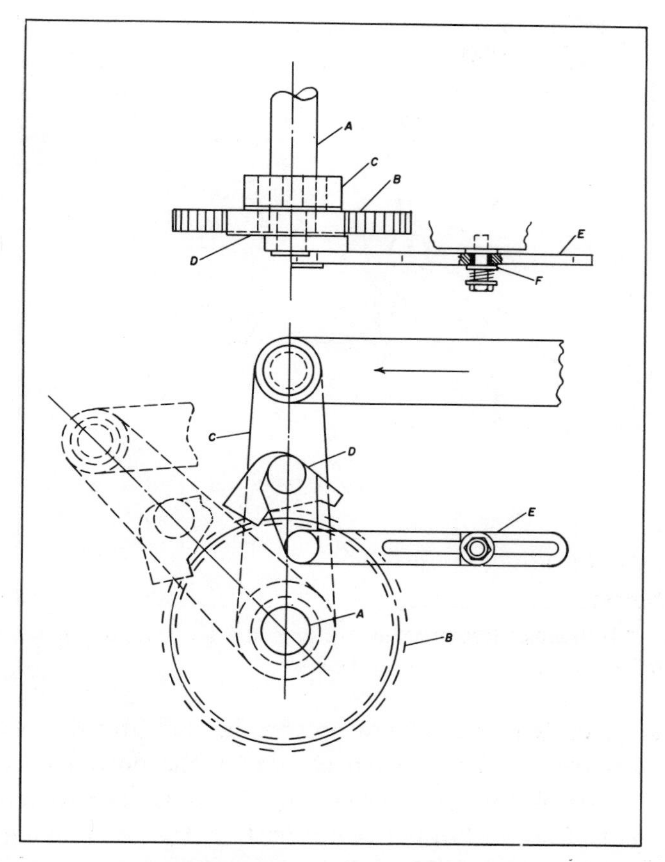

FIG. 5. Ratchet set for movement of slide-bar E to left, in a mechanism for forward and reverse movements.

D is connected to one end of slide-bar E, which is supported on a stud fixed to the machine. A light spring provides frictional resistance to the movement of the slide-bar during part of the cycle. The bar is of reduced thickness at the right-hand end and a flanged bronze bushing applies the frictional resistance under the action of the spring. The length of the bushing shank is such that friction is applied to the thicker section of bar E only, as can be seen in the top view.

Referring to Fig. 5, which shows the mechanism at the mid-point of the cycle, lever C is moving toward the left, and pawl

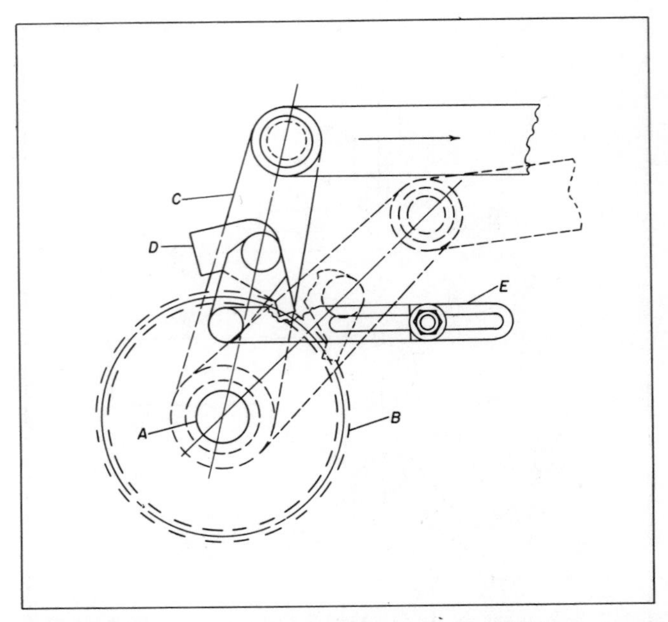

FIG. 6. Mechanism showing ratchet engaged for slide-bar movement to right.

D is engaged with one of the notches in the ratchet wheel. As lever *C* moves to the position shown by the dotted outline, it rotates shaft *A* through the action of the ratchet wheel. The movement of lever *C* draws bar *E* with it, the bar moving without any resistance at this point.

On the return stroke, one end of pawl *D* rides back over ratchet wheel *B* without transmitting motion to it. In Fig. 6, the mechanism is illustrated at the point where the heavier portion of bar *E* has contacted the bronze supporting bushing and the latter now can move only against the resistance created by the spring. This causes pawl *D* to reverse its position so that the other end engages a notch in the ratchet wheel, as shown. The reverse movement of shaft *A* begins at this point.

The resistance applied to the movement of bar *E* holds pawl *D* in engagement until lever *C* has reached the end of its travel to the right, as shown by the dotted outline. When lever *C* begins the forward stroke, the resistance applied to bar *E* causes pawl *D* to immediately reverse and engage another notch in ratchet wheel *B* for the forward stroke. This mechanism produces a forward rotation of shaft *A* equal to the movement of five ratchet teeth and a reverse movement equal to that of two teeth.

Reversing Linear Feed with Adjustable Stroke

A means was required for guiding wire onto reels by reciprocating a guide head between the flanges. In order to accommodate different flange widths, as well as varying distances between hubs, it was found necessary to provide an adjustment for the linear travel and relative location of the head. Further, to be able to wind different wire sizes on the reel, the linear speed of the guide head had to be variable. The mechanism shown in Figs. 7 and 8 was designed for the application.

The guide head *A*, Fig. 7, is free to travel back and forth on two nonrotating shafts *B* supported by two brackets *C*. A lead-screw *D* engages a full thread in the guide head and is supported

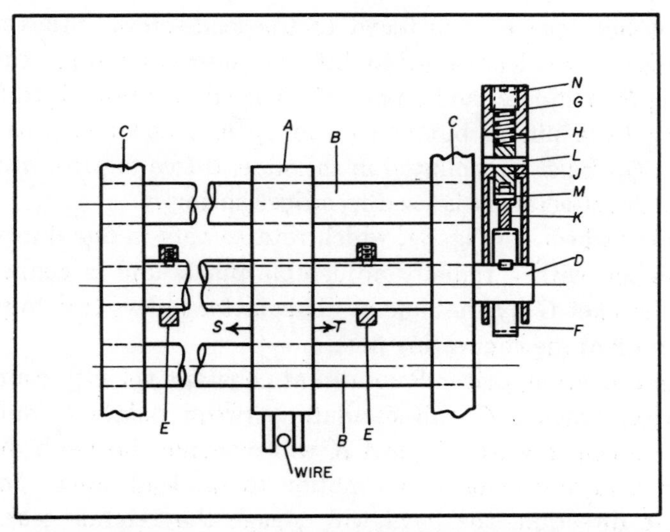

FIG. 7. Front view of feed device which provides a continuously reversing linear movement. Drive wheel and linkage, shown in Fig. 8, are omitted for clarity.

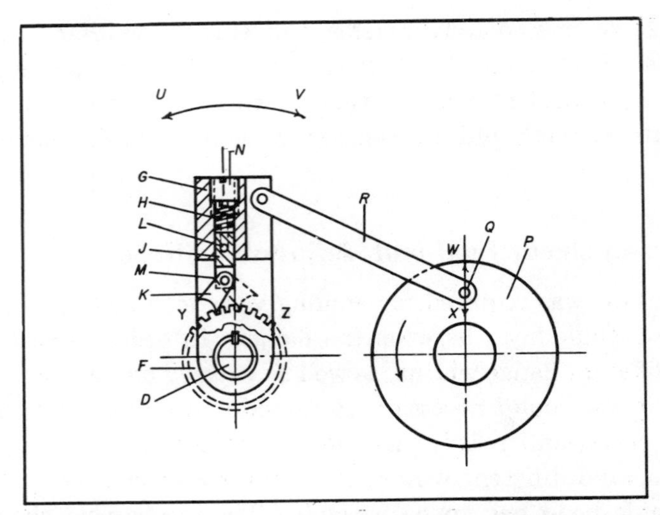

FIG. 8. Partial side view of mechanism shown in Fig. 7. Movement of guide head can be varied by adjusting the position of pin Q.

in journal bearings in brackets C. Stop collars E can be moved along the lead-screw and locked in any required position by means of set-screws.

A ratchet gear F is fastened to the lead-screw. Bracket G, supported on the lead-screw by integral journal bearings, carries a spring H, a slide J, and a pawl K. The slide is restrained from rotating by a pin L, but can move up and down in a slot in bracket G. Pawl K, mounted in the slide, is free to pivot around pin M. A set-screw N is used to adjust spring H.

Driving wheel P (Fig. 8), which rotates only in one direction, is equipped with a radially adjustable pin Q and is connected to the bracket G by the link R. This latter link is free to pivot about both of the connecting pins.

If, in operation, driver P rotates at constant velocity counterclockwise, bracket G will oscillate between points U and V. When moving toward V, pawl K will override the teeth of the ratchet gear and transmit no motion to the lead-screw. In the opposite direction, the pawl will engage the ratchet gear and rotate it counterclockwise. This, in turn, will move the guide head in direction S.

When the linear movement of the guide head is stopped by collar *E*, lead-screw *D* is unable to turn, as the guide head is retained by shafts *B*. If bracket *G* is still forced to move toward *U* and the lead-screw cannot rotate, pawl *K* will force slide *J* up and compress spring *H*. This will allow pawl *K* to flip over from position *Y* to *Z*. The oscillating movement of bracket *G* then will rotate the lead-screw clockwise and reverse the direction of travel of the guide head. When the guide head comes to the opposite stop collar *E*, the cycle is repeated.

By adjusting the position of pin *Q* in direction *X* or *W*, the oscillating bracket movement can be varied in magnitude. This, in turn, will vary the linear movement of the guide head to accommodate different wire sizes.

Machine Counter with Dwell Interval Operated Through Slotted Discs

On a warp machine used in the textile industry, the counter which keeps tabs on the amount of yarn drawn is operated through a device which incorporates a series of slotted discs to obtain a desired amount of lost motion. The reason for this is that the device must reverse its rotation several turns after drawing a certain length of yarn, then must rotate forward the same number of turns before more yarn is drawn, at which time the counter must pick up where it left off.

A drawing of the device appears in Fig. 9. The main element is a gear *A*. Within the gear bore is a series of eight discs *B*. Fastened to the right-hand face of the gear is a flanged plate *C*. On the left-hand face is a second flanged plate *D*. The latter is free to rotate, but is restricted from axial movement by a groove in its periphery which engages dog-point set-screws *E* in the gear. Plate *D* carries connecting-rod *F*, the top of which is joined to the counter (not shown). The entire assembly is bolted to the frame *G* of the machine.

One of the discs is shown in Fig. 10. Each has a large open area in the form of a curved slot *H*, and a hole *J*. The width of the slot is made greater than the diameter of the hole. This construction permits the large diameter of a shouldered pin *K*,

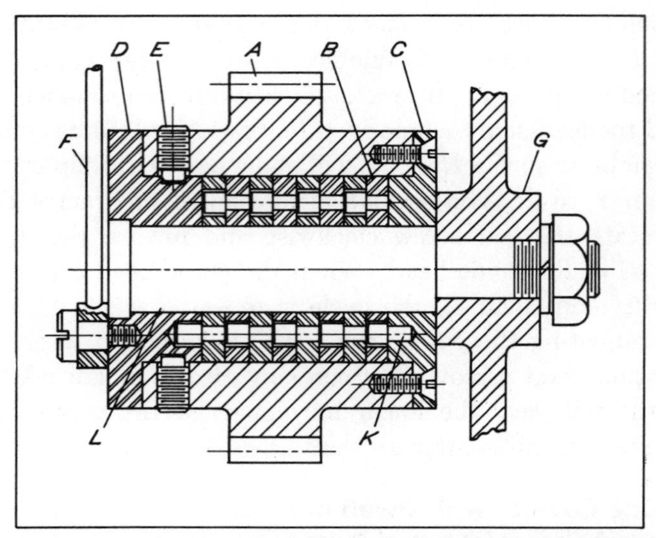

FIG. 9. When gear *A* is reversed, plate *D* remains fixed while the lost motion is absorbed.

Fig. 9, to bear in the slot of one disc, and the small diameter of the pin to fit the hole of the preceding disc. (For the first disc, the small diameter fits a hole in plate *C*. Similarly, the inner face of plate *D* has a hole accommodating the large diameter of the pin, the small diameter of which bears in the slot of the last disc.)

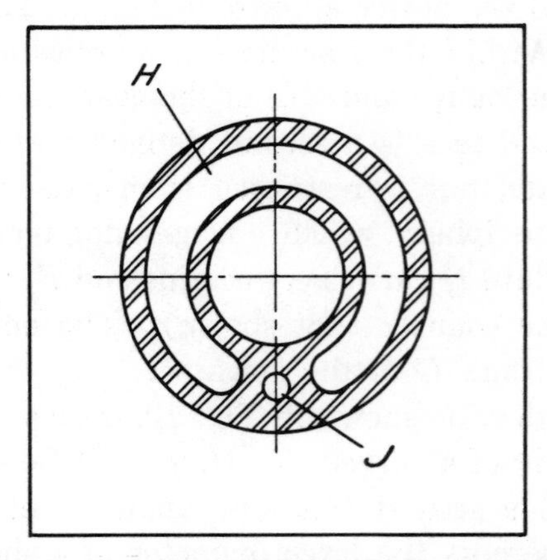

FIG. 10. Slot *H* in the disc accommodates the large diameter of one pin, and hole *J*, the small diameter of another pin.

When gear *A* is running forward and yarn is being drawn, it revolves as a unit with plate *D* around shaft *L*, and the connecting-rod operates the counter. The line of transmission extends from the gear to plate *C*, then to each shouldered pin and disc, and finally, to plate *D*. Then when the gear must be reversed for several revolutions, plate *D* remains fixed and the counter does not operate. The reason for this is that as the reversal is set up, the pin in plate *C* must be dragged from one end of the slot in the first disc to the other before the first disc joins in this reverse rotation. Similarly, there is a delay of almost a complete revolution before each of the following discs in turn starts its reverse rotation. Thus, a total of approximately seven revolutions of backward turning is available if required.

When the gear again runs forward, plate *D* remains fixed until all the pins in turn have been dragged to the opposite end of their respective disc slots. At that time, yarn again is drawn from the machine and plate *D* resumes its movement, and the counter again operates.

Direction-Changing Drive

A large number of machine tool components such as lathe saddles, milling and boring machine tables, and boring machine heads, incorporate subassemblies that move at right angles to each other. Usually, each of these subassemblies must be independently reversible. It is also frequently necessary to synchronize the movement of the subassemblies to obtain combined movement of 45 degrees. A patented, compact direction-changing drive, capable of performing all these functions through the actuation of a single control lever, is described.

The outside of the gear-box, from which the gears and control lever have been removed, may be seen at the left in Fig. 11.

Four gears, *C*, *D*, *E*, and *F*, comprise the train. Two of them, *C* and *F*, are sliding gears and are mounted on the two output shafts *G* and *H*. The remaining two are fixed gears: gear *D* is an intermediate gear and gear *E* is the driving gear, being mounted on input shaft *J*. Both sliding gears are moved by the action of a single control lever *K*, view B-B.

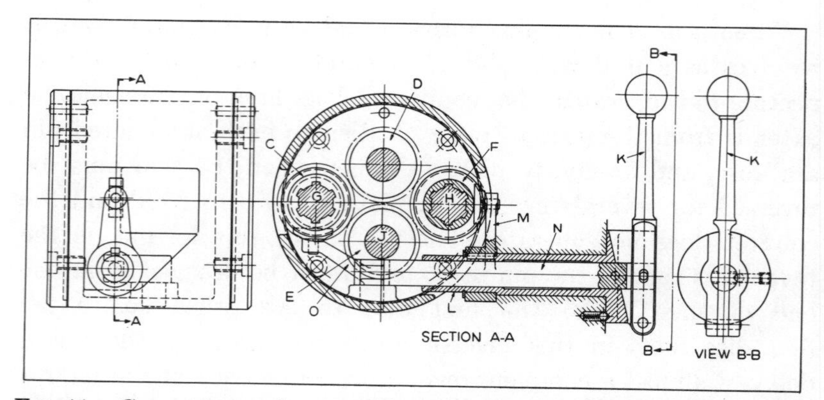

FIG. 11. Compact gear-box provides single-lever selection of eight directional output combinations of shafts *G* and *H* from a unidirectional input shaft *J*.

When the control lever is moved at right angles to the center line of sleeve *L*, sliding gear *F* will be shifted by means of arm *M*. This provides positive control over the engagement of the sliding gear with either intermediate gear *D* or driving gear *E*. This may be more clearly seen in the developed section of the gear train, Fig. 12. Movement of the control lever in a direction parallel to the center line of sleeve *L* will cause the shifting of sliding gear *C* through the action of push-pull rod *N* and offset bellcrank *O*. In this way, through the movement of only one control, eight drive motions can be called upon. A ninth, neutral position is represented by the gear arrangement shown in Fig. 12.

The position of the control lever, together with the positions assumed by the involved gears for each of the eight obtainable output motions, are shown in Fig. 13. In each of these cases input shaft *J* is rotating in a counterclockwise direction as viewed from the output side of the unit. When control lever *K* is pulled away from the gear-box, as seen at P, gear *C* is moved to the left into engagement with driving gear *E*. This causes output shaft *G* to rotate in a clockwise direction as indicated by the arrow. It may be noted that intermediate gear *D* and driving gear *E* are engaged at all times.

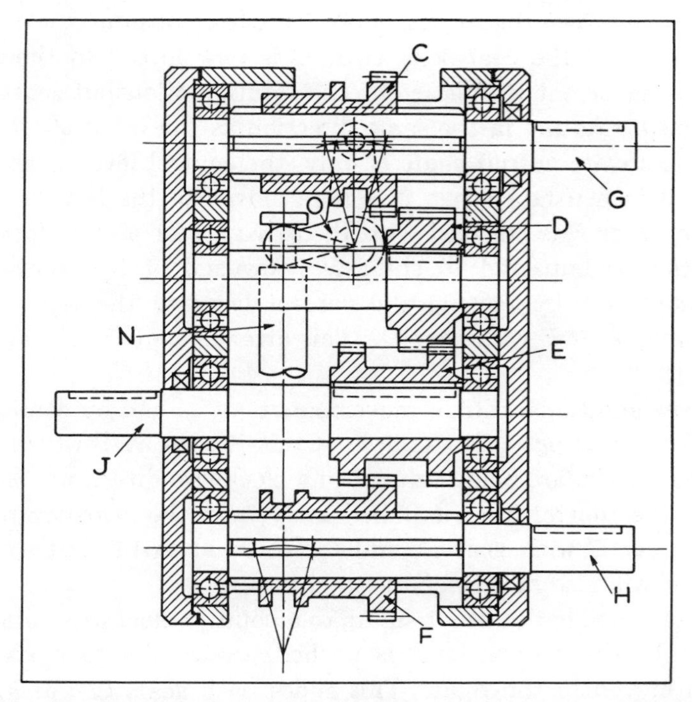

FIG. 12. Developed section of gear train showing types of gears and their position when the control lever is in a neutral position.

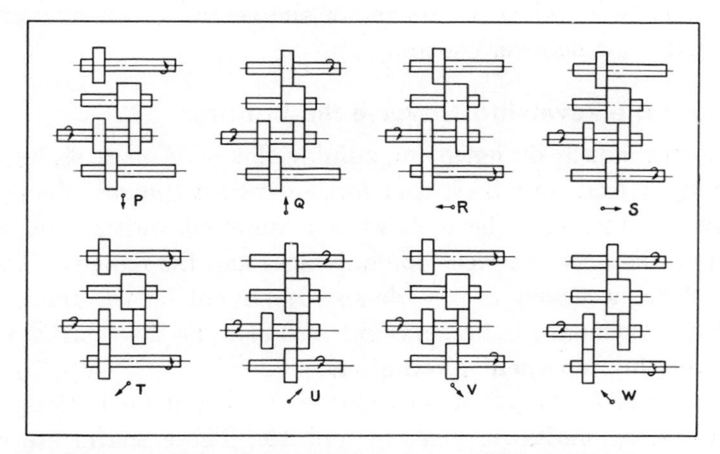

FIG. 13. Diagrammatic representation of the gear and lever positions for each of the eight output variations possible with the direction-changing drive.

At Q is shown the condition resulting from pushing the control lever toward the gear-box. Gear *C* is now forced to the right into engagement with gear *D*. The rotation of output shaft *G* is reversed, rotating in the same direction as the input shaft.

To activate output shaft *H* only, the control lever is first returned to neutral. When it is then moved to the left, as at R, sliding gear *F* is engaged with the driving gear and a clockwise rotation is imparted to shaft *H*. Reversal of this motion is accomplished by moving the control lever to the right, thus engaging gear *F* with gear *D*. This arrangement is seen at S in Fig. 13.

Four additional sets of movements may be had by combining the basic settings. An example of this is seen at T where both output shafts are being rotated in a clockwise direction. In this case, the control lever is pulled away from the gear-box to engage gear *C* with gear *E*, and then moved to the left so that gear *F* will also be engaged with gear *E*.

To reverse both output shafts to a counter-clockwise rotation, as at U, the control lever is pushed toward the gear-box and then moved to the right. This slides both gears *C* and *F* into engagement with gear *D*. The remaining two settings, shown at V and W, illustrate the two additional sets of movements that can be obtained by means of simple two-directional shifting of the single-lever control.

Adjustable Reversing Traverse Mechanism

A mechanism designed for guiding flat wire on reels by traversing a feed unit back and forth between the reel flanges is shown in Fig. 14. The reels were of different widths, and with various distances between the hub faces and the flanges. It was, therefore, necessary to provide an adjustment for the amount of linear movement of the feed unit and also for the relative location of the movement alternations.

The wire-feeding unit reciprocates back and forth along two nonrotating shafts *A*, Figs. 14 and 15. These shafts are supported by two brackets *B*, only one of which is shown. Bracket *C* (the wire-feeding unit) is free to slide on shafts *A*. This

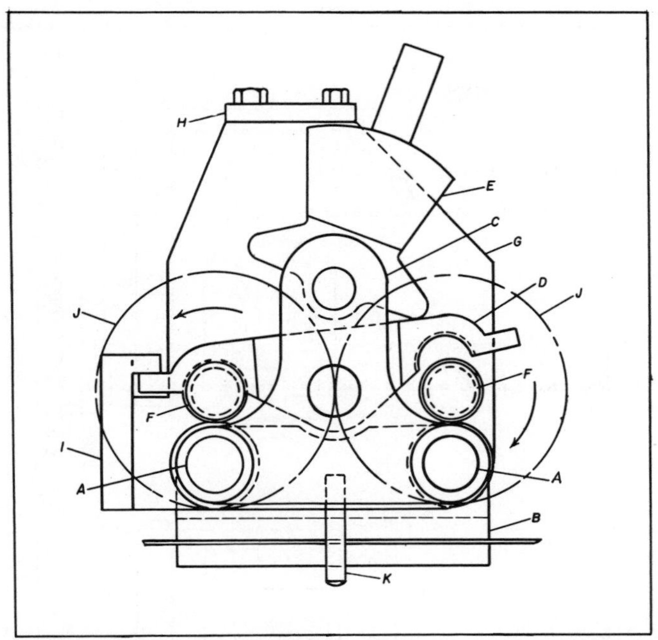

FIG. 14. End view of mechanism that proves an adjustable reversing traverse movement.

bracket carries a swinging lever *D*, which is provided with two half-thread arms that alternately engage one of two threaded portions of rotating shafts *F*. Either one of shafts *F* may be a driver to transmit rotation in the reverse direction through gears *J*. The threaded portions of shaft *F* may both be right-hand or both left-hand, depending upon the direction of rotation of gears *J*. Bracket *C* also carries a swinging weight *E* which has lower projections that alternately contact the upper surface of lever *D*.

Bracket *G* can be moved along shafts *A* and locked in any required position by means of a set-screw. This bracket carries a cam-plate *H* and a bar *I*. A similar bracket with parts *H* and *I* positioned in reverse positions, as shown in the dotted diagrams at the left in Fig. 15, is mounted on the other end of shafts *A*. Brackets *G* are positioned to control the distance by the feeding unit and the location of the movement relative to

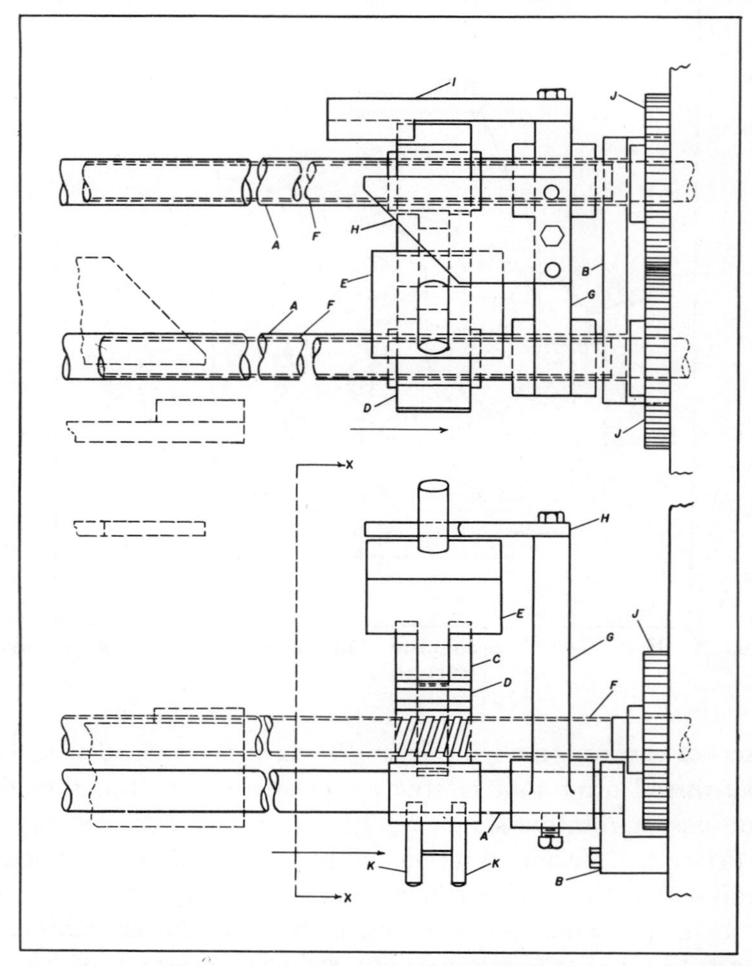

FIG. 15. Top and front views of mechanism that provides a reversing traverse movement which is adjustable for position and length of traverse.

the machine. The wire is guided between two pins K moved on bracket C.

Bracket C is moved in the direction indicated by the arrow when one of the half-threads on lever D engages one of the threaded shafts F. When this occurs the cylindrical projection of weight E has contacted plate H, causing weight E to be tipped off center, as shown in Fig. 14, so that one of its projections contacts the upper surface of lever *D*. Lever *D*, however, cannot pivot at this point because the projection on its left-hand end is retained by the projecting portion of bar *I*.

In the top view of Fig. 15 bracket *C* has moved to a point where the projecting end of lever *D* is about to emerge from under bar *I*. When bar *I* has been cleared, lever *D* will be permitted to swing under the action of weight *E*. This will cause disengagement of the half-thread from one shaft *F* and engagement with the other shaft so that the direction of the linear movement of the feeding unit is reversed.

The purpose of bar *I* is to closely control the timing of the mechanism reversal and to prevent the half-nuts from lifting out of the shaft threads while weight *E* is swung to the opposite side.

CHAPTER 7

Reciprocating Motions Derived from Cams, Gears, and Levers

Described here are mechanisms which give reciprocating motions derived from cams, gears and levers. Cams, gears and/or levers may be used to vary the stroke in some way or to impart a mechanical movement essential to meet a particular operating requirement.

Other reciprocating mechanisms, based on the action of gears, cams and levers are described in Chapter 9, Volumes I and II, and Chapter 7, Volume III of "Ingenious Mechanisms for Designers and Inventors."

Slow Reciprocation of Grinding Wheel Obtained from High-Speed Drive

Slow reciprocation of the abrasive wheel on a special centerless grinding machine is obtained from the high-speed rotary movement without the need for a reduction gear by means of the mechanism shown in Fig. 1. Reciprocation is accomplished by means of an axial cam drive and an additional pulley on the grinding wheel spindle.

Grinding wheel spindle *A* can rotate and slide in adjustable bearings *B*. Sleeve *C*, on which grinding wheel *D* is mounted, is pinned to the spindle. The wheel is rotated by belt *E*, which is mounted on a split pulley (*F* and *G*). Half-pulley *F* is screwed on sleeve *C*, while adjustable member *G* is secured to *F* by set-screws. By varying the axial location of member *G*, the effective diameter of the pulley and, consequently, its speed can be changed. A second pulley *H* — this one freely mounted on the spindle — is rotated by belt *J*. Mounted between this

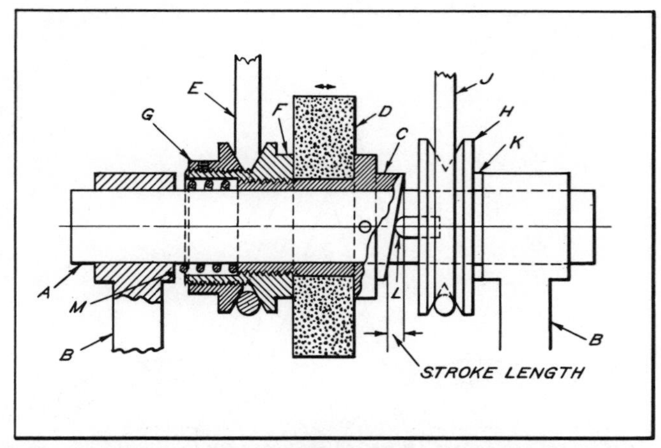

FIG. 1. Slow reciprocation of grinding wheel D is obtained by an axial cam drive mechanism. Freely mounted pulley H must rotate at a different speed than the wheel-driving split pulley F and G.

pulley and the right-hand bearing is a bronze washer K, and a pin L is pressed into the left-hand face of the pulley.

A helical cam surface provided on the right-hand end of sleeve C is kept in contact with pin L by spring M. Belts E and J can both be driven from the same shaft, but they must rotate the split pulley (F and G) and pulley H at different speeds in order to reciprocate the grinding wheel. If these pulleys were rotated at the same speed, there would be no relative motion between the pin and the contact point on the cam surface. However, as soon as there is a difference in the pulley speeds, the grinding wheel is reciprocated. The rate of reciprocation depends upon the difference in pulley speeds, and the length of stroke is controlled by the rise on the cam surface, as indicated on the drawing.

Geared Five-Bar Linkage for Straight-Line Motion

On an automatic machine where it was desired to guide a machine member along a straight line it was not possible to provide a guiding surface for the member. The solution is shown in Fig. 2.

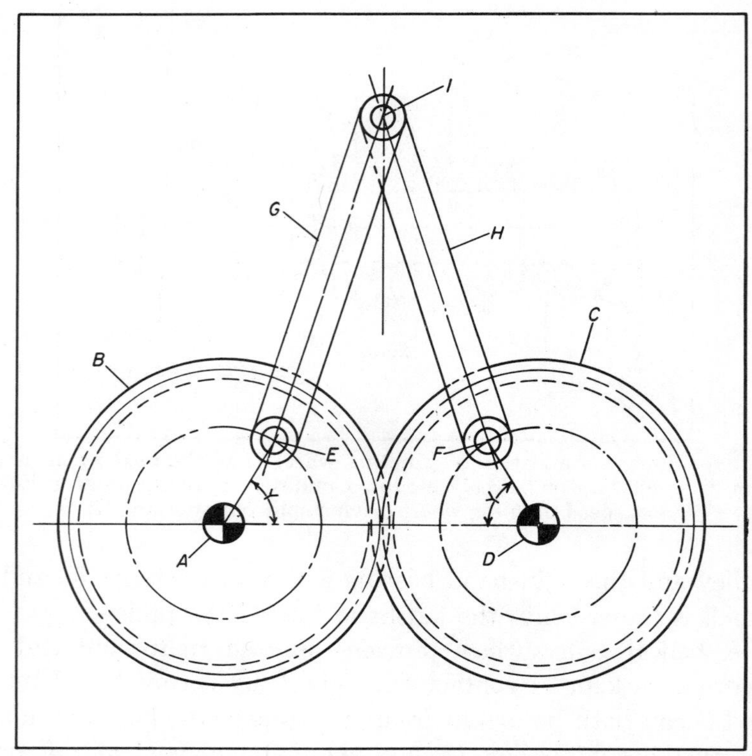

FIG. 2. Simple mechanism for obtaining straight-line motion through gearing.

In this movement driving shaft *A* carries gear *B* which is in mesh with gear *C*. Gear *B* is keyed to shaft *A*, and gear *C*, to shaft *D*. Both gears carry two studs *E* and *F* at equal radii from the centers of their respective shafts. They are also located so that angle *Y* is always the same for both studs. Links *G* and *H*, which are attached to the two studs, are of equal length. When gear *B* rotates, it drives gear *C*, and point *I* moves on a straight vertical line.

Reciprocating Cam with Half-Cycle Dwell

An irregular reciprocating motion was required on a machine used in the production of a formed wire part. This indicated the necessity of a cam. Due to the absence of a rotating part to which a conventional cam could be attached, it was decided to use a nearby reciprocating member to carry a straight cam. This, however, presented one drawback in that a reciprocating cam transmits motion to the follower on both the forward and the return strokes, the movement being reversed on the latter. As this was not permissible, the modified cam shown in Fig. 3 was designed.

Cam-slide *A* rides in a dovetailed groove machined in a stationary part *B* of the machine. Follower *C* contacts the cam surface of the slide, as shown at X. Adjacent to the cam-slide is a short groove in which rides a small slide *D*. Friction is maintained on the slide, which is slightly higher than the cam sur-

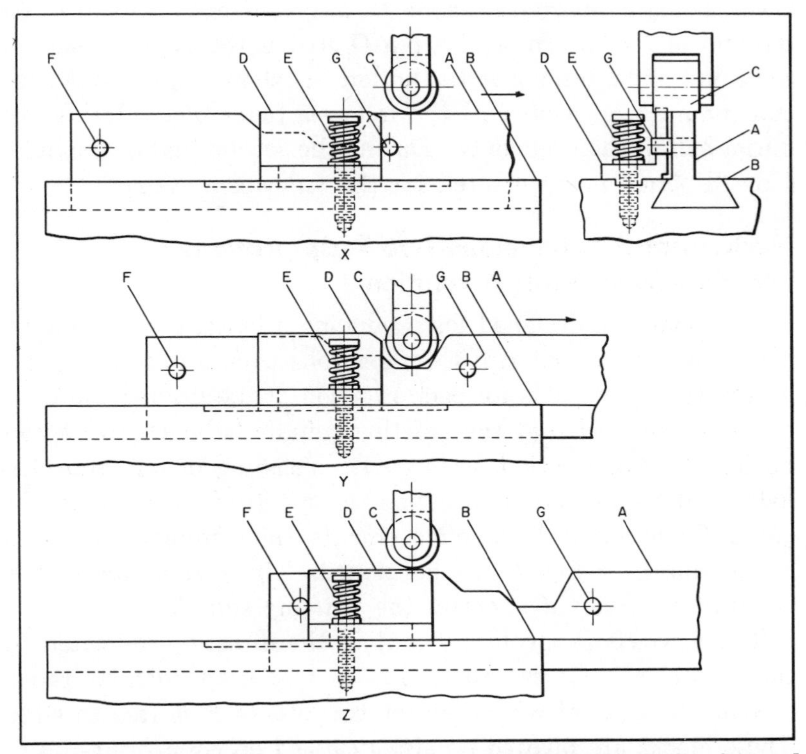

FIG. 3. Follower receives motion from reciprocating cam during forward stroke, but dwells during return stroke.

face, by spring *E*. The spring is retained by a stud that passes through a slot in the base of the slide. Two pins *F* and *G* protrude from the side of the cam.

The components can be seen at X in the position they assume at the beginning of a stroke. As the cam-slide moves in the direction of the arrow, follower *C* traces a path along the cam outline, as shown at Y. Continued movement of the cam permits pin *F* to contact slide *D* and force it under the follower, as illustrated in view Z. Due to the height of the small slide, the follower is lifted out of contact with the cam. This completes the first phase of the cycle of operation during which the desired reciprocating motion was transmitted to the follower linkage.

On the return stroke of cam *A*, pin *F* breaks contact with the small slide, which is held stationary by spring friction. The follower, being supported on slide *D*, also remains stationary. At the end of the return stroke, pin *G* strikes the opposite side of the small slide, forcing it to the left, as shown again at X. In this manner, movement of the follower takes place during the initial half of the cycle only. During the second half of the cycle a dwell period is substituted for the follower movement.

Mechanism that Develops Two Reciprocations with One Drive-Shaft Revolution

On a machine designed for producing a twisted wire product, it was necessary that two reciprocations back and forth be imparted to a wire guide for each rotation of the driving shaft. It was also required that the twisting spindle be given two cycles during the same period, each cycle consisting of one and one-half rotations in either direction, in synchronization with the guide. To accomplish these movements, the trammel gear mechanism shown in Fig. 4 was developed. The wire guide is indicated at *E*. Shaft *J* operates the twisting spindle.

The drive-shaft revolves disc *A* in the direction indicated by the arrow. Disc *A* is provided with two perpendicular radial grooves, in each of which one of two blocks *B* is free to slide. These blocks are pivoted on studs carried on connecting-rod *C*, which is attached to slide *D*. Slide *D* carries the wire guide

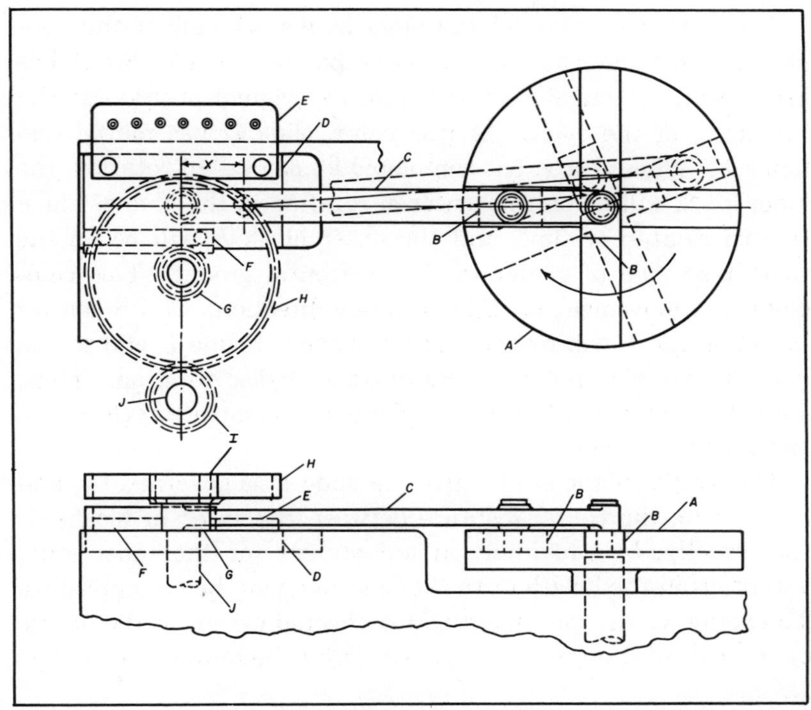

FIG. 4. Mechanism that incorporates a trammel gear for effecting reciprocating and oscillating motions.

and a rack *F* that meshes with gear *G*. Gears *G* and *H* are fastened together and rotate freely on the stud shaft on which they are mounted. Gear *H* meshes with gear *I*, which is attached to the twisting spindle shaft *J*.

Blocks *B*, being placed in slots perpendicular to each other, maintain their relative positions regardless of the circumferential position of disc *A*. As the disc revolves, the outer block *B* is caused to rise in its groove by the movement of the inner block *B*. The rotation of disc *A* acting on the outer block *B* causes the inner block to be drawn to the right as it rises. The combination of the two movements causes the blocks to move around the path of an ellipse which has its major axis on a center line extending between the center of disc *A* and the center of the connecting-rod stud on slide *D*.

The dotted outlines of the slots in disc *A* and of the connecting-rod *C* indicate their relative positions when disc *A* has been rotated about 60 degrees from its original position at the beginning of the cycle. At this point, slide *D* has moved distance *X*. When disc *A* has completed 90 degrees of rotation, the inner block will be on the center of rotation in the groove which is vertical at that time, and the outer block *B* will be on the right-hand side of center in the horizontal groove. This completes the movement of slide *D* in one direction. Continued rotation of disc *A* will reverse the movement of slide *D* and return it to the starting point at 180 degrees of disc rotation. Thus, a one-half rotation of disc *A* produces one complete cycle in the movement of slide *D*.

The length of the stroke given to slide *D* is governed by, and is equal to, the center distance between the studs in blocks *B*. Incidentally, by providing an adjustment for the inner stud, variations in the length of the slide stroke may be accomplished. The action of the movement is not affected except in the degree of shaft *J* rotation, which is governed by the movement of slide *D*.

Modified Eccentric Provides Rest Interval

In the fabrication of a wire product, a slide-actuated mechanism was required to transfer the part being made from a loading station to a work station, and then to an unloading station. The part was to be loaded on the forward stroke of the transfer slide, and unloaded on the return stroke. It appeared that a rotating eccentric disc offered the simplest method of moving the slide. This did not, however, prove entirely satisfactory because the velocity of the slide was too great when passing the unloading station to permit certain discharge of the part.

The difficulty was overcome by modifying the eccentric mechanism to provide a rest interval at the mid-point of the return stroke without otherwise affecting the work cycle. Figure 5 shows a mechanism that was satisfactory. Side views of the device are shown at X and Y and an end view at Z.

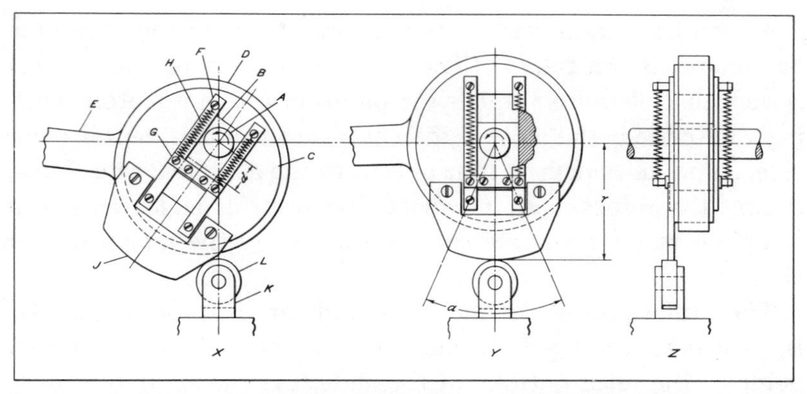

FIG. 5. By having the center of disc C momentarily coincide with the axis of shaft A, a rest interval is introduced in the action of this eccentric mechanism.

The shaft A which drives the mechanism is supported on bearings, not shown, and rotates in the direction indicated by the arrow. Keyed to this shaft is a block B fitting a rectangular slot cut through a disc C. Encircling the disc is a ring D integral with the rod E which extends to the transfer slide. Wear plates F support the disc. A bar G on block B carries studs that retain springs H. The opposite ends of springs H are anchored to studs attached to the wear plates.

A cam-plate J is fastened to one face of the disc. The greater part of the periphery of the cam-plate forms an arc that is concentric with the disc. Supported on a stationary bracket K is a roller L with which the cam-plate comes into intermittent contact during the functioning of the mechanism, and thus sets up a rest interval.

In view X, the disc center is offset from the shaft axis by a distance d, being in effect a conventional eccentric arrangement. The springs serve to keep the block at one end of the rectangular slot in the disc. In the position shown, the rise section of the cam-plate has just contacted the roller. Further rotation of the shaft causes the roller to act on this section, creating a movement of the disc center toward the axis of the shaft.

As can be seen in view Y, the curved section of the cam-plate has mounted the roller. (The vertical position of the bracket is such that distance *r* equals the radius of the cam-plate.) During this period, the axis of the shaft and the disc center coincide, so no movement is transmitted to the eccentric ring. Thus, during the rotation of the shaft, the transfer slide remains a complete rest for an interval determined by the magnitude of angle *a*.

The radial position of the block and cam-plate *J* on the disc is, of course, arranged to time the rest interval to the desired point in the return stroke of the slide. As the shaft continues to rotate, the contact of the cam-plate with the roller is terminated, and until contact is again made, the mechanism operates in the manner of an ordinary eccentric.

Two Slides Reciprocated Intermittently from One Source

Figure 6 shows a device used on a machine for producing a woven wire product. Its purpose is to carry strands of wire into the required positions in proper sequence.

Three views show the mechanism at different points in its operating cycle. Referring to view X, spur gear *A* rotates freely on a stud that is carried by a long bar *B*, moved by a cam (not shown).

Gear *A* meshes with two rack slides *C* and *D*, one above and one below. Rack *C* is free to slide and is retained by plate *E*. Two pins *F* protrude from this plate and act as stops.

Rack *D* is retained by plate *G*. This plate, however, is held against it under the pressure of springs held by a series of studs *H*, so that the rack must overcome a degree of frictional resistance before it can move. Both racks carry rods *J*, which are flattened and drilled on their outer ends to form eyes for guiding the wire strands.

View X presents the mechanism at the beginning of the cycle. At this point, rack slide *C* is in contact with right-hand stop-pin *F*. When bar *B* is moved to the left, rack *D* resists movement due to spring-loaded plate *G*. As rack *C* is not subject to such pressure, the gear is caused to rotate in a counterclockwise direction, rolling on the lower rack while carrying the upper rack with it.

This movement continues until, at the midpoint of the forward stroke, rack *C* is halted by left-hand stop-pin *F* — the device then being in the position shown in view Y. At this point, bar *B* has advanced through distance *V*, while rod *J* has traveled twice that distance.

Rack *C* is now restrained, so that continued movement of the main bar will overcome the friction of rack *D*. Now, rotation of the gear is reversed as it rolls on the upper rack while carrying the lower rack with it. This movement continues until the end of the forward portion of the cycle, leaving the mechanism as

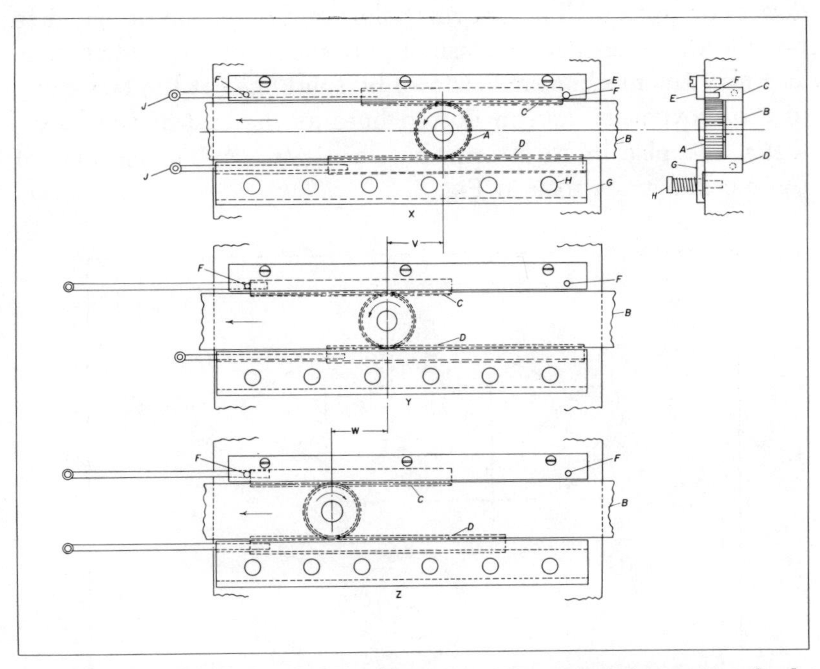

Fig. 6. Eyelet rods, *J*, are advanced and retracted intermittently by rotation of gear *A* on reciprocating bar *B*.

shown in view Z. The gear has advanced through distance *W* from its position in view Y while the lower rod *J* moved twice as far, bringing the eyes of the two rods into alignment in the forward position.

During the return half of the cycle, rack *C* is drawn back to the right-hand stop and rack *D* follows. In this manner, the upper rod is the first to move forward and the first to return. The operating cam is designed to provide a period of rest at each position so that work can be performed on the wire strands.

Mechanism Which Increases Movement by Levers and Chain

On a packaging machine it was necessary to provide a transfer movement for operating a slide. However, the movement of the slide had to be much greater than could be obtained with a direct connection. The drawings show a mechanism designed to provide the necessary increase of movement for the added slide. In Fig. 7 the mechanism is shown by solid lines at the beginning of the movement, and in dotted lines at the end of the movement, and also in an intermediate position. A bottom view of the mechanism is seen in Fig. 8.

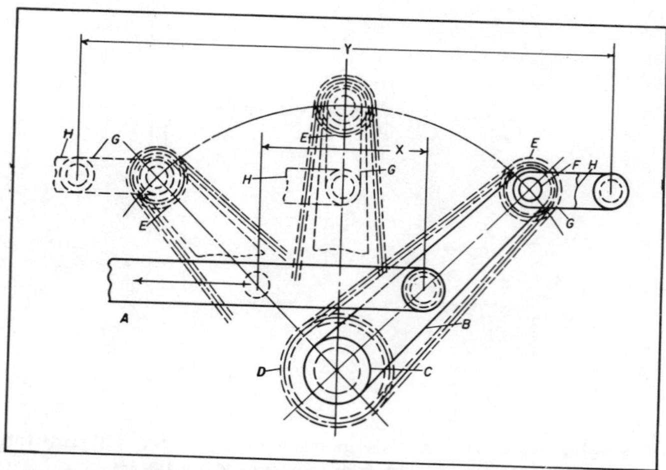

FIG. 7. Lever and chain mechanism designed to impart a traverse much greater than obtainable through a direct connection.

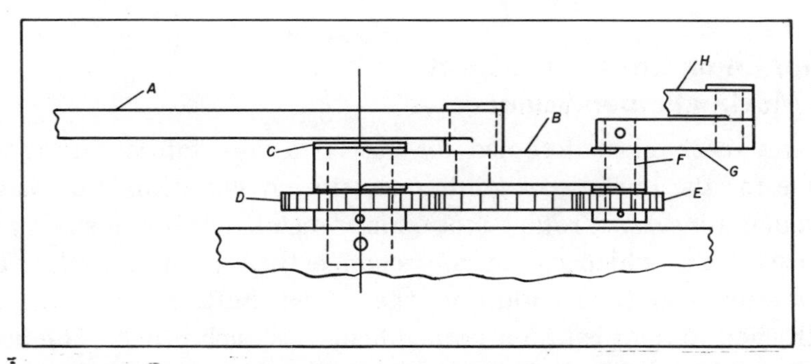

FIG. 8. Bottom view of mechanism which increases movement by levers and chain.

Bar *A* transmits motion from the existing slide to lever *B* of the added mechanism. Lever *B* is free to pivot in the same direction on fixed stud *C*. Sprocket *D* is locked on stud *C*. The sprocket is connected by a roller chain to sprocket *E* on shaft *F*. The shaft is free to turn in its bearing on the upper end of lever *B*. Lever *G* is attached to shaft *F* and carries a stud to which bar *H* is attached. Bar *H* transmits motion to the added slide.

When bar *A* is moved in the direction indicated by the arrow, lever *B* pivots in the same direction. As this occurs there is a change in the wrap-around position of the chain on sprocket *D*. This causes lever *G* to swing on the sprocket axis. Rotation continues until lever *B* reaches the end of its movement, at which position lever *G* has made a half-rotation. Its position is horizontal, as at the beginning of the cycle. However, lever *G* has been reversed relative to lever *B*. Thus, double the length of lever *G* has been added to the horizontal movement of shaft *F*.

Positions of lever *G* at the ends of the movement are controlled by the ratio of sprockets *D* and *E* relative to movement of lever *B*. The increase in the movement of bar *H* over that of bar *A* is indicated by the difference in the dimensions *X* and *Y*. Increase or decrease in the movements of bar *H* may be accomplished by a change in the length of lever *G* or, if adjustment is desired, lever *G* may be slotted to permit relocation of its stud which carries bar *H*.

Bead Chain Drive Turns Shaft on Moving Perpendicular Axes

In a mechanism designed for use on a wire fabricating machine for the purpose of guiding slowly moving strands of wire through a machine, rotary motion is transmitted from a shaft to two elements which continually change their positions relative to the drive-shaft. In addition, the driven shafts are positioned with their axes of rotation perpendicular to each other. The intention is to place the wire strands at various angles off their normal position at selected points in the cycle as determined by the wire pattern requirements. Figure 9 is a plan view of the mechanism; Fig. 10 is a front view; and Fig. 11 is a view of Fig. 9 from the right.

In the sketches the drive shaft *A* carries a bead chain sprocket *B* and may rotate in either direction. Bead chain idler sprocket *C* rotates on a stud mounted to the bed of the machine. Slide

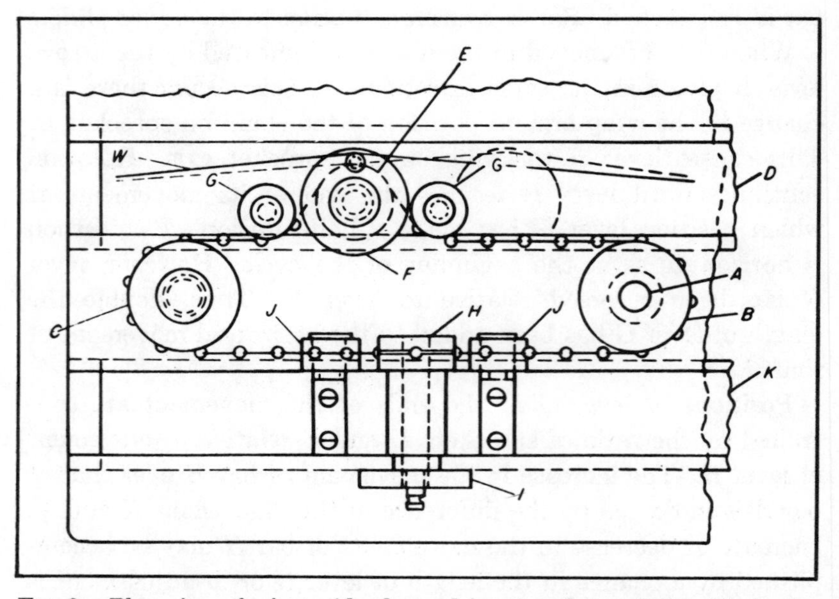

FIG. 9. Plan view of wire guide shows drive *A* and its relationship to the system of motions of the wire *W* as controlled by movement of slides *D* and *K*.

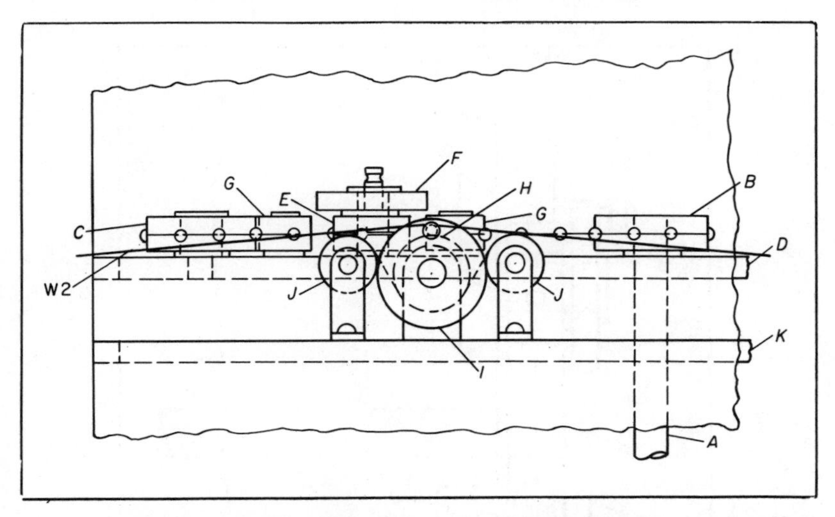

FIG. 10. Bead chain drive of discs *F* and *I* requires but little torque. This is a front view of the mechanism.

D, which is cam-operated, carries the sprocket *E* and disc *F*, keyed together and rotating freely on a stud. Two idler sprockets *G* are applied to maintain chain contact. Disc *F* carries a grooved pin which engages one strand of wire *W* to periodically place it in an angular position relative to its original straight-through position. Slide *K*, also cam-operated, carries a pillow block and shaft on which sprocket *H* and disc *I* are mounted. Disc *I*, like *F*, also has a grooved pin. Idler sprockets *J* are mounted on brackets carried by slide *K*.

In operation, slides *D* and *K* are moved intermittently by individual cams which are designed to provide the motion as programmed by the pattern to be produced in the product. In Fig. 9 the wire *W* is shown moved from its normally straight path. Figure 10 shows all parts in the same relationship as in Fig. 9, but another wire, *W2*, has also moved from its normally straight path. If the pattern required, either or both slides *D* and *K* could remain stationary while the discs *F* and *I* are given a number of rotations. In some wire design sequences, one slide may remain permanently in position while the other slide is moved in a required sequence. Because drive-shaft *A* rotates

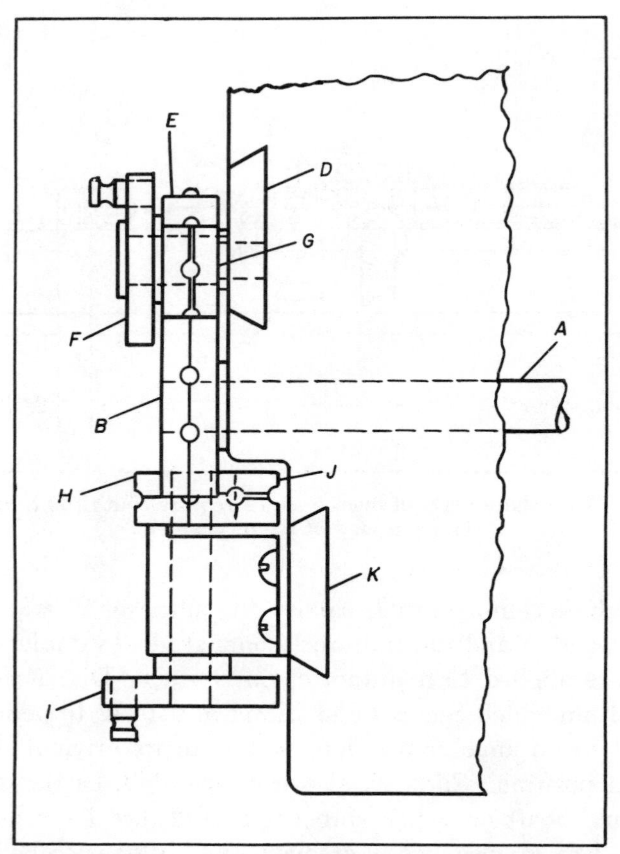

FIG. 11. Continuous rotation of wire guides on discs F and I sets up a "lay" pattern woven in a wire fabric, as shown by this right-hand view of the machine component plan seen in Fig. 9.

continuously, discs F and I also rotate continuously, regardless of the positions or movements of the slides. Any movement of slides D and K varies the angularity of the wires. In Fig. 9, disc F is shown dotted in its extreme position to the right, and indicates the change in the position of the wire as compared with the position represented by the solid outlines.

This mechanism operates at low speed under light load; therefore, a bead chain is ideally suited for transmitting motion from the drive to the perpendicular planes of rotation.

Reciprocating Drive Functions Around Roller Chain

During the development of a new product, the need arose for a simple reciprocating drive capable of operating under heavy loads. A stroke of 24 inches in length was required, but space considerations ruled out the use of conventional crank and lever type drives.

To meet these conditions, the illustrated sprocket and roller chain drive was developed (see Fig. 12). Either one of the two sprockets *A* or *B* can be used as the driving member, the remaining sprocket serving as an adjustable idler. Roller chain *C* is of standard design with one exception, one of the link rivets has been replaced by a long pin *D*. On each end of this pin is a roller *E*, held in place by cotter-pins.

In operation, the rollers are located between two follower-plates *F*, and also drive against them. The followers fit closely within slots in housing *G* and are held in place by cotter-pins on both the top and bottom. A gap is cut in one leg of the follower-plates to provide clearance over the sprocket hubs when the housing is at either end of the stroke. Bracket *H* is welded to

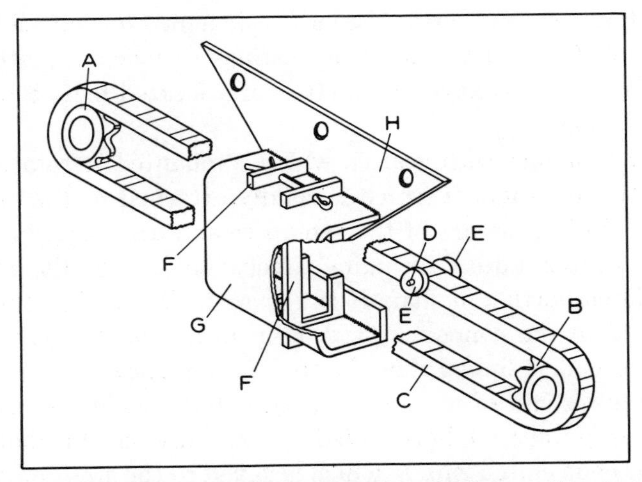

FIG. 12. Roller chain imparts reciprocating driving motion to machine slide (not shown). Rollers *E* engage follower-plates *F* which carry slide bracket *H* with them in both directions of travel.

housing *G* and bolted to the machine slide to be reciprocated (not shown).

As the roller chain moves with the two rollers *E*, located between the follower-plates, linear motion is imparted to the housing and then to the machine slide. When the chain link supporting the rollers reaches one of the sprockets, it descends, changing direction and returning on the lower span of the chain. Remaining between the two followers, the rollers now drive the housing and the machine slide in the opposite direction, providing the desired reciprocating motion.

Gear and Clutch Mechanism for Variable Operating Conditions

The various slides of multiple-slide wire-bending machines must be operated at different lengths of stroke, speeds of travel, and dwell periods at points of reversal because of differences in the thickness, springiness, and shape of the material.

The principal operating features embodied in the mechanism are shown in Fig. 13. A short connecting-rod *A* has one end affixed to an eccentric (not shown), attached to a continuously revolving shaft. The other end of the rod is attached to lever *B* by means of pin *C*. Lever *B* is fastened to shaft *D* by key *E*, the shaft being mounted in bearings supported by the machine frame. Also keyed to shaft *D* and located behind lever *B*, is a spur gear *F*.

Gear *F* meshes with gear *G*, which is mounted to rotate freely on stationary shaft *H* which is rigidly attached to the machine frame. Both gears are of the same size and are located on horizontal center lines. Swinging freely on a slightly reduced shouldered portion of shaft *H* is a lever *I*. One end of this lever is linked to the connecting-rod *J* by means of pin *K*, and the opposite end is linked directly to a former-slide.

The lower end of lever *I* has a large boss *W*, in the center of which is a tapered hole. Fitting smoothly in this hole is a circular cone clutch ring *L*, which is keyed to the stationary shaft *H* by key *M*. Shaft *H* projects beyond the clutch ring, the extension being externally threaded for lock-nuts *N*. By adjusting

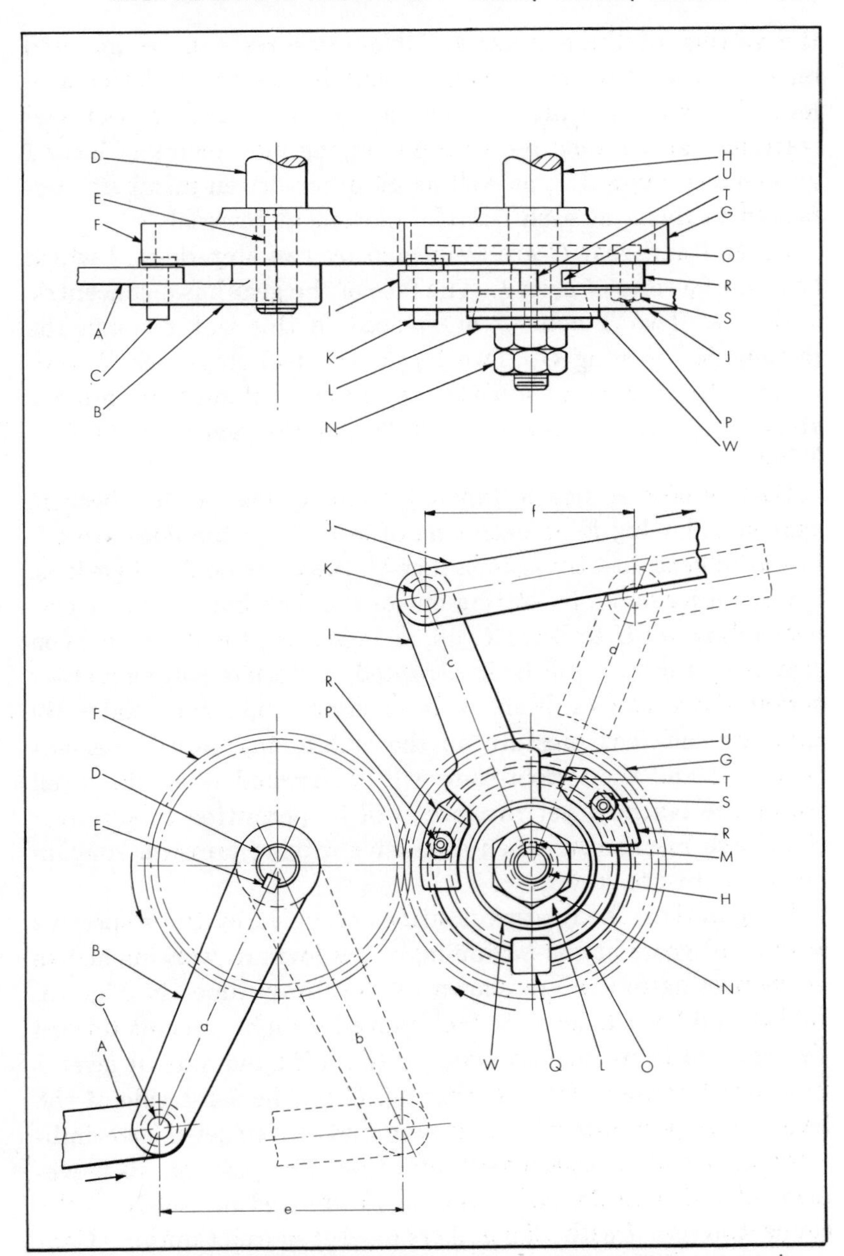

Fig. 13. Gear and clutch mechanism which produces a reciprocating motion with dwell periods. The driving and driven members are connecting-rods *A* and *J*, respectively.

the setting of these nuts, variable pressures can be imposed on the clutch ring in its engagement in the tapered recess of lever *I*. With this arrangement a certain degree of frictional restraint can be imparted to the swinging movements of lever *I* and connecting-rod *J*, as well as all other driven members connected to them, at each point of movement reversal.

Lever *I* and gear *G* are connected by two stop-dogs *R* which bear on the sides of lever *I*. The face of the gear has a concentric T-slot *O*. Two T-bolts *P* are placed in this slot through the rectangular opening *Q*. Two hardened steel stop-dogs *R*, provided with tongues to suit slot *O*, are circumferentially adjustable. They may be locked to the face of the gear by nuts *S* on T-bolts *P*.

Each stop-dog has a tapered contact nose *T* for bearing against a flat lug *U* on each side of lever *I*. If the dogs are adjusted to bear directly against lugs *U*, there are no dwell periods. On the other hand, by setting one of the dogs back from its corresponding lug *U* to leave a gap, as indicated, the drive between gear *G* and lever *I* will be interrupted, and there will be a short period of inaction, or dwell, of the lever and the former-slide. By applying sufficient pressure on the clutch ring, any movement of lever *I* and the former-slide will be arrested when the dwell points are reached, yet the lever will be permitted to slip over the clutch mechanism when a positive driving pressure is again imparted to the lever.

In operation, the relative positions occupied by the respective levers and gears at the beginning of the forward working stroke of connecting-rod *A* are shown by the heavy lines in Fig. 13. Rod *A* and lever *B* have, in fact, moved a slight amount toward the right to bring the stop-dog on the left-hand side of lever *I* into direct contact with bearing lug *U* on the same side of the lever. The positions of the driving and driven levers are indicated at *a* and *c*, respectively, and from this position all movement of rod *A* to the right will be transmitted positively to the connecting-rod *J* with all members moving in exact unison. Connecting-rod *J* and the former-slide attached to it will travel at precisely the same speed as connecting-rod *A*, since gears *F* and

G have a one-to-one ratio and levers *B* and *I* are the same length.

When rod *A* reaches the extreme right-hand limit of its movement, as shown by the dotted lines at *b*, rod *J* and lever *I* will have reached position *d*. With the return movement of rod *A*, however, the driven members will remain stationary for a certain period, or until the stop-dog *R* on the right-hand side of lever *I* is brought to bear against lug *U* on that side of the lever. When this contact takes place, the lever will be drawn toward the left in unison with the movement of rod *A*. At the termination of this return stroke, when rod *A* and lever *B* have been drawn slightly to the left of the positions indicated at *a*, a second dwell period will occur, which lasts until the left-hand dog *R* bears against the adjacent lug of handle *I*.

The dwell periods may be varied by setting the dogs in different positions on the driven gear. Respective movements of levers *B* and *I* will then be different; the latter will have a shorter length of travel. The different stroke lengths are shown at *e* and *f*.

A modification to produce a wider range of stroke and dwell adjustments is illustrated in Fig. 14. This is accomplished by using gears of other than a one-to-one ratio, and by providing a series of connected holes in the end of the driving lever *B* to permit varying the radial setting of rod *A*. The fifteen-tooth intermediate gear shown reverses the stroke direction.

Linear Movement Reduced by Differential Chain Drive Mechanism

A machine producing a woven wire product was required to have two strands of wire traversed across it simultaneously at different rates of travel and over different distances but with the same time cycle. Figure 15 shows a mechanism designed to perform this task.

Bar *A* is slidably dovetail-mounted in a stationary part of the machine, and is caused to reciprocate by a cam, not shown. Bar *B* is slidably dovetail-mounted in bar *A*, and carries two sprockets *C* which are free to rotate on their studs. Block *E* is at-

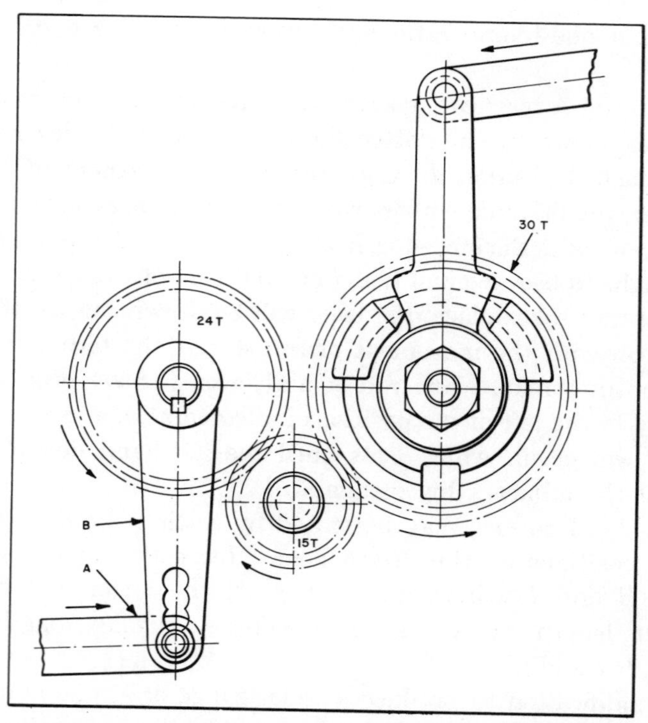

FIG. 14. Modification of mechanism shown in Fig. 13, which provides reversal of stroke direction and various stroke lengths and dwell periods.

tached to a stationary part of the machine. Two chains *D* engage sprockets *C*, one end of each chain being attached to block *E*, the other end being attached to bar *A*. The guides that direct the path of the wire are not shown. One of these guides is attached to bar *A*, and the other to bar *B*.

In the upper view, the assembly is shown at the mid-point of the traverse movement. In the lower view, the assembly is seen at its extreme left-hand position. In operation, the bar *A*, in moving toward the left, carries with it the two lower ends of the chains *D*. This movement results in an increase in the tension of the chain on the right, and a decrease in the tension of the chain on the left, so that motion is transmitted to the bar *B* through the right-hand sprocket *C*. As the left-hand sprocket C travels with bar B, it is impossible for slack to develop in the left-hand chain D.

Thus, due to the fact that the upper ends of the chains D are fixed in position and that the sprockets are movable, a differential motion is produced which results in a 50 per cent reduction in the length of the travel movement of the bar B, as indicated by the distances X and Y, in which X represents the movement of bar A, and Y represents the movement of bar B. The bar B therefore follows the same motion pattern as bar A, but to a reduced magnitude.

Variable Reciprocating Motion Derived from Uniform Rotation

On a machine used in making a formed wire product, the partly completed piece is transferred to various positions for the different operations. One of these transfer movements is accomplished by the use of an eccentric operating a reciprocating member. Owing to variations in the size of the work-pieces, it is necessary that the distance through which the part travels

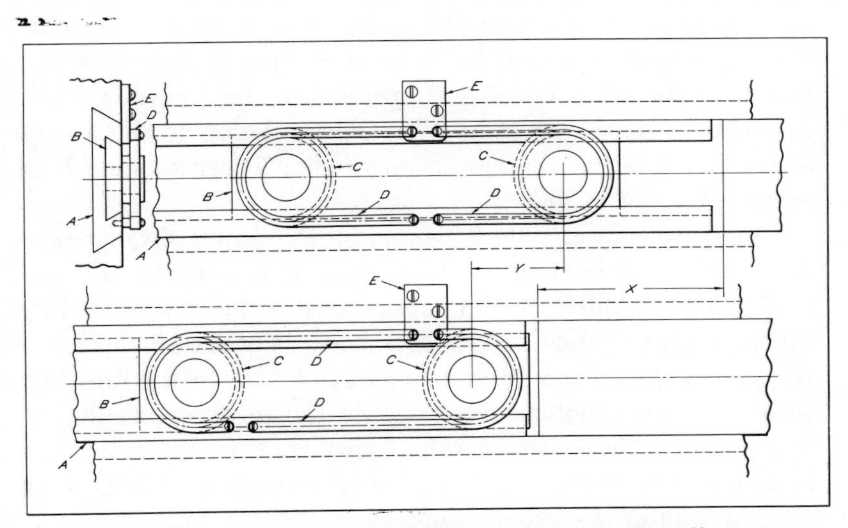

FIG. 15. Differential chain mechanism designed to reduce linear movement of slide.

from the point of transfer be varied without stopping the machine. It is also necessary that the loading position, which is the location at which the cycle is started, remain unchanged, regardless of the adjustment in length of stroke.

Figure 16 shows a mechanism that was designed to obtain the desired variable reciprocating motion from uniform rotation of an eccentric. Eccentric *A*, which rotates at a uniform rate to operate the mechanism, transmits reciprocating motion to slide *B* through the follower *C*. This slide is supported in machine frame *E*, and carries another slide *F*, which is free to reciprocate in the dovetailed slot in slide *B*. Slide *G*, also supported in machine frame *E*, carries a roller *H* which engages an angular slot in slide *F*. At one end of slide *F* is mounted another roller *P*, which engages a groove in the adjustable block *J*. This block is keyed to the vertical shaft *K*, which is supported in bracket *L*. Shaft *K* also carries worm-gear *M* which meshes with worm *N*. The worm, mounted on bracket *L*, is rotated by handwheel *O* to adjust angle of block *J*.

In operation, the rotation of eccentric *A*, which is shown in the plan view (center sketch) with follower *C* in its extreme right-hand position, causes slide *B* to move to the left. As slide *G* is connected to slide *B* by roller *H* in slide *F*, slide *G* is also moved to the left. As roller *P* operates in the groove in block *J* (which is shown set at an angle with relation to the line of motion of slide *B*), the movement of slide *B* causes slide *F* to move in the slot in slide *B*.

This movement of slide *F*, caused by the action of roller *H* in the angular slot in slide *F*, results in motion being transmitted to slide *G*, in addition to the motion that is directly applied by the movement of slide *B*. However, because of the angularity of the slot in slide *F*, any movement of this slide in the direction shown produces motion of slide *G* in the reverse direction, or to the right. This reverse motion subtracts from the movement transmitted to slide *G* by slide *B*, the movement of slide *G* being the resultant of the two motions.

In the partial plan view (top sketch), from which some of the parts have been omitted for clarity, slide *B* has completed

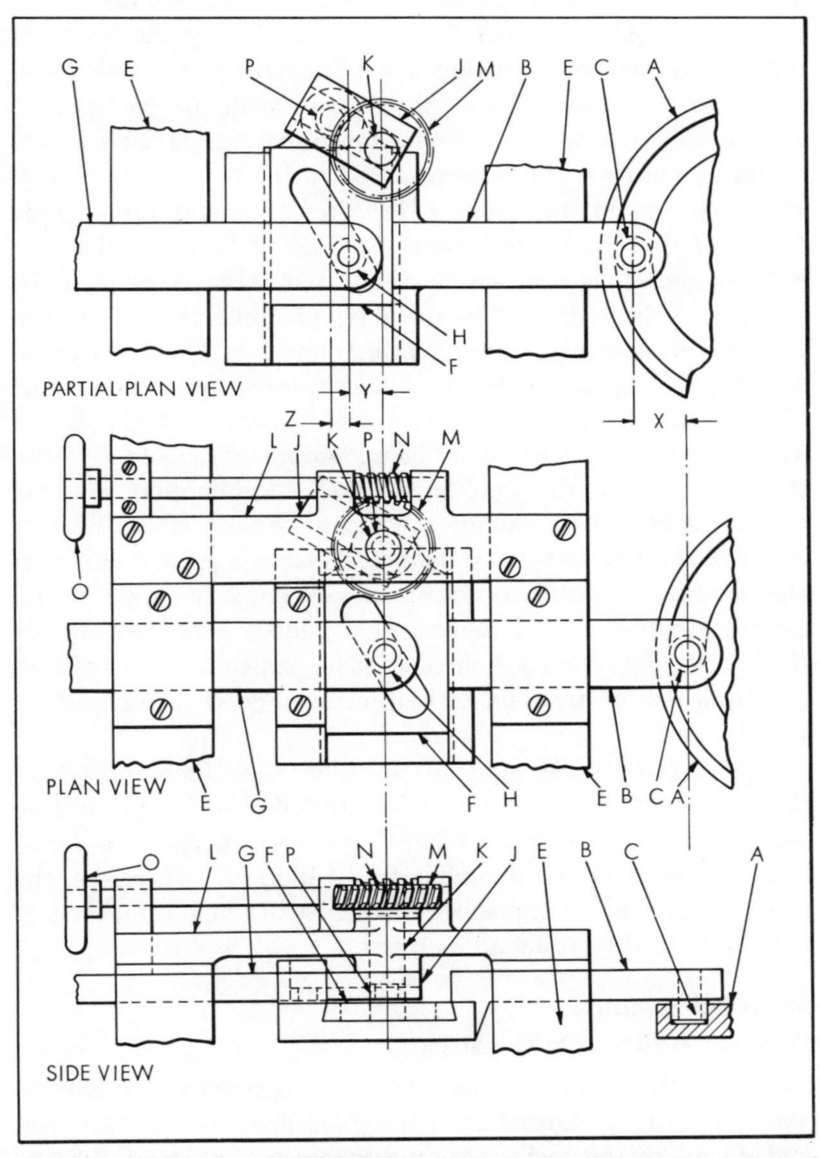

Fig. 16. Mechanism designed to obtain variable reciprocating motion of work-transfer slide *G* on a wire-forming machine from an eccentric *A* that is rotated at a uniform rate. The length of stroke of slide *G* is controlled by the angular setting of block *J*, which can be varied by rotating handwheel *O*.

its movement to the left, roller *C* having reached the high point on eccentric *A*. Also, slide *F* has been moved by the action of roller *P* in the groove in block *J*. Comparing the position of roller *H* in the plan view with its position in the partial plan view, it will be noted that its horizontal position relative to the center line of slide *F* has been moved to the right a distance *Z*, which represents the distance lost in the movement of slide *G* due to the upward movement of slide *F*. The actual movement of slide *G* is distance *Y*, which is equal to the rise of the eccentric *A* (as indicated by distance *X*) minus the distance *Z*.

As previously mentioned, the angularity of block *J* can be adjusted by the handwheel *O* through worm *N* and worm-wheel *M*. If block *J* were set with its groove parallel with the line of motion of slide *B*, there would be no movement of slide *F* within slide *B*. As a result, there would be no lost motion, and the movement of slide *G* would be equal to the movement of slide *B*. If block *J* were rotated counter-clockwise beyond this parallel setting, the resultant downward movement of slide *F* would cause an increase in the movement of slide *G* over that of slide *B*. The setting of block *J* should not be perpendicular or almost perpendicular to the line of motion of slide *B*, since binding would occur.

It should be noted that in the plan view (center sketch) the axes of shaft *K* and roller *P* exactly coincide. This condition will exist at all times when roller *C* is at the low point on eccentric *A*, regardless of the angularity of block *J*. Therefore, the loading point of the transfer movement will remain unchanged, regardless of the stroke adjustment.

Modified Eccentric Provides Adjustable Die Stroke

On a machine employed for producing stamped wire products, one of the dies is carried on a reciprocating slide. In the original design of the reciprocating mechanism, shown at the top in Fig. 17, the slide *C* on which the die was mounted fitted into a dovetail in the bed of the machine, and was reciprocated by an eccentric *B* carried on a rotating shaft *A*. Despite the fact

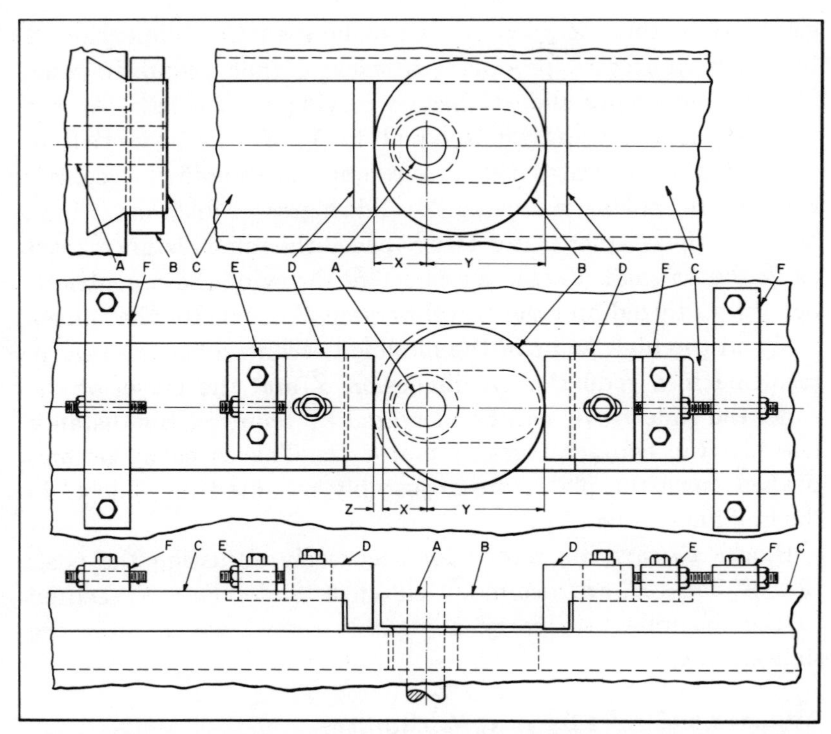

FIG. 17. In original design of reciprocating mechanism (top), the high point of eccentric *A* wore rapidly. In modified design (center and bottom), contact plates *D* are adjustably mounted on slide *C* to provide take-up for wear.

that the contact surfaces of the eccentric and plates *D* were hardened, wear on the high point of the eccentric soon resulted in lost motion.

Since the only means of correcting the lost motion was to refinish the contact surfaces of the eccentric and the wear plates and then to apply shims behind the plates, it was decided to redesign the mechanism to provide take-up for wear. This was accomplished by means of the modifications shown in the center and bottom views of Fig. 17. The redesigned mechanism also provides a means for varying the length of stroke of slide *C*.

In the center view, the outline of the original eccentric is shown in broken lines, while the modified eccentric is shown solid. With this modification it can be seen that dimension X has been shortened, producing a space Z that results in some lost motion before slide C begins moving to the left. As the "throw" of the eccentric is equal to $Y-X$, the reduction in dimension X increases the throw, and the modified eccentric is therefore capable of moving the slide a greater distance. Since, however, the contact plates D are spaced the same distance apart as in the original design, the effective throw of the eccentric is not fully utilized and the travel of slide C is exactly the same.

When the high point of the modified eccentric becomes worn, resulting in a reduction in dimension Y and the travel of the slide, the slide travel can be increased by reducing the distance between the contact faces of plates D. This is easily accomplished since the plates are adjustably mounted on slide C in the modified design.

Blocks E, carrying screws for accurately adjusting the position of plates D, are also mounted on the slide. Bars F, secured to the machine bed, carry adjustable screws which limit the slide stroke.

Adjustable-Stroke Driving Mechanism Reciprocating Slide

The unusual lever driving mechanism shown in Fig. 18 was designed to replace an ordinary connecting-rod arrangement used to impart a reciprocating motion to a machine slide. This slide, indicated at H, was an essential part of a wrapping machine and had to be driven by a constantly revolving shaft in the machine. Adjustments in the position of the reciprocating stroke were frequently necessary.

When making adjustments with the original equipment, it was necessary to stop the machine, sometimes three or four times in succession, before obtaining the correct amount of slide travel or proper positioning of the reciprocating slide in its guideways. With the new drive, provision was made for obtaining adjustments without interfering with the machine drive in any way. The mechanism for the control was designed to accomplish its purpose in a quick and safe manner.

Referring to Fig. 18, member *A* is the driving shaft which rotates in a horizontal bearing *B*. Shaft *A* is driven at a constant speed of rotation in a clockwise direction by a simple spur gear drive (not shown) at the rear of bearing *B*.

To the front end of shaft *A* is keyed a cast-iron circular disc *C*, which carries the projecting crankpin *D* fastened in place by a lock-nut at the rear of the disc flange. The crankpin is immovable in the disc and cannot be adjusted to obtain variations in the length of stroke as in the original drive. Mounted on the crankpin is the connecting-rod *E*, this being retained in the correct endwise position by the collar *F*.

A connecting-rod *G*, having the same effective length as *E*, is linked at its right-hand end to the reciprocating slide *H* by pin *K*.

Slide *H* moves in the horizontal plane within the stationary dovetail guide track *L*. The remaining end of links *E* and *G* and

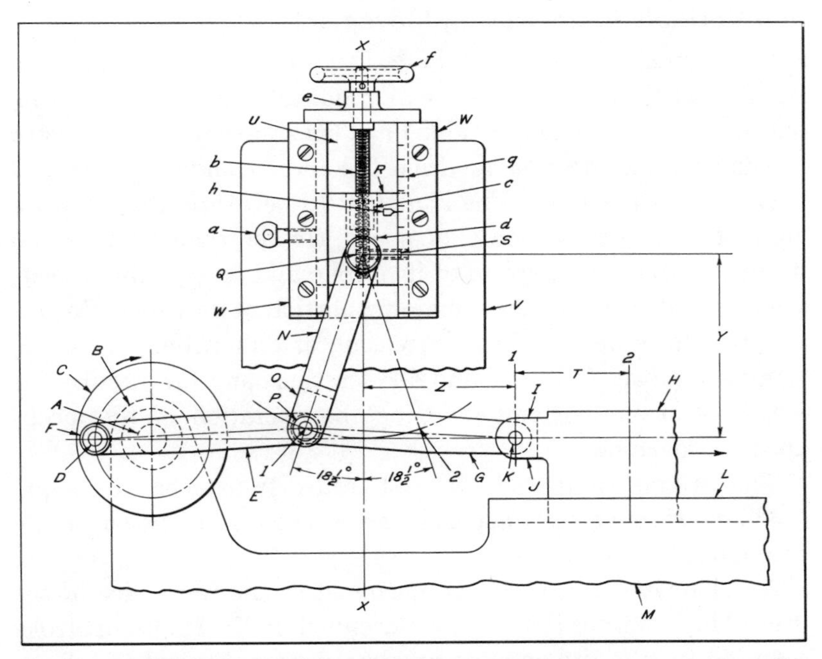

FIG. 18. Adjustable-stroke driving mechanism for reciprocating slide set for minimum stroke length of slide *H*.

the lower end of link *N* are linked by and pivot about stud *P*. The upper end of arm *N* is attached to the front of the adjustable slide *R* by a fulcrum stud *Q*.

Slide *R* is actuated by a fine-pitch lead-screw *b*. The upper end of the lead-screw is guided in the fixed bearing plate *e* fastened to the top of the guide bracket. A handwheel *f* is keyed to the projecting upper end of the lead-screw, so that it may easily be rotated. The top surface of the retaining plate *W* at the right-hand side has the accurately measured gradations *g*. The small pointer *h*, fastened on top of slide *R*, is set adjacent to the graduation marks to facilitate setting the slide.

The position of the stroke of *H* is adjusted by rotating handwheel *F* to change the position of stud *Q*. It should be noted that as the length of link *N* is not infinite, the length of stroke will change a small amount as its position is changed.

Converting Oscillating Circular Motion into Variable Reciprocating Movement

An unusual lever mechanism designed to transform a uniform circular oscillating motion into a periodically varying reciprocating movement is shown in Fig. 19. This simple and inexpensive mechanism was devised to drive the former slide on a special wire-forming machine. The slide had to be reciprocated with a gradually accelerating rate of speed (faster than that of the driving shaft) for a portion of its stroke and an equally decelerating rate of speed for the remaining portion of its stroke. Means also had to be provided for increasing or diminishing the duration of these accelerations and decelerations and for altering the length of stroke without stopping the operation of the mechanism or machine.

The driving shaft and the cast iron disc attached to it are oscillated through 120 degrees by an eccentric, not shown, acting through a short link arm.

An elongated radial slot is machined in the front face of the disc. The length of the slot is determined by the length of stroke required for the driven slide and the permissible outside diameter of the disc. A phosphor-bronze slider block, fitted into this

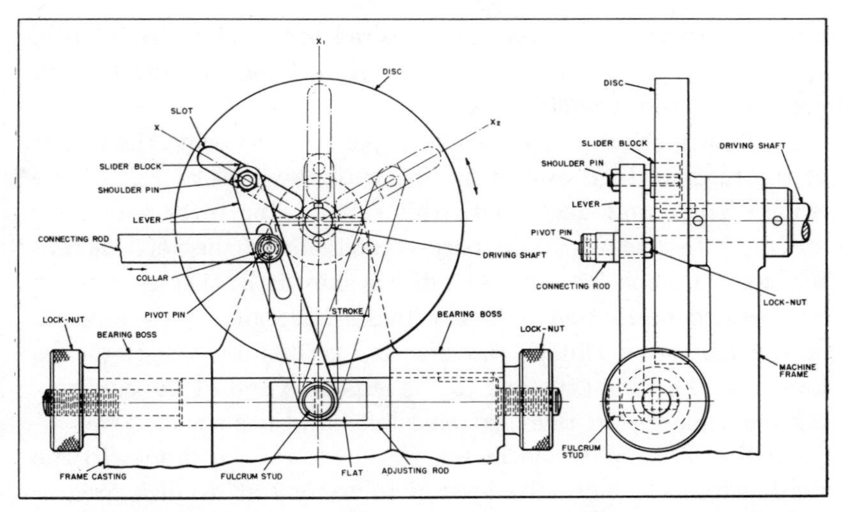

FIG. 19. Mechanism designed to drive a former slide on a special wire-forming machine.

slot, is secured to the end of a lever by means of a threaded shoulder-pin and lock-nut.

The lever is linked to a connecting-rod by means of a pivot pin, located in a slot machined in the center of the lever. A collar pinned to the projecting end of the pivot pin retains the connecting-rod, but allows it to swivel about the pin. The pivot pin may be adjusted to any desired position along the slot, and is locked in place by a nut at the rear of the lever.

The opposite end of the connecting-rod is attached directly to the driven slide (not shown), which has to reciprocate horizontally. The lower end of the lever is mounted on a stationary fulcrum stud that is fixed to the side of a cylindrical adjusting rod. This horizontal rod is carried in two bearing bosses on the machine frame. The lever, thus mounted, can swivel freely upon the stud. A flat on one side of the adjusting rod provides a bearing surface for the location of the stud and the side of the lever.

Each end of the rod is reduced and threaded, and is provided with a knurled lock-nut which bears against the end face of the bearing boss. By adjusting the two lock-nuts, the rod can be moved horizontally to give any desired setting for the fulcrum stud. A key is provided within the right-hand bearing boss to prevent the rod from rotating.

As shown, the rod has been adjusted so as to bring the center of the fulcrum stud exactly in line with the vertical axis of the driving shaft and disc. With this setting, the lever will move an equal distance — 60 degrees — each side of the vertical center line of the mechanism. The disc is shown about to commence its forward oscillation in a clockwise direction. This position of the elongated slot in the disc is denoted by X. Both the lever and the slide attached to the left-hand end of the connecting-rod thus are in their extreme left-hand positions.

As the disc moves clockwise, the lever will be drawn to the right. However, since the lever is fulcrumed at a much greater radius than that of the shoulder-pin on the disc, the slider block will gradually move down the slot and approach the center of the disc. Throughout that portion of the stroke of the disc between positions X and X_1, the connecting-rod and the attached former slide will travel at an increasingly faster rate than the driving shaft.

When the lever has passed position X_1, however, the slider block will gradually move outward in the disc slot, and the driven slide will be correspondingly slowed down. This deceleration will be at exactly the same rate as an acceleration occurring in the first portion of the stroke. When point X_2 has been reached, reversal occurs.

Throughout the return stroke from X_2 to X_1, the speed of the lever and the driven member will be increased; their speed will be decreased when the lever passes point X_1. Thus, for each complete oscillation of the disc and its driving shaft, the driven slide will receive two accelerations and two decelerations of equal duration.

The maximum stroke is obtainable by adjusting the pivot pin so that the connecting-rod is set at the highest point within the slot in the lever. By moving the pin to the opposite end of the slot, so that it lies at the shortest possible distance from the center of the fulcrum stud, the minimum stroke is obtained.

Variations of stroke length ranging between these maximum and minimum values can be readily obtained by changing the position of the fulcrum stud relative to the vertical axis of the mechanism. Such changes can be effected while the mechanism is running merely by adjusting the lock-nuts and the lateral position of the adjusting rod in its bearings.

Unique Pumping Mechanism Applied to a Homogenizer

A mechanism designed to impart pulsations to three plungers has been incorporated in a homogenizer. Hydraulic suction and pressure are alternately obtained in this equipment by rubber pulsators *R* through the action of pistons *P* (see Fig. 20).

The pumping portion of the mechanism, with ball suction and discharge valves and a pressure relief value, is shown at the right-hand end of the drawing. The driving mechanism consists of a camshaft *D*, on which are mounted three cams C_1, C_2, and C_3, which actuate pistons *P*. There are three sets of parallel pistons, plungers, etc., which extend toward and through the pump housing.

Camshaft *D* is geared to drive-shaft *B*. Pistons *P*, their cylinders, and part of their operating mechanism are constantly immersed in the actuating fluid, which is oil.

Each pulsator *R* is made of synthetic rubber and bonded to a sleeve *E*. A metal support *S* inside the pulsator is connected to cylinder *F*. Each piston is provided with a collar *H* that transmits the pressure of spring *J*.

Rocker arms *A* pivot on shaft *T*. Each cam rides against a ball-bearing roller *G* which is mounted on a stub shaft in a corresponding arm *A*. A second ball-bearing roller on this arm contacts the left end of the piston. In the position shown in the illustration, cam C_1 has pushed piston *P* into the discharge position, against the pressure of spring *J*. When the camshaft makes a half revolution, cam C_1 will have moved 180 degrees to permit piston *P* to move toward the left and uncover ports *O* to compensate for any slight oil leakage from the cylinder and prepare for the discharge stroke.

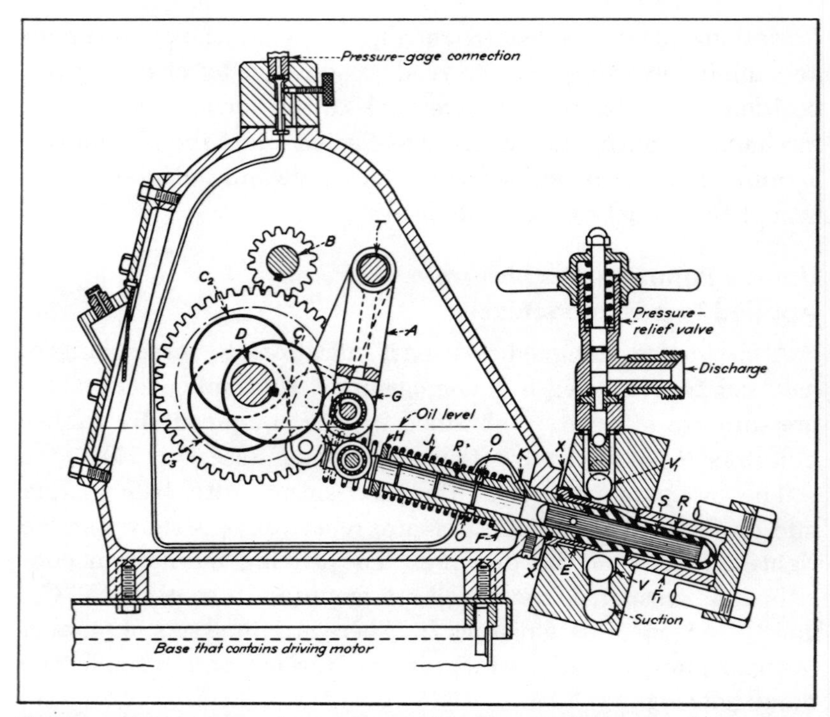

FIG. 20. Triplex type of pumping mechanism which imparts three pulsations to a homogenizer with each revolution of a camshaft.

Pressure inside the rubber pulsator is relieved at this time, and it contracts to create a vacuum inside cylinder F_1. Liquid then flows into this cylinder through the ball suction valve V. When cam C_1 makes another half revolution, arm A moves piston P into the position shown, and forces oil into the rubber pulsator. The pulsator then expands and forces liquid out of cylinder F_1, through the ball discharge valve V_1 and out through the discharge connection.

The pump, being of a triplex design, makes three pulsations for each revolution of the cam-shaft, which insures a relatively constant flow of oil. The discharge pressure is measured by a gage connected to the cylinder at K.

CHAPTER 8

Crank Actuated Reciprocating Mechanisms

The special designs of crank mechanisms described in this chapter are for transmitting motion to slides or other parts having a reciprocating action. These drives may be arranged to produce some special movement, such as, for example, arresting the motion of the slide momentarily during some part of the stroke or providing a quick return movement to reduce the idle period; or the design may be special in that certain parts of the mechanism are to be operated at constant or varying velocities.

Other crank-actuated reciprocating mechanisms are described in Chapter 9, Volume I and Chapter 8, Volumes II and III of "Ingenious Mechanisms for Designers and Inventors."

Oscillating Shaft Driven with Simple Harmonic Motion

In a printing machine it was desired that a shaft be rotated through 180 degrees, and at the same time that the angular motion of the shaft be simple harmonic.

A mechanism that fulfills the requirements is shown in Fig. 1. Its mode of operation is as follows: To the input shaft *A* is fastened gear *B*, which is in mesh with gears *C* and *D* (both having the same diameter). Gears *C* and *D* rotate discs *G* and *H* through the shafts *E* and *F*, in the same direction and with the same angular velocity. Discs *G* and *H* carry crankpins *I* and *K*. They, in turn, carry the rack *L*, which is in mesh with gear *M*.

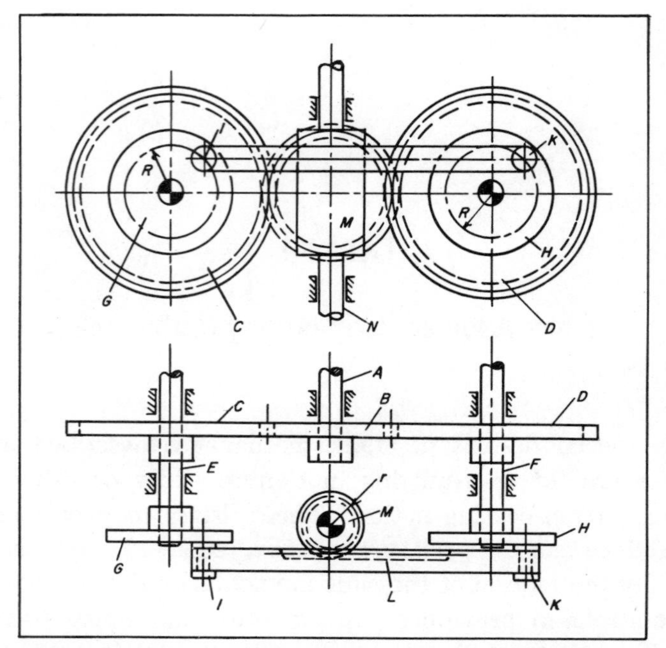

FIG. 1. With drive-shaft A in constant rotation, this mechanism causes output shaft N to oscillate with simple harmonic motion through 180 degrees when proportioned according to the stated formula.

When discs G and H rotate they will impart a simple harmonic motion to gear M and output shaft N. The angle ϕ through which N rotates is determined by

$$\frac{2\pi r \times \phi}{360} = 2R \text{ or } \phi = \frac{360\,R}{\pi r}$$

With geometrical proportions shown in the sketch, $\phi = 180$ degrees.

Rapid Return and Dwell Period Provided by Spring-Loaded Eccentric

A conventional eccentric and connecting-rod were used to apply holding pressure on a wire product during a machine-forming operation. Although this setup was satisfactory for the job at hand, a product design change made it necessary to modify the clamping motion slightly to furnish a quicker release of the wire part. This revised mechanism is shown in Fig. 2.

Eccentric *A* is free to rotate on hub *B* of carrier dog *C*. The carrier dog is, in turn, keyed to shaft *D*, which rotates in the direction indicated by the arrow. Pin *E* is the only connection between the eccentric and the dog. The pin, passing through eccentric *A*, contacts carrier dog *C* during part of the operating cycle.

The end of pin *E* that protrudes from the front of the eccentric supports bushing *F*. An external annular groove cut into the bushing carries one end of tension spring *G*. A pin *H* and a bushing *J*, supported by the machine frame, carry the opposite end of the spring. The machine-slide to be reciprocated (not shown) is attached to the upper end of connecting-rod *K*.

When the mechanism is functioning, carrier dog *C* rotates with shaft *D* in a counterclockwise direction until the dog contacts the end of pin *E*. Further movement of shaft *D* will cause eccentric *A* to rotate, at the same time extending spring *G*. The eccentric, as shown at X, is in its uppermost position, or point of greatest pressure application to the work. Pin *E*, at this point, is slightly above the center of shaft *D*. Continued rotation

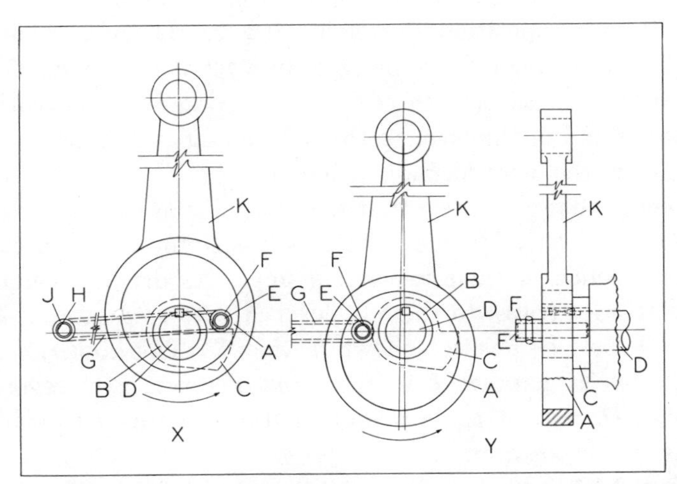

FIG. 2. Modified eccentric provides rapid return of machine-slide followed by a dwell period.

of the eccentric will cause the high-pressure point to be passed, at which time spring *G* will pull the eccentric sharply in its normal direction of rotation. As a result of this movement, connecting-rod *K* and its attached machine-slide undergo a rapid partial return.

At Y, in Fig. 2, can be seen the position assumed by eccentric *A* and pin *E* at the end of their spring-induced rotation. The tension of spring *G* prevents additional rotation of the eccentric until carrier dog *C* once again contacts pin *E*. Thus, in addition to a rapid partial return of the machine-slide, a dwell period is provided.

Generating Two Reciprocating Strokes for One

An arrangement that produces two strokes for each stroke of a reciprocating driving member is shown in Fig. 3. Alternate movements generated in each direction are of different lengths. This mechanism was designed and used to position a forming head on a machine for processing a wire product.

Lever *A* is mounted on two studs *B* and *C* which are located on a stationary part of the machine. Two curved slots permit movement of the lever of these studs. Each slot is machined to a radius centered on the opposing stud (*B* or *C*) when the lever is in the position shown in view X. Driving member *D* and driven member *E* are pivoted on lever *A* by means of studs *G* and *H*, respectively. In addition, a spring *F* resists motion of the lower end of the lever to the left. Member *D* is actuated by a cam on the wire-forming machine.

View X shows the device at the midpoint of the movement of bar *D* to the left. In this position studs *B* and *C* are in contact with the ends of their respective slots. As driving member *D* completes its travel to the left, lever *A* pivots on stud *C* in the lower slot to the position shown in view Y. This causes member *E* to move a distance *L* in the same direction. On reversal of member *D*, lever *A* again pivots on stud *C* returning member *E* to the center position.

After driving member *D* passes the midpoint, continued movement to the right causes lever *A* to pivot on stud *B*, again mov-

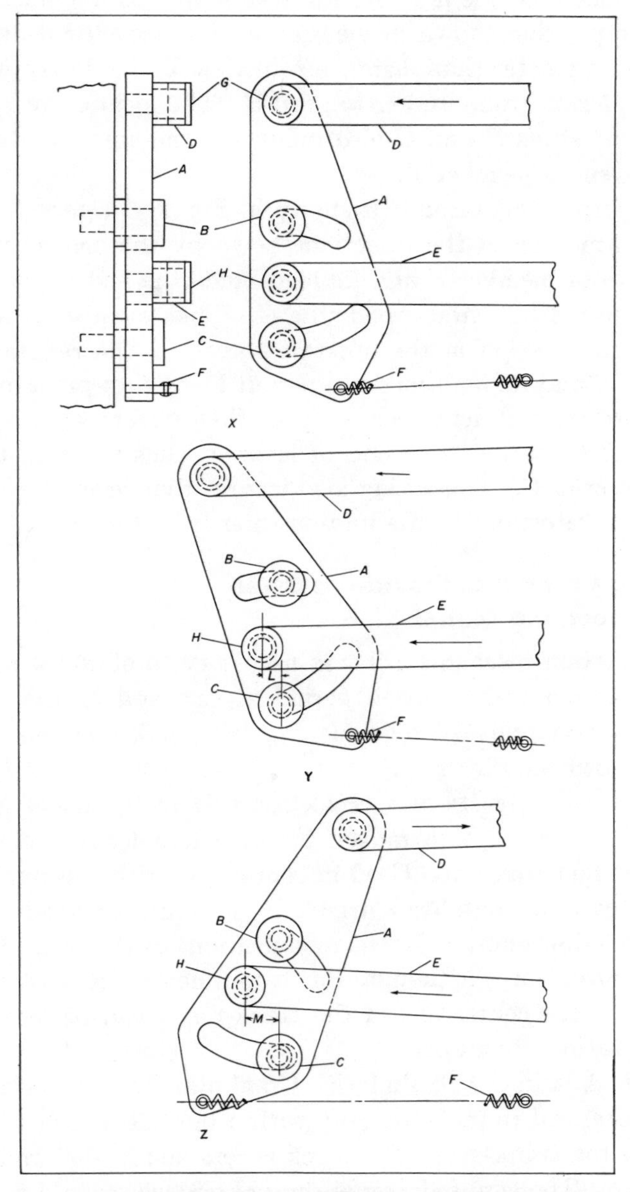

FIG. 3. Lever arrangement that provides two strokes for every stroke of a reciprocating drive rod *D*. Alternate strokes in each direction are of different lengths.

ing member E to the left. At the end of this stroke, member E is in the position shown in view Z and has traveled a distance M. This is greater than distance L in view Y, due to the change in the fulcrum from stud C to stud B. The location of stud H relative to studs B and C also influences the amount of movement given to member E.

The purpose of spring F is shown in Fig. 3. In view X member D is moving in the direction shown by the arrow and the resistance of member E and its load holds the end of the lower slot in lever A in contact with stud C. When member D returns to the mid-position in the opposite direction, the resistance of member E and its load tends to permit lever A to pivot on stud H. However, sufficient resistance to this movement is applied by spring F to the lower end of lever A, thus insuring that it pivot on stud C. The spring also insures that lever A pivot on stud B in returning to the mid-position from the right.

Rotating Crank that Provides a Dwell in Reciprocating Motion

In a certain mechanism it was necessary to obtain a reciprocating motion with a dwell period at one end of the stroke. This was accomplished by designing the crank and cam mechanism shown in Fig. 4.

The device consists of a crank rotating in the usual fashion but with its crankpin arranged to move radially in a slot provided in the crank-arm. Used in connection with a suitable stationary cam, a cross-head driven by a connecting-rod can be made to follow many different modifications of the reciprocating crank movement. Although the cam-follower in this case is located on the center line of the crankpin, it can be located to suit operating conditions.

Crank A is keyed to shaft B. Crankpin C and cam-follower D are arranged to move radially with a block E which slides in a slot in the crank-arm. The block is retained in the crank-arm by gibs F. The cam-follower and crankpin are secured to block E by means of a screw and nut, as shown in cross-section X-X of the assembly.

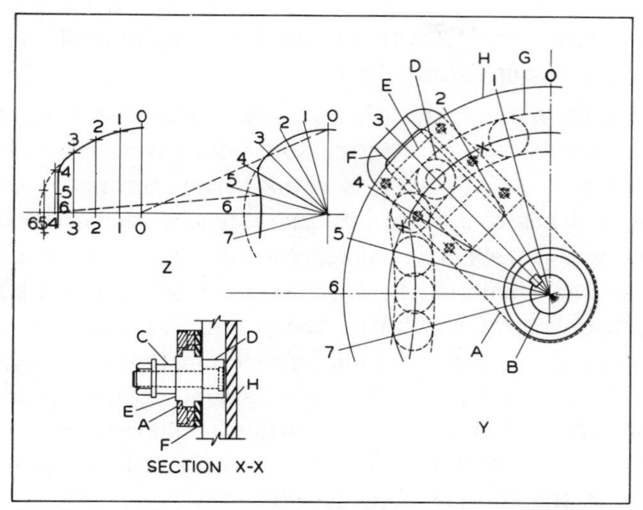

FIG. 4. Rotating crank and cam mechanism produces dwell in reciprocating movement.

The reciprocating motion with a dwell is obtained when cam-follower *D* moves within a recessed cam-groove *G* in a stationary plate *H* during the swing of the crank, as indicated in view Y. By referring to the diagrams showing the movement of the crank device, view Z, it will be apparent that when the follower *D* passes points numbered 0, 1, 2, and 3, the driven member will follow the normal crank-arm arc. However, since the path through points numbered 5 and 6 is concentric with the point at which the connecting-rod is attached to the cross-head or other reciprocating member, no horizontal movement can occur as the cam-follower moves through this portion of its path. In diagram Z, the diagonal dotted line between points 0-0 and the dotted line between points 5-6 represent two positions of the connecting-rod.

Since part of the horizontal movement of the driven member is lost at the dwell portion of the cycle, the length of the crank-arm radius must be suitable for the desired total movement. The transition from movement to dwell period can be varied to suit conditions.

Crank-Driven Plate Obtains Near-Uniform Velocities Through Compensating Cam

In a certain manufacturing process, sheets arriving on one conveyor section are transferred to another section by means of a suction plate. This plate is crank-driven through a connecting-rod and has a straight-line movement between the ends of the sections. Sheets arrive at regular intervals, and it is essential that the movement of the suction plate by synchronized and approximately uniform during the pick-up and deposit of each sheet, since the transfer cannot be made instantaneously.

The design of the transfer mechanism created a problem, because simple linkage would obviously convert the constant angular velocity of the crank to a continuously varying linear velocity of the suction plate.

The time in the cycling of the plate when a uniform movement was wanted corresponded to 10 degrees of crank rotation during the pick-up of the sheet, and another 10 degrees during its deposit.

A compensating cam solved the problem. As can be seen in Fig. 5, suction plate *A* is joined by connecting-rod *B* and link *C* to crank *D*. The crank is keyed to drive-shaft *E* which rotates at a constant speed. Compensating cam *F*, having a drop *G* and a rise *H*, is rigidly attached to the frame.

At one end, link *C* carries a roll *J* which is spring-loaded against the cam. The drive-shaft rotates clockwise. When the roll contacts the drop on the cam, the effective length of the crank is decreased, and, thus, the velocity of suction plate is reduced. Conversely, the velocity of the suction plate is raised when the roll is on the rise on the cam.

The drop and rise areas are so located on the cam as to produce the near-uniform velocities at the desired points in the cycle of the suction plate. In one instance, an unwanted acceleration of the plate is cancelled by a drop in the cam, and in the other instance, an unwanted deceleration is cancelled by a rise in the cam.

Winding Head for Skeins of Embroidery Floss

Skeins of embroidery floss are formed by winding the thread around two stationary bars. As each skein is formed, it is moved along the bars, away from the winding head, to be cut and labeled. The path of a winding arm swinging around the outside of the two bars must not be restricted by any type bracket, such as might be required for retaining the bars in a stationary position. The illustrated mechanism, utilizing a Scotch yoke, is being used in such a machine.

Floss to be wound passes through a hole drilled in the center of shaft *A* (see Fig. 6). The shaft is supported in bearing bracket *B* and is retained by collar *C*. From there it continues through hollow winding arm *D*. The winding arm is revolved at high speed, wrapping the floss about the two stationary arms *E* mounted in bracket *F*.

Support for this bracket is obtained from pins *G* and *H*, which can slide in holes drilled through integral bracket ears. Two

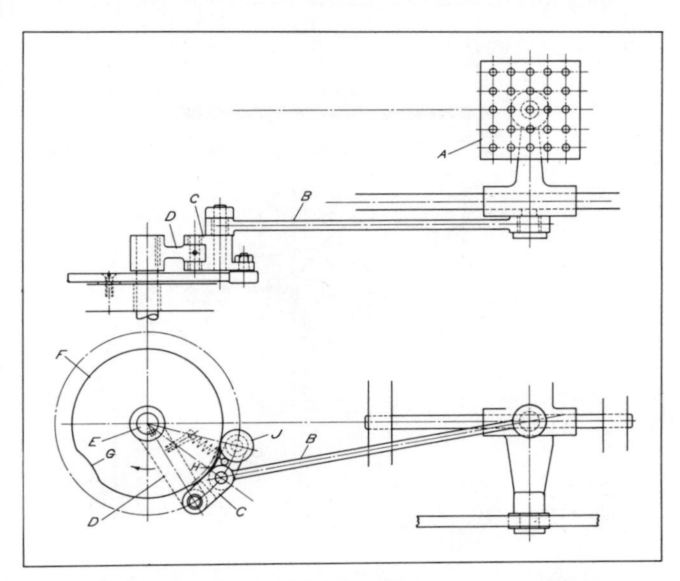

FIG. 5. The net effect of compensating cam *F* is to produce near-uniform velocities at two points in cycle of suction plate *A*.

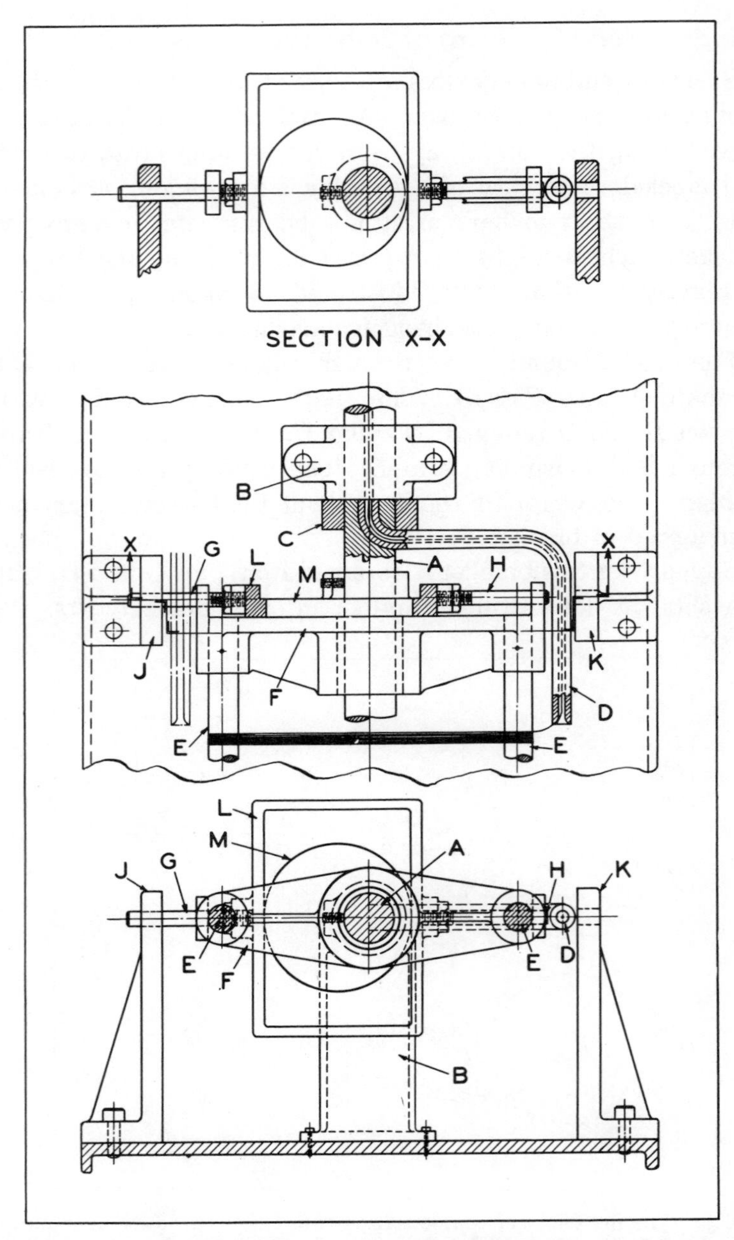

FIG. 6. Head for skein-winding machine features Scotch yoke for reciprocating movable bracket-support pins.

identical angle supports *J* and *K*, screwed to the machine base, are drilled to receive the outer ends of the pins. The inner ends of the two pins are threaded into bosses on either side of Scotch yoke *L*. Cam *M*, mounted on shaft *A*, rides within the yoke.

The cam is timed to move pin *G* into engagement with support *J* when the winding arm is passing between bracket *F* and support *K* as illustrated. As the winding arm approaches the left-hand side of the unit, the yoke and attached pins move to the right. This advances pin *H* into engagement with support *K*, while retracting pin *G* from the opposite support. A path is thus opened between bracket *F* and support *J* to permit unrestricted passage of the winding arm.

Bracket *F* is supported at all times. During the time that winding arm *D* is approaching either vertical position, and just after it leaves that position, both pins are engaged simultaneously. The appropriate pin is completely disengaged only during the instant necessary for the winding arm to pass. The relationship of the working members to each other may be more clearly seen in section X-X.

Oscillating Drive-Shaft Actuates Slide at Variable Speed and Stroke

A slide to move light packages across a wrapping table had to be driven by an oscillating shaft and lever. The position, length, and amplitude of the lever arm were fixed, since the lever also actuated another mechanism synchronized with the motion of the slide. It was not possible to attach the lever directly to the slide because of certain peculiarities of the slide motion.

First, the slide stroke had to have a range of adjustment, from a 9-inch minimum to a 10½-inch maximum (direct attachment of the lever and slide would have produced a 13-inch fixed stroke). Second, in its forward movement the slide initially had to travel in unison with the lever for a distance of 2 inches, then gradually decelerate; and in its return movement, the slide initially had to accelerate gradually to a point 2 inches from the end of the stroke, then travel in unison with the lever. Third, provision had to be made to vary within small limits the distance that the slide traveled in unison with the lever. It was required that the distance of unison slide vary from 2 inches at the maximum slide stroke to 3¼ inches at the minimum stroke.

The mechanism that was devised is shown in Fig. 7. The upper and lower views show, respectively, the position of the elements at the beginning and end of the forward stroke. Slide *A* is dovetailed to and reciprocates along guideway *B* formed in the vertical side of frame *C*. Oscillating drive-shaft *D* and lever *E* are supported on the same frame, but at a considerable distance from the slide.

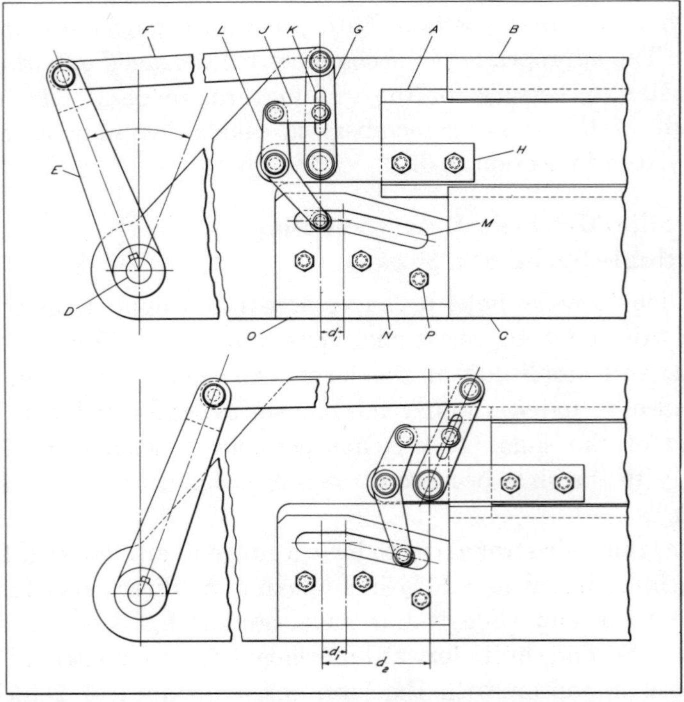

FIG. 7. Converting the oscillation of drive-shaft *D* to a reciprocation of slide *A* through this mechanism permits the stroke to be of adjustable length and made at varying velocity.

The top of lever *E* has a slot through which is pinned a long rod *F*. At one end, rod *F* is joined to arm *G*, which, in turn, pivots on bar *H* fastened to slide *A*. Link *J* pivots around an adjustable point in slot *K* of arm *G*, and the left-hand end of the link is pinned to the upper limb of bellcrank *L*. The bell-crank swings on bar *H*, and its lower limb carries a roller fitting slot *M* in plate *N*. It will be noted that one section of the slot is shorter and parallel to guideway *B*, and the other section is longer and at a decline. The extent of these two sections, as well as the angle of the decline, is carefully determined in accordance with the desired total length of slide travel and range of speed variation.

Providing bell crank *L* is prevented from rotating, arm *F* will push slide *A* in unison. At the start of the stroke, bell crank *L* is prevented from rotating by keeping the distance between the attached roller in *M* and the slide *A* constant. When the roller enters the inclined part of slot *M*, the distance increases causing bell crank *L* to rotate clockwise. The clockwise rotation of *L* causes *A* to move to the left with respect to arm *F*, thus causing deceleration. If the stud connecting *J* and *G* is lowered, rotation of *L* will increase deceleration as it can cause greater rotation of *G*. Lowering of the stud will also change the distance of unison travel. However, the distance of unison travel can be readjusted by changing the position of *N* in slot *O*. Elevating the position of the stud between *J* and *G* will decrease the amount of deceleration.

Drives for Reciprocating Members with Varying Relative Motion

When designing special-purpose machinery for packaging articles of various kinds, it is often necessary to have two elements move in unison for a period of time and then have one travel forward at an accelerated rate. Generally, there is insufficient space to enable each movement to be obtained by the usual method with cams, links, and multiple slides. Such an arrangement may also result in excessive overhang or undue remote operation.

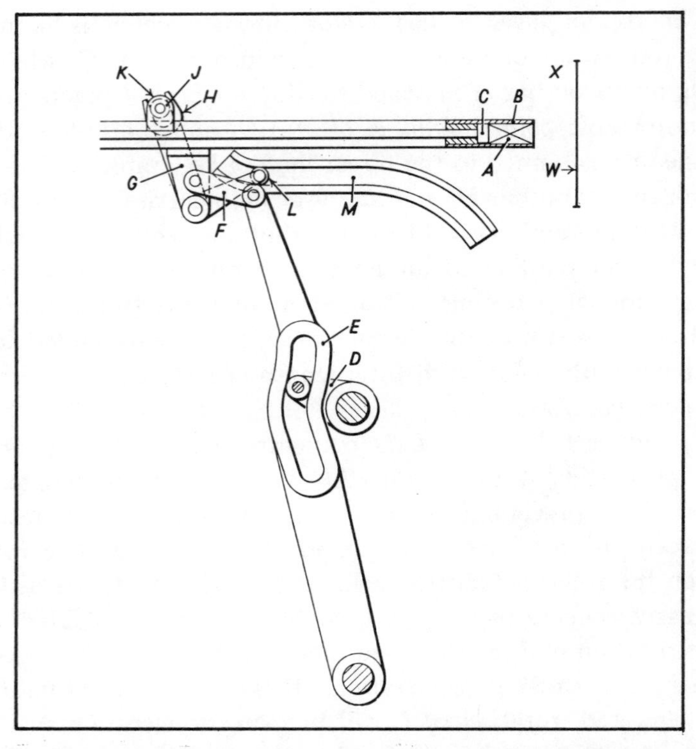

FIG. 8. Transfer device in which an ejector is made to travel at a faster speed during part of the cycle.

A packaging machine may be required to hold the article to be wrapped, transfer it into its wrapper while performing some of the folding operations, and finally eject it into some other receptacle. Alternatively, the machine may be required to apply a hold-down plate to an article while it is being elevated and enveloped on three sides by the wrapper. After this, the wrapper is folded onto the base of the article and then onto the end faces. In the next movement, the article is transferred over a folding plate to complete the wrapping, and the package is carried to the following work station.

The requirements of the first method are met by the arrangement shown in Fig. 8. An article *A* is temporarily held between a pair of blades *B*, at the rear of which there is a sliding plunger

C. These components are carried through the various movements by means of a crank *D*, driving a slotted lever *E*. A link *F* connects lever *E* to a sliding carriage *G*. For actuating plunger *C* separately, there is a bellcrank lever *H*, one end of which is forked to engage a roller *J*. This roller is carried by a stud supported on a bearing bracket *K* attached to the bar carrying plunger *C*. The other arm of bellcrank lever *H* is fitted with a roller *L*, which engages the track of a stationary cam *M*. The pivot stud of lever *H* is secured to sliding carriage *G*.

When the article *A* has been introduced between plates *B*, the carriage assembly moves to the right. As the article reaches work station X, it is thrust into the wrapper *W*, since plunger *C* has by this time moved the trailing edge of the article to the leading edge of blades *B*, through the action of cam *M* on bellcrank lever *H*. At this point, blades *B* have almost reached the end of their travel, and because of the form of cam *M*, the plunger is traveling at an increased rate. As a result, the article is pushed from between the blades, and is deposited ready for the following operation.

A mechanism designed for the second set of conditions is illustrated in Fig. 9. It is shown at the completion of its stroke and about to return to the starting position, indicated by the phantom view. With the mechanism in the starting position, the article *A* is pushed up through well *B*, together with the wrapper so that the latter is folded into an inverted U-form. The top of the article is held in contact with stop plate *C* by a platform (not shown). Plate *C* rests on the article for the duration of the period that the latter is being enveloped by the wrapper, and travels with it until the partially enclosed article has been positioned at the next work station in the packing sequences.

The plate is mounted on a sliding carriage *D* to which is also attached a pusher *E*. The latter is fitted with wrapper-folding members (not shown) that make the two rear side folds on the package. Carriage *D* is reciprocated on guide ways *F* by means of a link arm *G*, pivoted on lever *H*. Member *H*, in turn, is driven by a crank *J* by means of a connecting-rod *K*.

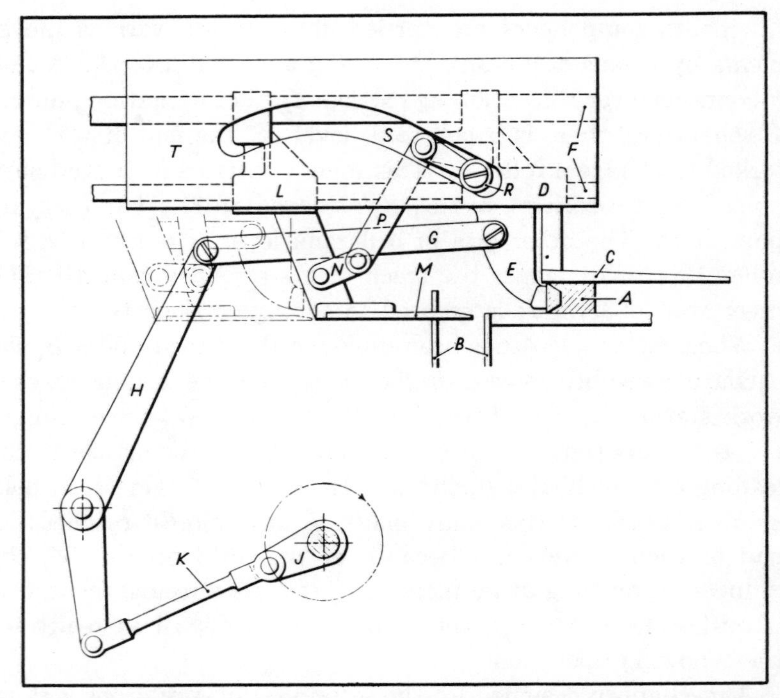

FIG. 9. Crank-driven mechanism with a stationary cam to vary the speed of member M for wrapper folding.

A second sliding carriage L, to which is attached a bottom folder M, is moved on the same guide ways as carriage D by means of link N. This link is pivoted on the long arm of bellcrank lever P. The latter, in turn, is pivoted on a stud on carriage D. The short arm of bellcrank P carries a roller R which engages a cam slot S in a stationary plate T.

As the stop plate C and the plunger E are being moved to the right to engage the rear of the partially wrapped article A, the bottom folder M initially moves forward at a faster rate because of the upward curvature of the cam slot S. Thereafter the rate is reduced, and the folder finally dwells while pusher E moves the article on to the next work station. At this stage, the article has been enclosed on four surfaces by its wrapper and an inward fold has been made at the rear of each end of the package.

CHAPTER 9

Variable Stroke Reciprocating Mechanisms

Means of adjusting the length, speed or timing of the reciprocating stroke are described in this chapter. Other variable stroke reciprocating mechanisms are described in Chapter 9, Volume III of "Ingenious Mechanisms for Designers and Inventors."

Fixed Stroke Converted into Variable Stroke

The demand for varying widths of a flat wire product where previously only one standard width was processed made it necessary to provide existing machines with a simple mechanism to insure even winding onto spools. Even winding was accomplished without modifying the width of the spools by providing an adjustment for the stroke of the guiding member.

As can be noted in Fig. 1, when double screw *A* is rotated, it transmits reciprocating motion to the slide *C* through the forked, swinging follower *B*.

To provide for adjustment in the length of travel, a few modifications had to be made. A rack-and-pinion and "Scotch yoke" arrangement was devised which consisted of a gear, rack, roller, slide-bar, and support block. Gear *D* rotates freely on a stud in slide *C* and meshes with rack *E* which is attached to a stationary part of the machine. Any linear movement of slide *C* produces rotation of gear *D*, the pitch diameter of the gear *D* being such that it will be given a complete half-revolution as the slide *C* covers its range of travel. The hub *I* of gear *D* has a T-slot containing a T-bolt on which roller *F* rotates freely.

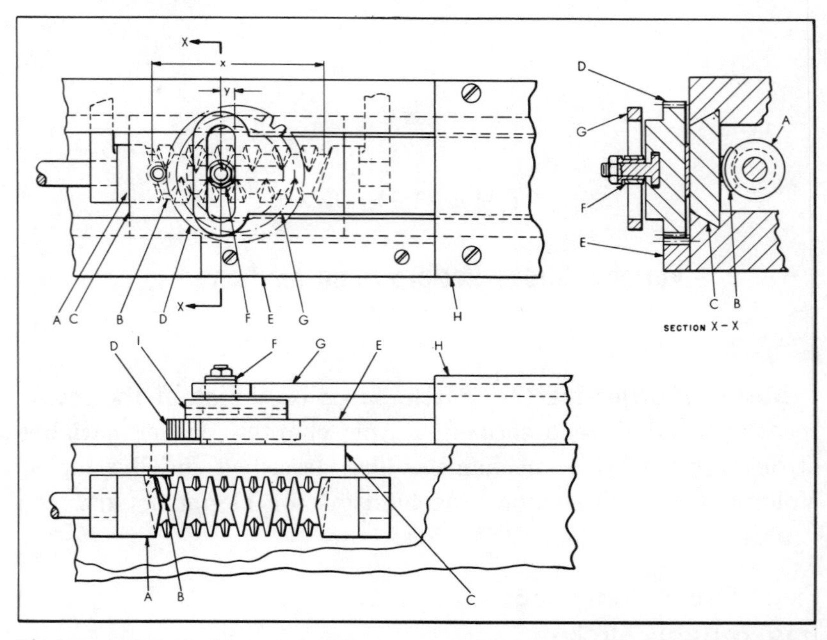

FIG. 1. This mechanism, which originally consisted of a double screw and slide arrangement to give a fixed stroke, was modified by the addition of a rack-and-pinion and a "Scotch yoke" to provide an adjustable stroke.

Movement of the T-bolt in the T-slot permits adjustment of the position of roller F relative to the axis of gear D. Roller F engages a slot in slide-bar G which, in turn, is supported in block H. To the opposite end of bar G is fastened the guide which delivers the product to the spools.

The center of roller F is offset from the center of rotation of gear D, a distance y. As gear D rotates, the center of roller F describes a semi-circle, with radius y, about the axis of gear D. Roller F acts in the slot of slide G, and a half-revolution of gear D rotates to the opposite side of the center of gear D. The relative movement between the bar and the gear is $2y$. Since the center of gear D moves through a distance x due to the linear movement of slide C, the combination of the two movements results in a total length of travel of slide G equal to x plus $2y$.

If the roller *F* were placed on the opposite side of the center of gear *D*, with the slide *C* in the position shown, the total movement of slide *G* would equal x minus $2y$. Furthermore, if roller *F* were placed exactly over the center of gear *D*, roller *F* would merely rotate on its center, and the total movement of slide *G* would be the same as that of slide *C*.

Although the added motion is harmonic it is small enough not to be objectionable.

Variable Straight-line Reciprocation from Uniform Rotary Motion

On a machine producing a woven-wire product, some of the strands of wire are displaced varying amounts to produce a random decorative pattern. This is done by converting a uniformly rotating input to a straight-line reciprocation which can either change length or be the same length from action to action.

Shaft *A* (see Fig. 2), mounted in a stationary part of the machine, rotates at a uniform rate in the direction indicated by the arrow. The shaft carries and is keyed to a disc *B*, which in turn carries a pin *C* that extends from one side.

Ring *D* is grooved to varying depths on equally spaced radial lines, each of which is numbered in the drawing. The ring is mounted on plate *E*, and is free to rotate. Two guide bars *F* are carried by the plate, and slide freely in channels, being retained by strips *G*. These guide bars are wide enough to support the ring in the back.

Plate *E* carries an extension on its top, to which the guide for the wire is attached. Stop-pin *H*, at the bottom, serves to hold the sliding parts in the starting position between movements.

In the position of pin *C* shown, it is entering the No. 1 groove in the ring, and the ring thus has to rotate in the direction indicated by the arrow. Until the pin contacts the bottom of the groove, the movement of the ring is rotative only. But when the pin reaches the bottom of the groove, it transmits to the ring (and thus, to the plate) a linear movement as well.

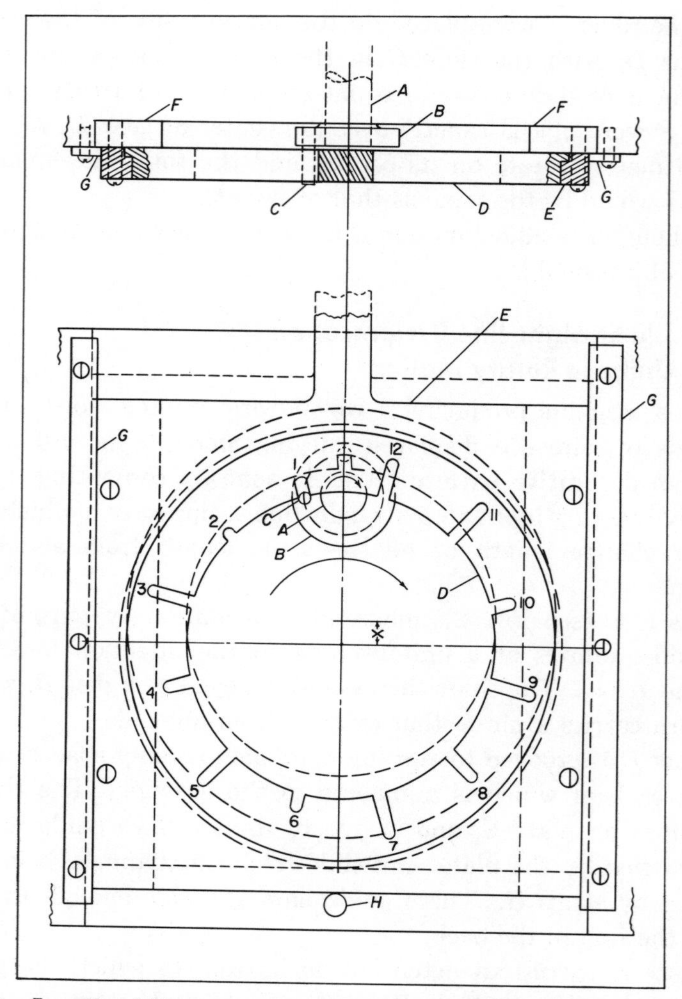

FIG. 2. Length of linear movement of plate *E* is controlled by the respective depths of the twelve grooves in ring *D*.

By the time the pin reaches the vertical center line of the mechanism, both the ring and plate *E* have moved upward a distance equal to *x*. Continued movement of the pin transfers the No. 1 groove to the position previously occupied by the No. 12 groove, each rotation of disc *B* indexing the ring to the next groove.

The amount of linear movement given the plate is governed by the depths of the grooves in the ring. Since, in this instance, the No. 2 groove is the shallowest, it will produce the greatest amount of plate movement. Grooves Nos. 3, 5, 7, 9, and 11 are deep enough so that the pin cannot contact their bottoms. Thus, when the pin is engaged in any of these grooves, the ring rotates without any accompanying linear movement of the plate, as is desired.

On the machine to which this mechanism is applied, the axes of the rotating parts are in a horizontal plane, so that the weight of the plate will return it to contact stop-pin *H* after each reciprocation. In an application where the axes are vertical, a return spring will be required.

Adjustable Eccentric

In the design of various types of machines, an eccentric motion is often needed to operate certain mechanisms. A relatively simple arrangement for obtaining such eccentric movement from a rotary drive is shown in Fig. 3. The resultant drive is positive and can be varied quickly and easily, thus avoiding time-consuming dismantling and reassembly.

In construction, adapter *A*, made from a rectangular piece of stock, is turned on one end. In addition, a hole is bored

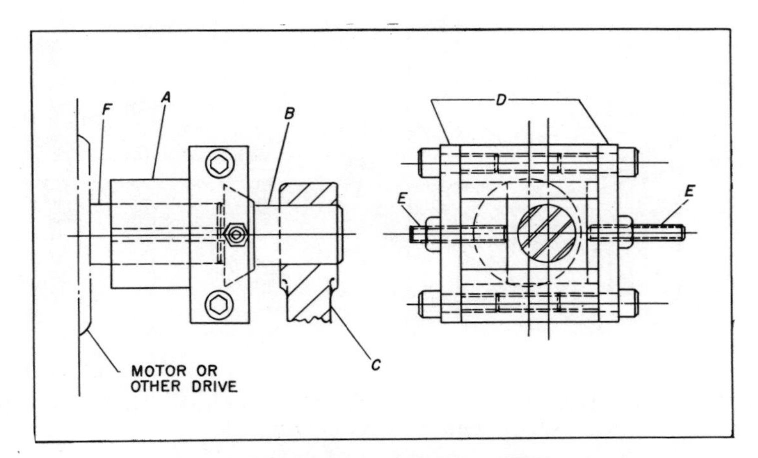

FIG. 3. Simply constructed adjustable eccentric.

through the part and a keyway is provided to fit the driving shaft. A female dovetail is milled in the rectangular section at the opposite end. From a square piece of stock, a male dovetail slide *B* is then machined to fit the adapter. One end of this member is turned to fit a connecting-link arm *C*. Two side-plates *D*, made from flat stock, are bolted to the sides of the adapter with socket-head cap-screws, as shown. Each plate also has a hole tapped to receive headless set-screws *E*, which are used for adjusting the amount of eccentricity and for locking the slide *B* securely in place during operation of the connected drive.

The slide and adapter may be scribed with gage lines, if desired, in case minute adjustment is needed.

Eccentric Driving Mechanism Permits Stroke Adjustment During Operation

Small variations in the stroke length of a reciprocating slide can be made while it is operating, by means of the mechanism shown in Fig. 4. Driving disc *A* has an integral shank revolving in fixed bearing *B*, where it is retained by bearing cap *C*. Driving gear *D*, keyed to the shank, revolves continuously.

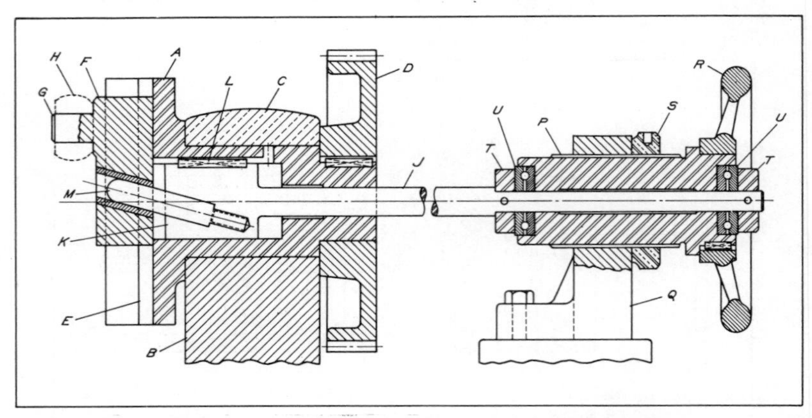

FIG. 4. By revolving handwheel *R*, the throw of crank pin *G* can be varied without stopping the slide.

The disc face contains a dovetail *E* milled across its diameter. Crankpin block *F*, fitting the dovetail, has an integral crankpin *G*, over which is fitted one end of a connecting-rod *H*. At its other end, the rod is attached to the reciprocating slide (not shown).

Rod *J*, by which the device is adjusted, has a shouldered section *K* fitting the bore of disc *A*. Key *L* causes the rod to revolve with the disc yet permits a short axial movement of the rod along the bore. A hard pin *M* is pressed at an angle into the end of section *K*. This pin engages a hole that is drilled in the crankpin block.

Rod *J* extends from any convenient distance to a control point. At its right end, the rod is reduced in diameter and can rotate in sleeve *P*. An external thread on the sleeve engages a threaded hole in angle-bracket *Q*. By revolving handwheel *R*, keyed to the sleeve, the sleeve can be adjusted axially. Threaded ring *S* locks the sleeve, once it has been adjusted. Rod *J* moves axially in unison with the sleeve, by means of the stop collars *T* and thrust bearings *U*.

If the stroke of the slide has to be lengthened, ring *S* is released, and the handwheel revolved counterclockwise. This movement is transmitted to the rod, and pin *M* is retracted a corresponding amount from the crankpin block, causing the block to move radially outward in disc *A*. Thus, crankpin *G* has a greater throw. By revolving the handwheel clockwise, the throw of the crankpin is similarly decreased.

Mechanism for Effecting a Varying Reciprocating Motion

On a wire-forming machine, a reciprocating slide was required to move at a uniform rate of speed during part of the cycle and, at a predetermined point, to increase in speed for the remainder of the cycle. The motive power for operating this slide was taken from a uniformly reciprocating slide and then transformed into varying reciprocating motion by the mechanism here shown.

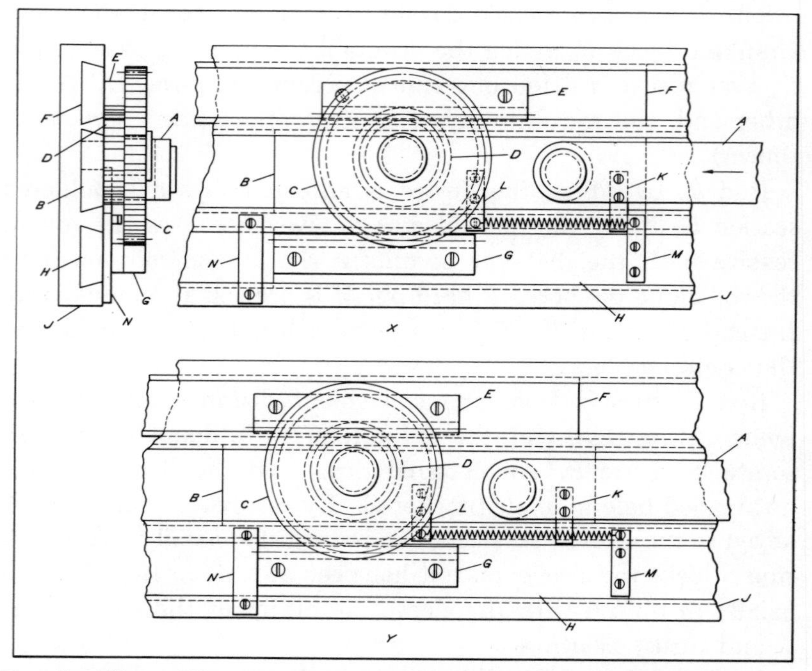

FIG. 5. Three-slide arrangement, utilizing gears in the capacity of a lever, varies motion supplied by a uniformly reciprocating source.

Rod *A* (see Fig. 5), reciprocates at a uniform rate of speed and drives slide *B*. This slide carries gears *C* and *D* which are keyed together, yet rotate freely on a stud. Gear *D* meshes with the rack *E* which is fastened to a second slide *F*. The larger gear meshes with rack *G* on a third slide *H*. All three slides are dovetailed in the stationary slide bracket *J*.

Dog *K* and spring retainer *L* are attached to slide *B*, while spring retainer *M* is secured to slide *H*. A stop *N* is fastened across the lower dovetailed slot.

In operation the uniform reciprocation is obtained from rod *A* which moves slide *B* to the left. As long as contact between parts *M* and *K* is maintained by means of the tension spring, the central and lower slides will move in unison. The gears cannot rotate at this point as there is no change in the relative positions of slides *B* and *H*. Therefore the three slides will move together during this phase.

It will be noted that, at Y in the diagram, rack *G* has contacted stop *N*, thus preventing further movement of slide *H*. As slide *B* continues its movement, gear *C* is forced to rotate due to the relative movement between it and the now stationary rack *G*. Gear *D*, which rotates with the larger gear, transmits this motion to the upper slide through rack *E*. Slide *F* now moves at a greater speed than slide *B*. The ratio of gears *C* and *D* governs the extent of the increased speed of slide *F* as compared with that of slide *B*.

On the return stroke of rod *A*, the upper and central slides will move to the right in the same speed ratio with respect to each other as they did on the forward stroke. This will continue until dog *K* contacts spring retainer *M*, at which time all three slides will once again move together.

Adjustable Eccentric Produces a Variable Throw

Rotary movement of a driving shaft must be converted frequently into rectilinear movement for the actuation of such members as slides or levers. One of the most common and effective means for accomplishing this conversion is the combination of an eccentric and connecting-rod. In its conventional form, however, an eccentric is applicable only when a fixed length of stroke is desired. If, on a particular machine, it is necessary to vary the stroke length imparted to the driven members, it is advantageous to employ an adjustable eccentric drive mechanism.

Figure 6 shows the design of such an adjustable eccentric. Outer eccentric member *A* has a running fit within the head of a connecting-rod (not shown) which couples the eccentric to the reciprocating machine member. A large hole is bored through the outer eccentric member, being offset distance *X* which is determined by the amount of throw required.

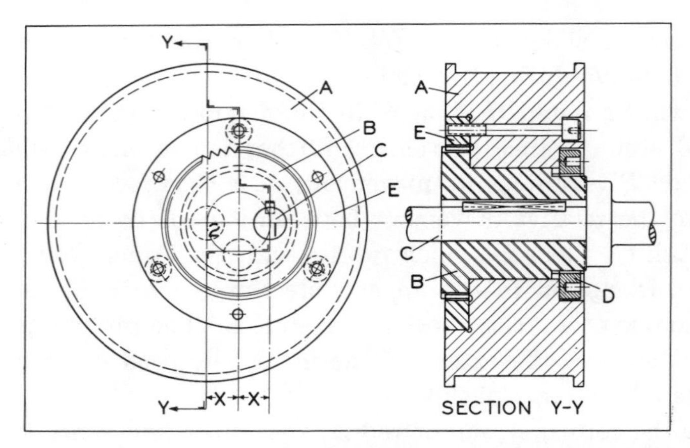

FIG. 6. In this drive, throw of outer eccentric A is adjusted by rotating sleeve B carrying drive-shaft C.

Mounted in the bored hole is flanged sleeve B. Sleeve B is bored to receive, and is keyed to, drive-shaft C. The shaft hole is located eccentrically in the sleeve, the amount of offset provided in the illustrated unit being X. Circular lock-nut D is threaded on the right-hand end of the flanged sleeve to retain it in place. Spanner holes are drilled in the face of the lock-nut to facilitate removal and replacement.

The flanged end of sleeve B has a series of accurately spaced vee serrations machined on its periphery. The serrations mesh with similar internal serrations provided on ring E. Cap-screws and dowels, passing through eccentric member A from the opposite side, locate and retain ring E in the recess provided. The dowels must be of sufficient size in order to bear the main driving load.

With sleeve B situated so that shaft C is on the horizontal center line in the position shown at 1, maximum throw of the eccentric is obtained. If lock-nut D is backed off, and sleeve B is rotated 180 degrees in relation to the outer member, shaft C will be coincident with the center of the eccentric, Position 2. At this setting, no motion will be delivered to the machine slide. Sleeve B may be located at any intermediate position between the neutral and maximum throw settings.

Steplessly Variable Stroke Movement

An arrangement employed on an optical profile-grinding machine that enables the length of stroke of a horizontal reciprocating slide to be steplessly varied is shown in Fig. 7. Of compact design, the unit has provision for bolting directly to a speed reducer.

Body *A* is keyed to the shaft of the reducer and houses a slide *B*. This slide incorporates a crankpin and is retained by a cover plate (not shown in the elevation). From the pin, motion is transmitted to the machine slide by a connecting-rod.

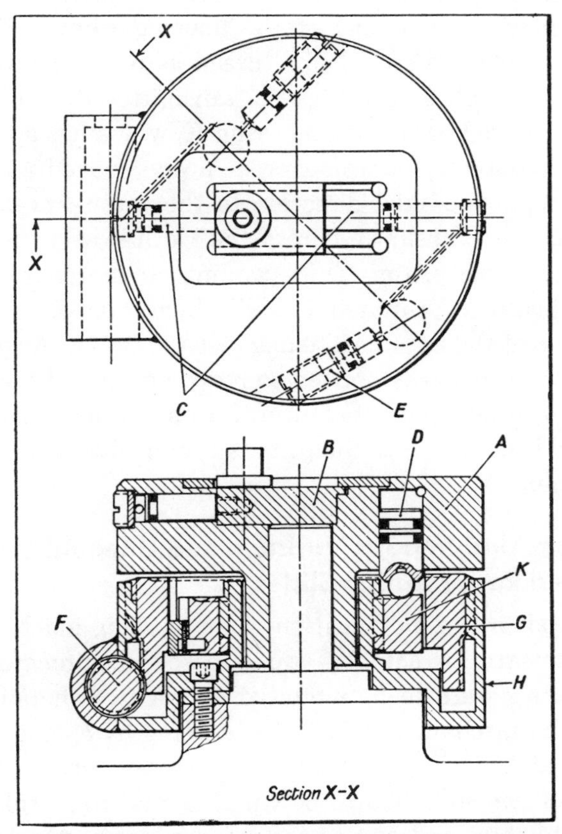

FIG. 7. Device for steplessly varying stroke of a reciprocating drive. Stroke can be changed while drive is in motion.

Two pistons *C*, forming part of a closed hydraulic system, determine the position of the slide *B* and, consequently, the throw of the crankpin and the stroke of the machine slide. The pistons *C* are controlled by pistons *D*, the cylinder bores being connected by transverse ports as seen in the elevation. To facilitate charging and setting, a piston *E* with screw adjustment is provided for each branch of the hydraulic circuit in this mechanism.

Adjustment of the stroke can be effected while the machine is running by turning a handwheel connected to the worm *F*. This member meshes with a worm-wheel *G* which in turn is threaded into a stationary housing *H*, permitting the worm-wheel to move into or out of the housing when the worm is rotated. A left-hand stub Acme thread is used for this purpose. From the worm-wheel, drive is also transmitted through a sliding phosphor-bronze key to the inner ring *K*, which has a right-hand thread. In consequence, this ring is always moved in the housing in the opposite axial direction to the worm-wheel.

Movement is transmitted from the worm-wheel or the inner ring to one of the pistons *D*, depending on the direction of the stroke adjustment, through a ball which runs in a track on the end face of the appropriate member (*K* or *G*). As one piston is raised, the other is permitted to move downward by reason of the right- and left-hand thread arrangement, and a corresponding motion is imparted to the pistons *C* and slide *B* which carries the crankpin.

Mechanism That Imparts Variable and Unequal Strokes to Opposed Reciprocating Slides

Incidental to the modification of an existing machine, it was found necessary to actuate two opposed reciprocating slides. Forming tools were to be mounted on each of the slides. The drive for the mechanism had to be powered by a constant-speed shaft that also motivated other machine movements. The length of stroke of one slide had to be variable, and the stroke of each slide had to start and end at the same instant. Also, the slides

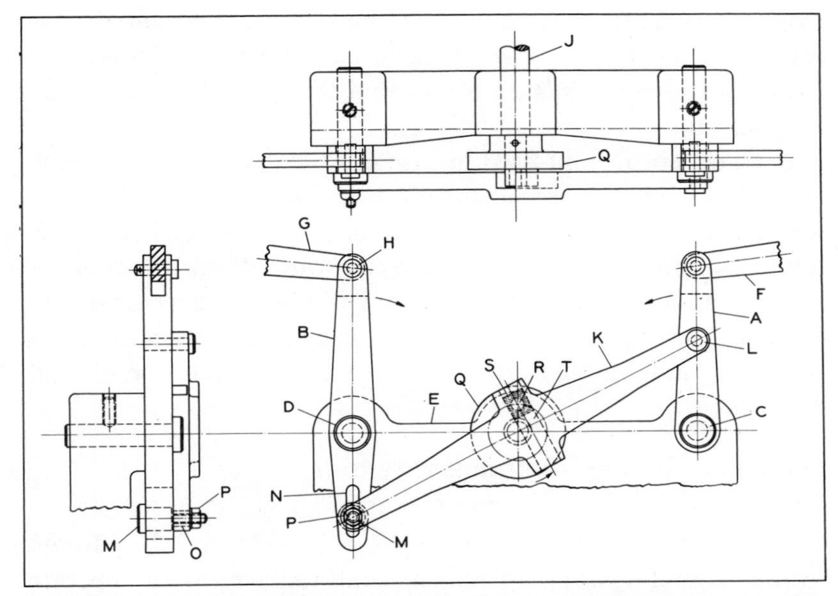

FIG. 8. Drive mechanism for opposed reciprocating slides that permits stroke adjustment of one slide.

were always to move in opposite directions. The mechanism illustrated in Fig. 8 was designed to satisfy these conditions.

The two levers *A* and *B* are mounted to pivot freely on headed studs *C* and *D*, respectively. The top half of lever *B* and lever *A* are of the same length. The levers are mounted along the same horizontal center line in machine frame *E*. The upper end of each lever is slotted for short connecting links *F* and *G*. Each link pivots on a headed stud *H*. The opposite end (not shown) of each of these connecting links is coupled to a reciprocating slide.

The two levers are mounted at approximately the same distance each side of the driving shaft *J* which extends from a bossed portion of the machine frame. The levers are constrained to pivot in opposite directions by means of the long steel connecting-rod *K*. The right-hand end of the connecting-rod is coupled to lever *A* by the headed stud *L* secured to the lever at a fixed center distance from the fulcrum stud *C*. The connecting-rod pivots on stud *L*. The left-hand end of connecting-rod *K* is attached to lever *B* in such a way that its point of pivoting may be changed.

A crank pin *R* is pressed in disc *Q* which is keyed to driving shaft *J*. A slide *S* pivots about pin *R* and slides in slot *T* in connecting-rod *K*. The rotation of crank pin *R* slides *S* in slot *T* and reciprocates *K*. *K*, in turn, reciprocates *B* and *A* in opposite directions. The amount of throw of pin *R* is such that link *F* moves the required distance. The size of the reciprocation of *G* can be varied in any instance by altering the position of stud *M* in slot *N*.

Lever Type Driving Mechanism Permits Stroke and Dwell Adjustments

The lever type mechanism shown in Fig. 9 was incorporated in the drive of a wrapping machine, and was required to operate a transfer slide which transported wrapped packages from the machine to a conveyor belt alongside. A drive shaft with an oscillating rotary movement of 35 degrees was the source of motion for the driving mechanism. Another factor involved in the reciprocation of the transfer slide was that it had to be readily adjustable to suit the various sizes of packages normally handled by the machine.

Drive shaft *A*, shown at X in Fig. 9, is mounted horizontally within a bearing hole through the small upright boss *B*.

Securely keyed to the drive-shaft are two identical levers *D*, one being situated at each side of boss *B*. They are retained in position by means of the headed end *E* of the drive-shaft at the right-hand side, and collar *F*, cross-pinned to the shaft, on the left-hand side.

Pin *G*, which supports lever-arm *H*, passes through in-line holes in the upper end of the twin levers *D*, and is retained by collar *J*. The upper limb of the lever-arm *H* pivots about pin *K* within the forked end of connecting-rod *L*. Connecting-rod *L* is linked directly to the transfer slide of the wrapping machine.

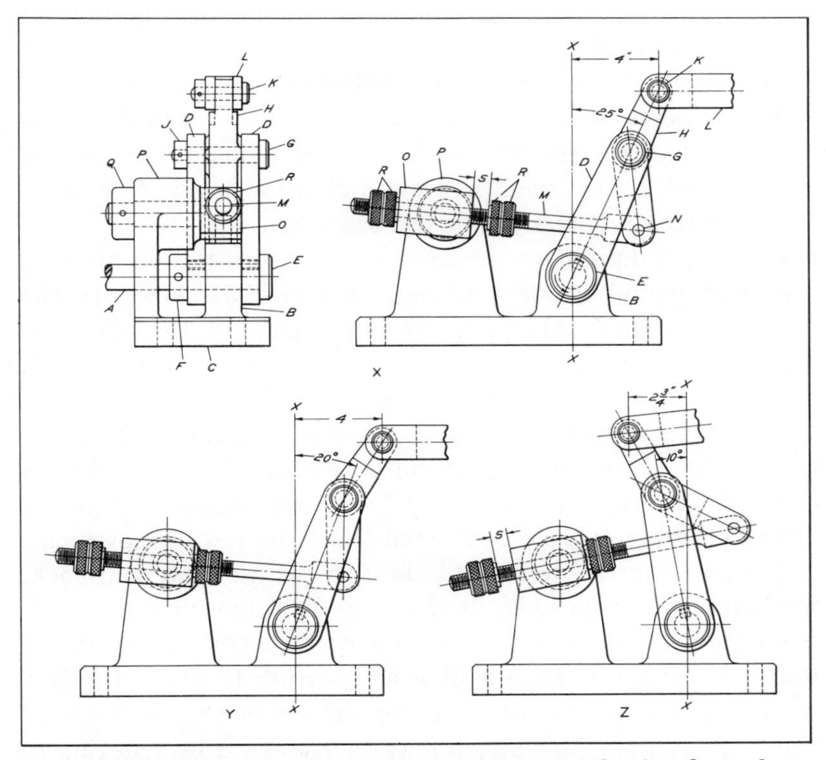

Fig. 9. Knurled nuts of the lever type driving mechanism here shown may be adjusted during operation to vary the dwell duration and also the points of connecting-rod reversal.

The lower limb of the lever-arm lies at a slight angle to the upper limb, and is slotted to receive the end of rod *M*. A pin *N* joins these two members so that they are free to pivot.

The opposite end of rod *M* can slide within trunnion block *O*. A shaft, machined on one side of the block, rides in a bearing hole in a large boss *P*. Collar *Q* is pinned to the shaft to retain the trunnion block in position.

For the major portion of its length, rod *M* is threaded to receive four knurled nuts *R*, which are located as shown. The nuts are adjustable, and are normally secured in any desired setting by simply locking them together in pairs.

The diagram at the right-hand side of X shows the relative positions assumed by the various components as drive-shaft *A* starts to move in a counterclockwise direction.

At this point in the cycle, connecting-rod *L* is in its retracted position, that is, situated 4 inches to the right of vertical axis X-X. It will be observed that levers *D* are inclined at an angle of 25 degrees to the right of the same axis.

Distance *S* must be calculated according to the duration of the dwell period required at each end of the stroke. In the illustrated example, this distance is regulated so that the connecting-rod and the transfer-slide will remain inactive at each end of the stroke during a 5-degree travel of shaft *A*.

As the mechanism moves from its starting point in the direction of the arrow, connecting-rod *L* remains stationary due to the resistance of the transfer slide. *L* will remain stationary until the levers *D* have completed the 5 degrees of movement, whereupon knurled nuts *R*, at the right of the trunnion block, will contact that member, as shown at Y in the illustration.

As the drive-shaft continues its movement beyond this point, rod *M* is prevented from sliding any further to the left. Thus connecting-rod *L* receives the combined movements of levers *D* and lever arm *H*. This action may be visualized by referring to diagram Z in the illustration. The diagram shows the relative positions of the components when drive shaft *A* has reached the end of its forward oscillation. At this stage, levers *D* are inclined 10 degrees to the left of vertical axis X-X, while the connecting-rod has moved 2¾ inches to the left of the same axis, a total travel of 6¾ inches.

As the return stroke commences, the connecting-rod again remains stationary, due to the resistance of the transfer slide, for the first 5 degrees of drive shaft travel. After this distance has been covered, the knurled nuts *R*, on the left-hand side of the trunnion block, reach their limit of movement, and motion is once again transmitted through the connecting-rod to the transfer slide of the machine.

Duration of the dwells at each end of the stroke may be altered by adjusting the setting of the knurled nuts on rod *M* while the machine is operating at slow speeds. The nut-setting may also be altered to effect variations in the points of reversal of the stroke. By setting both pairs of nuts in contact with the end faces of the trunnion, the dwell periods will be eliminated and the connecting-rod will then have its maximum length of stroke.

Common Drive for Two Slides with Partially Synchronized Travel

A lever drive mechanism installed in a wire-bending machine now allows a high degree of versatility in the equipment. The mechanism has permitted two existing tool-slides to operate with partially synchronized travel, so that different sizes and forms of wire can be handled, and a greater variety of bent shapes can be produced.

Both tool-slides reciprocate in the same horizontal plane, but are a considerable distance apart in the machine. The mechanism transmits the drive from a common shaft, centrally located, which oscillates through an arc of 40 degrees. The first tool-slide has a short, fixed travel; the other, a considerably longer and adjustable travel. Although both tool-slides start and stop together, the second moves in unison with the first only during the initial portion of the forward travel and the final portion of the return travel. In the interim the second tool-slide increases in speed so as to compensate for its greater length of travel.

A full-scale diagram of the mechanism appears in Fig. 10. The oscillating drive-shaft *A* is carried in the bearing bracket *B*. Keyed to the drive-shaft and oscillating with it is a lever *C*. Through a slot in the top of the lever is cross-pinned a connecting link *D*. The opposite (left-hand) end of this link, not shown, is joined to the first tool-slide, which has the short, fixed travel.

A bellcrank *E* fulcrums on a stud *F* extending from one side of the lever. The second tool-slide with the long, adjustable travel, is joined to another connecting link *G* cross-pinned in a slot in the upper arm of the bellcrank *E*.

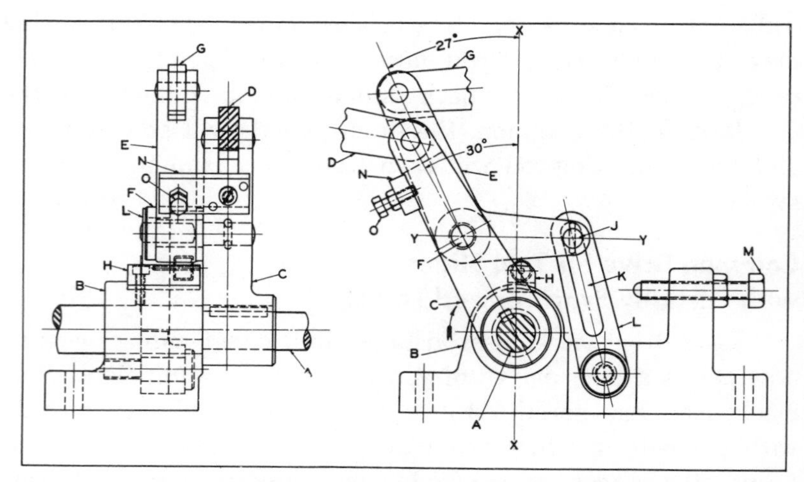

FIG. 10. At the start of the forward stroke, the bellcrank E moves in unison with lever C.

The lower arm of the bellcrank extends laterally, being supported by a roller assembly H mounted on the top of the bracket. The height of the assembly is designed so that the arm is in light contact with the roller when the mechanism is in its starting position, as shown in Fig. 10. The vertical center line of the roller is offset about 1/8 inch to the right of the vertical center line X-X of the drive shaft.

The ratio of the radii of the arms of the bellcrank determines the stroke-length range of the second tool-slide. In this particular instance, the radius of the lower arm is approximately two-thirds that of the upper arm.

At its extremity, the lower arm of the bellcrank contains a short dowel pin J which has a sliding fit in the slot K of a link L. This link pivots in a channel in the bearing bracket, and is a slight distance behind the lower arm of the bellcrank. A stroke adjusting screw M is located in an integral rib in one side of the bearing bracket, to the right of the bellcrank. As will be explained, this screw controls the extent that the second tool-slide travels in unison with the first tool-slide.

To the left side of lever C is doweled and screwed a short rectangular plate N. This plate extends across the adjacent side of the bellcrank, and serves to transmit the forward motion of the lever to the bellcrank. Because of the slightly different angularity of the bellcrank, its side is somewhat clear of the plate. So that both tool-slides can start in unison, a set-screw *O* in the plate is adjusted to bear on the bellcrank.

When the drive-shaft starts its forward, or clockwise, oscillation — indicated by the arrow — lever *C* lies at an angle of 30 degrees to the left of the line X-X, and the upper arm of the bellcrank lies at an angle of 27 degrees; and pin *J* is on the horizontal center line Y-Y of stud *F*. Movements of the lever and bellcrank continue in unison until the lever is vertical, as in Fig. 11. At this point, link *L* has also pivoted clockwise into contact with screw *M*, and pin *J* has descended part way down slot *K*.

Since no further pivoting of the link is possible, continued movement of the lever is not transmitted uniformly to the bellcrank. Instead, the bellcrank fulcrums on stud *F* for the final 10 degrees of forward oscillation, Fig. 12; pin *J* descends to the bottom of the slot, and the link pivots to the left. Thus, the upper arm of the bellcrank, with its combined movements, travels in a

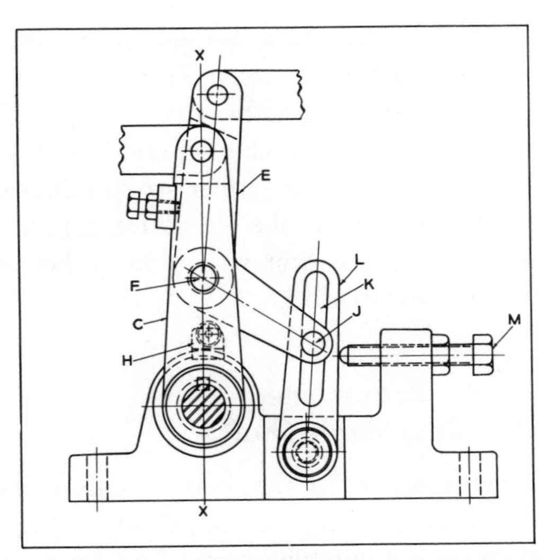

FIG. 11. When link *L* abuts screw *M* the bellcrank *E* starts to fulcrum on stud *F*.

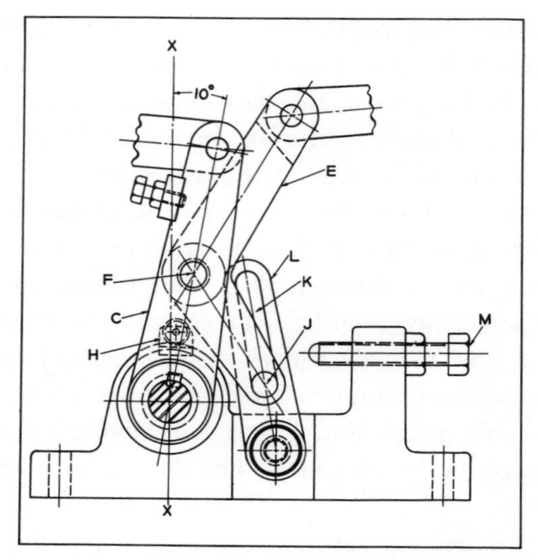

FIG. 12. At the end of the forward stroke, pin *J* has descended to the bottom of slot *K*.

considerably longer arc than the lever, causing the second tool-slide to travel a correspondingly greater distance and at a greater speed than the first tool-slide.

As counterclockwise rotation begins, the lower arm of the bellcrank rides over the roller assembly *H*, forcing pin *J* to rise in the slot. At the same time, the link pivots to the right into contact with screw *M*. Then, the bellcrank and lever move in unison for the balance of the return stroke. Thus, during the return stroke, the second tool-slide starts rapidly, then slows down to the speed of the first tool-slide — the reverse of the action during the forward stroke.

Two Rotary Slides Reciprocated in Synchronism on a Single Shaft

Two machine slides were required to rotate on a common shaft and, at the same time, to reciprocate along the shaft in opposite directions. A simple means had to be provided for altering the stroke length of both slides so that they could be adjusted to travel either equal or unequal distances with their reversal points occurring simultaneously. A mechanism incorporating these features is shown in Fig. 13.

Shaft *A*, shown at V, rotates slowly and carries the two machine slides *B* and *C*. The slides are forced to rotate with the shaft by means of keys *D* and *E* which are fastened to their respective members. A keyway *F* is machined along the length of the shaft to provide a sliding fit with the keys.

Reciprocating movements are imparted to slide *B* by T-shaped lever *G*, which pivots on stationary headed stud *H*. Connecting-rod *J*, providing the main source of motion, is free to pivot on stud *K* which connects it to the elongated slot at the lower end of the T-shaped lever. The length of the elongated slot is determined by the desired variation in stroke length of member *B*.

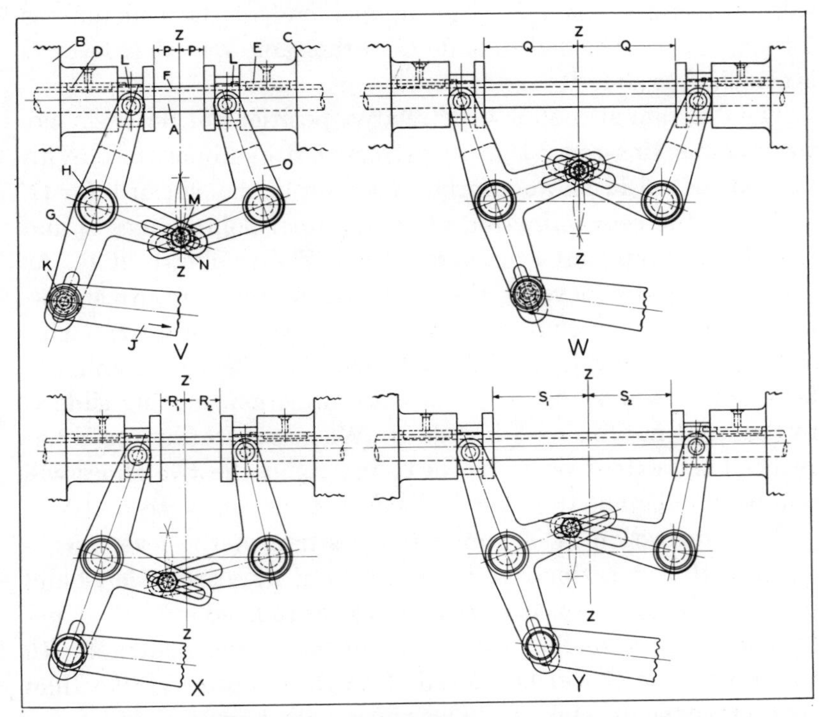

FIG. 13. Adjustable lever type mechanism provides synchronous reciprocation of two opposed, rotating slides.

The upper end of lever *G* is forked to straddle a cylindrical extension boss on the end of slide *B*. Each arm of the fork is linked with an annular groove in the extension boss by means of a hardened steel trunnion block *L*. In this way, the machine slide is free to rotate at the same time that it is being reciprocated.

Centrally located with respect to stud *H*, in the integral right-hand, or short, limb of lever *G*, is an elongated slot. Mounted in this slot is headed stud *M* whose opposite sides are flattened slightly to prevent it from turning in the slot, yet allowing it to slide freely. The stud is secured to the limb by lock-nut *N*.

A larger diameter of stud *M* has a sliding fit within an elongated slot cut along the lower limb of bellcrank *O*. The bellcrank pivots on a stud similar to *H*. Both of these studs are located the same distance from shaft *A*, and the same distance each side of vertical axis Z-Z. The uppper end of the bellcrank is forked, and is attached to slide *C* in the same way that lever *G* is attached to slide *B*.

The diagram at V shows the relative positions of the members when connecting-rod *J* is at its extreme left-hand position. With stud *M* locked in an appropriate position in the slot of lever *G*, and with the connecting-rod at its terminal point, slides *B* and *C* will be equidistant from vertical axis Z-Z as shown at *P*. As connecting-rod *J* moves to the right (arrow) on its return stroke, lever *G* will move counterclockwise, pivoting on stud *H*. This motion causes revolving slide *B* to move to the left. Similarly, bellcrank *O* will swivel in a clockwise direction, moving slide *C* to the right in unison with slide *B*. When connecting-rod *J* has reached the extent of its travel to the right, the two slides will still be equidistant from vertical axis Z-Z, as can be seen at W.

The illustration at X shows the position of the lever mechanism with connecting-rod *J* once again at its extreme left-hand position. In this case, stud *M* is set closer to fixed stud *H*, thereby conveying a smaller radial movement to the bellcrank. In this way, slide *C* will be moved through a shorter stroke than that imparted to slide *B*. The slides will, however, still move in unison, ending their sliding movements at the same instant,

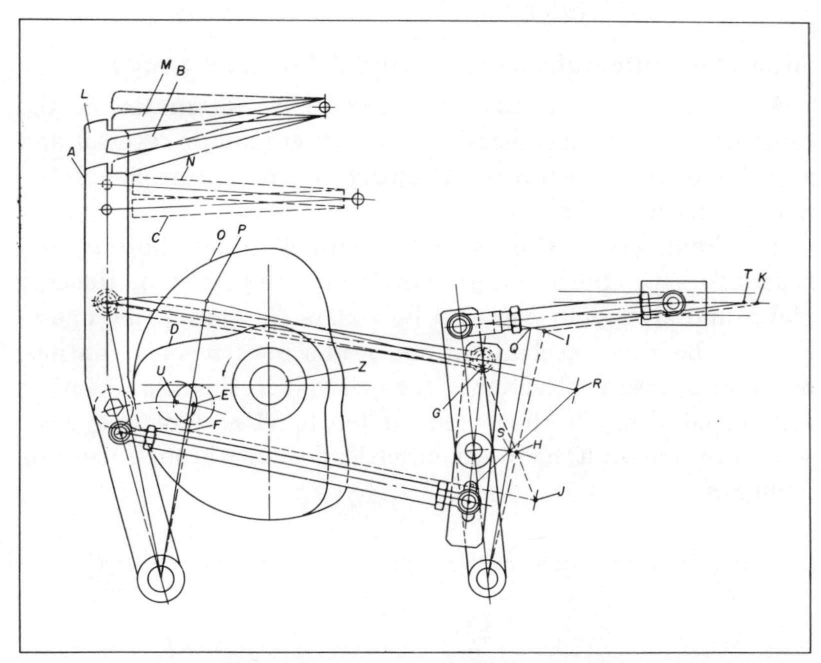

FIG. 14. The output motion to the slide, right, is changed in its stroke length by alternating the use of latches *B* and *M*.

although at unequal distances from vertical axis Z-Z, as seen at Y.

Slide Motion Differential

The object of the device is to change the length of slide stroke to position *T* or *K* by selection of latches *M* and *B* (see Fig. 14). The drive is rotating shaft *Z*. Levers *A* and *L* follow open track cams *D* and *O*. This view shows lever *L* blocked out by latch *M*. Lever *A* follows cam *D* under tension from spring *C*. Connecting-rod *F* is driven to the right, carrying the intermediate lever to successive positions *G*, *H*, *J*. The standing lever is thus carried to position *I*.

When lever *A* is latched out and lever *L* is allowed to follow cam *O* there is the same mode of operation but a different motion. The standing lever moves only to point *Q*, and the slide goes only as far as *T*.

Slide with Automatic Reversal and Adjustable Stroke

A half-nut that engages two lead-screws alternately is the heart of a slide mechanism having both automatic-reversal and adjustable-stroke features. Drawings of the principle of the device appear in Fig. 15.

The frame consists of two end plates *A*, square bars *B*, and round bars *C*. Both lead-screws *D* are supported in the end plates and are axially retained by collars *E*. During the operation of the slide mechanism, the lead-screws revolve continuously in opposite directions. The drive for this movement is introduced through the extended left-hand end of one lead-screw and transmitted to the other lead-screw through meshing pinions *F*.

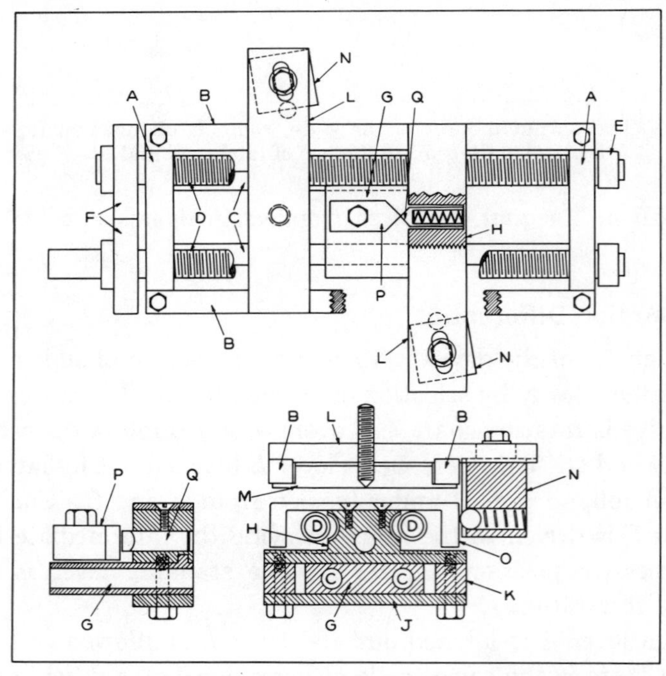

FIG. 15. Half-nut *H* is disengaged from the lead-screw *D* by the thrust of the trip block *N*.

Cross-head *G*, the member required to reciprocate, has a sliding fit on the round bars *C*. The half-nut *H* is splined to the top of the cross-head and is able to move transversely. This half-nut, which has the shape of an inverted T, has two thread sections, one adjacent to each lead-screw.

A keeper plate *J* prevents the half-nut from rising from the cross-head. This plate is joined to the base of the half-nut by cap-screws which run through spacing blocks *K*. Actually, the half-nut never remains in the neutral position illustrated but is engaged to one lead-screw or the other.

To set the stroke area as well as the length of stroke, two stops *L* are fixed along the frame where desired. These stops straddle the square bars *B*, to which they are clamped by keeper plates *M*. One end of each stop extends over an opposite side of the frame, and supports a trip block *N*. The working face of each trip block is set at an angle of 10 degrees to the center line of the device. In the working face is a spring-loaded ball detent.

A shift *O*, with a 10 degree bevel at either end, causes change of engagement of *H*.

In operation, one of the lead-screws moves the half-nut until the shift bar depresses the ball detent in the corresponding trip block, which then thrusts the half-nut out of engagement until it is slightly past a mid-center position over the cross-head.

The completion of the traverse movement of the half-nut *H* is caused by the action of detent *Q* on *P*.

Two Opposed Slides Driven with Rapid Variable Strokes

One phase of the operation of a particular wrapping machine involves the transfer of cartons across a stationary table. Each carton is gripped on two opposite sides by a pair of slides. Figure 16 shows a simple lever type driving mechanism designed to actuate the two slides with a synchronous movement in opposite directions.

Slides *A* and *B* may move freely within their respective dovetailed guide ways which are provided across the top of a cast-

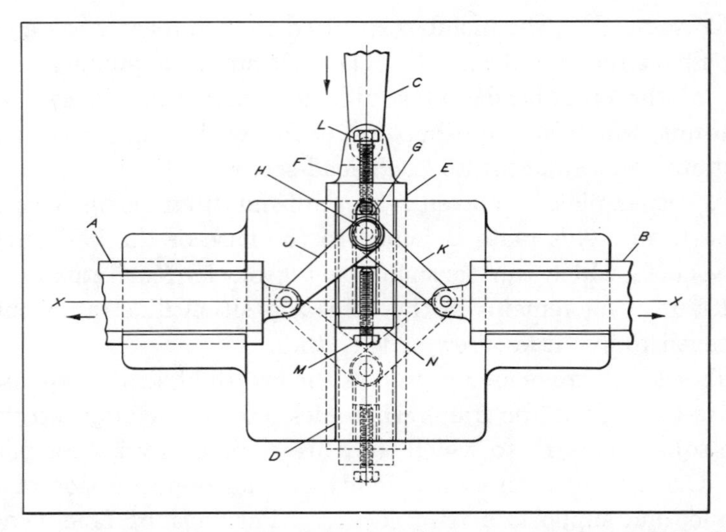

FIG. 16. An adjustable mechanism which drives two opposed slides from a single reciprocating rod.

iron baseplate. This baseplate is bolted to the frame of the machine in line with driving arm *C* which is a source of reciprocating motion.

Across the center portion of the baseplate is machined a third dovetailed guide way *D* situated at right angles to the first two. Riding within it is slide *E*. Lug *F*, which is integral with the third slide, is slotted to receive the end of the driving arm.

Machined along the top of slide *E* is a narrow T-shaped slot into which is fitted a steel sliding block *G*. This member is drilled to receive shoulder stud *H*. Mounted on the projecting portion of this stud are the ends of two identical levers *J* and *K*. The opposite ends of these two levers are connected to slides *A* and *B* respectively, as shown.

The sliding block *G* is locked in position by means of clamp screws *L* and *M*. Screw *L* is threaded through the rear end wall of slide *E*, while screw *M* passes through a small plate *N*. This plate is secured to the front face of the slide. When the lever mechanism is set for normal working conditions, the clamp screws *L* and *M*. Screw *L* is threaded through the rear end wall.

Operation of this mechanism and the manner of its adjustment will be clearly understood by referring to Fig. 16. The solid lines show the driven slides in their innermost position which occurs when slide *E* is fully retracted. To obtain the most efficient operation of the mechanism, the angle subtended by levers *J* and *K* in this position must never be less than 90 degrees. In practice it will be advisable to limit the terminal positions of these levers to a minimum included angle of 100 degrees.

As driving arm *C* moves forward forcing slide *E* ahead of it, levers *J* and *K* will straighten out. This action will force the driven slides apart equally. The maximum forward stroke imparted to these driven members will occur when the center of shoulder stud *H* lies on axis X-X along which the slides travel.

Further travel of slide *E* beyond axis X-X will cause the driven slides to retract in unison until they reach their innermost positions as shown by the light broken lines. A return stroke of the driving arm will produce a duplication of these movements.

Alterations in stroke length of the driven slides are obtained by adjusting the throw of the crankpin on the driving shaft (not shown). Variations in the working positions of the driven slides are obtained by adjusting the position of sliding block *G*. To facilitate setting of this block, the top surface of slide *E* may be graduated and the baseplate marked with a zero line.

Dwell periods may be procured at each terminal point in the movements of slides *A* and *B* if desired. By allowing block *G* to have a certain amount of independent sliding movement within slot in slide *E*, a corresponding motion will be subtracted from the stroke of the driven slides, thus imparting a dwell at the retracted positions of slides *A* and *B*. This type of setting is easily possible by merely adjusting clamp screws *L* and *M*. Each of these screws should be provided with a simple lock-nut arrangement. Provided slide *E* is located at a 90-degree angle to axis X-X, and levers *J* and *K* are the same size, then driven slides *A* and *B* will move in synchronism through identical distances in opposite directions.

Two Slides Operated Longitudinally and One Also Crosswise

For application on a wire fabricating machine, it was necessary to design a mechanism that would guide two strands of wire uniformly back and forth, and at the same time, reciprocate the one wire in a perpendicular plane. The wire to which the two movements are imparted must be stationary during a portion of the cycle. A view of the mechanism at the beginning of its motion, Fig. 17, is a front elevation, while Fig. 18 shows a plan view at an intermediate point of the cycle.

The mechanism includes a slide *A* which is mounted on a stationary part of the machine. The slide receives a uniform motion from a grooved face-cam (not shown) located at its left end. The slide carries a block *B* which is dovetailed to receive

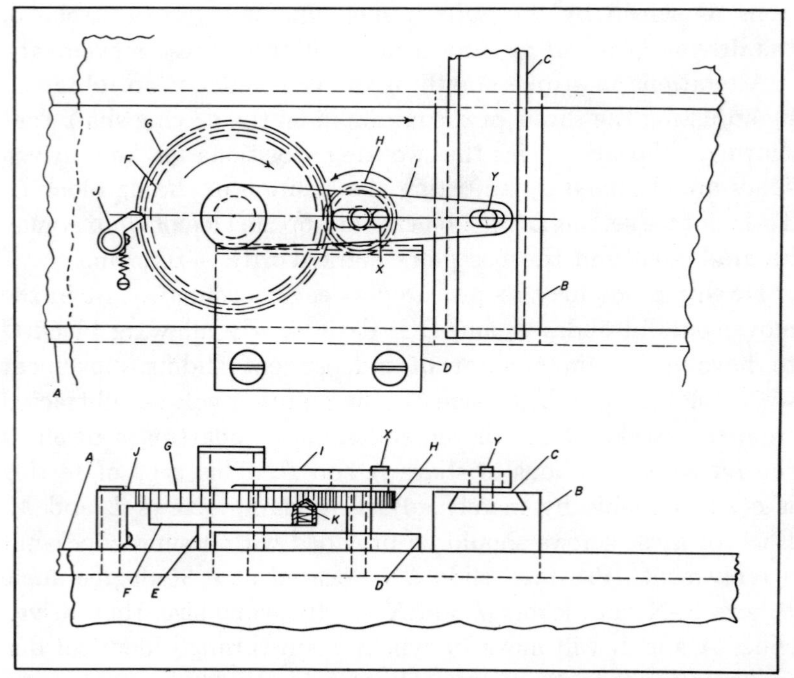

Fig. 17. Mechanism for wire fabricating machine which moves two slides lengthwise and one of them crosswise.

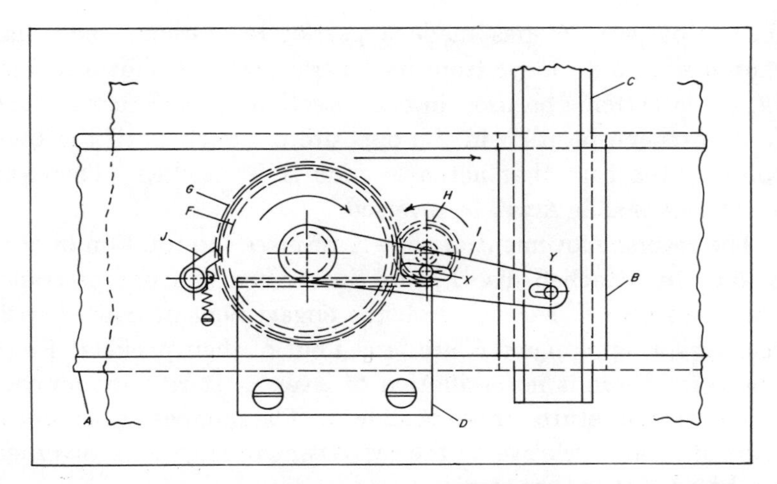

FIG. 18. Mechanism for wire fabricating machine when one-quarter of slide movement has been completed.

slide *C*. Rack plate *D*, also attached to a stationary part of the machine, meshes with pinion *E*. This pinion is attached to disc *F*. Both of them rotate freely on a stud mounted on slide *A*.

Disc *F* carries a series of spring plungers which engage pockets in the bottom side of gear *G* and transmits motion to it. This gear is also free on the stud. It meshes with gear *H*, which carries a pin *X* that engages a slot in lever *I*. The lever is also free on the stud. The end of the lever has a second slot at the right-hand end which engages a pin *Y* on slide *C*. A spring-loaded pawl *J* mounted on slide *A* engages the teeth of gear *G*, permitting rotation of the gear in one direction only.

When slide *A* begins its movement to the right from the position seen in Fig. 17, gear *E* will move with it and rotate in the direction indicated by the arrow, due to its engagement with rack *D*. Disc *F* and gear *E* will rotate as a unit. Gear *H*, meshing with gear *G*, is also caused to rotate. This causes pin *X* to impart an oscillating motion to lever *I*, which is transmitted to slide *C* through *Y*. Slides *A* and *C* carry the wire guides at their outer ends in parallel lines.

In Fig. 18, slide *A* has moved through a part of its motion and has caused the gears to reciprocate slide *C* through lever *I*. At this point, gear *H* has made a partial revolution, which has caused slide *C* to move from its central position, shown in Fig. 17, to its extreme position in one direction. It will now reverse to its extreme position in the opposite direction until the high point of the cam that actuates slide *A* is reached. Then the movement of slide *A* will be reversed.

The reverse movement of slide *A* reverses the rotation of gear *E* and the attached disc *F*, but the motion will not be transmitted to gear *G* because with the engagement of pawl *J* with gear *G*, spring plungers *K* are forced out of their pockets. From this point there is no oscillation of lever *I*. It remains immovable until the return stroke of slide *A*. The number of reciprocations of slide *C* relative to the movement of slide *A* is governed by the ratio of the gear train.

CHAPTER 10

Mechanisms Which Provide Oscillating Motion

Mechanisms which provide oscillating motion are described here. Similar mechanisms are also described in Chapter 10, Volume III of "Ingenious Mechanisms for Designers and Inventors."

Sprocket Operated Geneva Drive Provides Wide Design Possibilities

The normal Geneva drive in which the roller is moved around a circle has limitations which make the mechanism useless for certain purposes. The most severe limitation is that the sum of the time consumed in indexing and the time of dwell corresponds to the time that it takes for the roller to move one revolution around the axis of the driving component of the device.

Instead of moving the roller around a circle, it is possible to move it along a different path and thereby obtain more freedom of design. Figure 1 shows one possible divergence from conventional design. Sprockets *A*, *B*, *C*, and *D* are driven by a chain *E* on which there is fastened roller *F*. Indexing disc *G* has two slots spaced 180 degrees apart.

The roller is shown in a position where it is just about to enter a slot of the indexing disc. With proportions as shown on the drawing, the disc *G* is accelerated during 30 degrees of movement, then moved with a constant velocity through 120 degrees, and decelerated during 30 degrees. The disc then remains in its dwelling position until roller *F* enters the next slot.

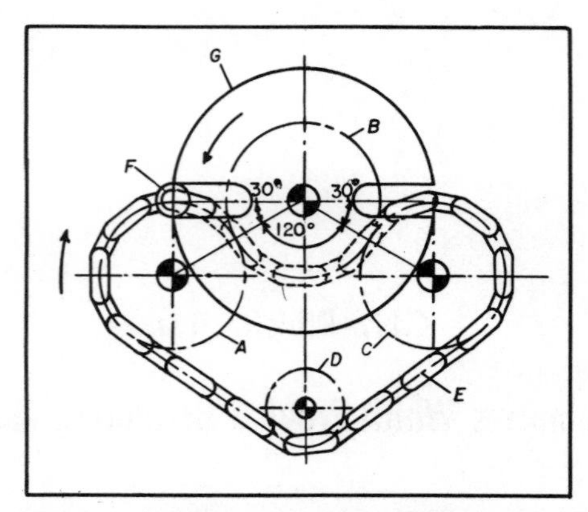

FIG. 1. Geneva drive in which actuating roller does not need to follow circular path.

Flexibility is a feature of this design because the number of slots, the number of rollers, and the length of the chain can be changed to suit various purposes.

Constant Pivot Linkages Replace Ball Bearings

Radar antennas are often constructed so that they rotate about a horizontal axis of no more than 180 degrees (from horizon to horizon). It is necessary to have an unobstructed area around the center of rotation so that the radar beams can pass freely. Therefore, many radar antennas are mounted on ball bearings. One such ball bearing is known to be 13 feet in diameter.

These large, expensive ball bearings can now be replaced by relatively low-cost constant pivot linkages, two forms of which are shown in the sketches. Figure 2 shows a parallelogram linkage. A and B are fixed pivot points and A_1 and B_1 move through circular arms having the same radius. Therefore C_1, too, will move in a circular arc. The same holds true for C_2. Thus, the circular element will rotate around C as center when either arm, A_1A_2 or B_1B_2 is moved.

The system shown in Fig. 3 has more links than the foregoing design but is more compact. In the drawing, B and C are fixed

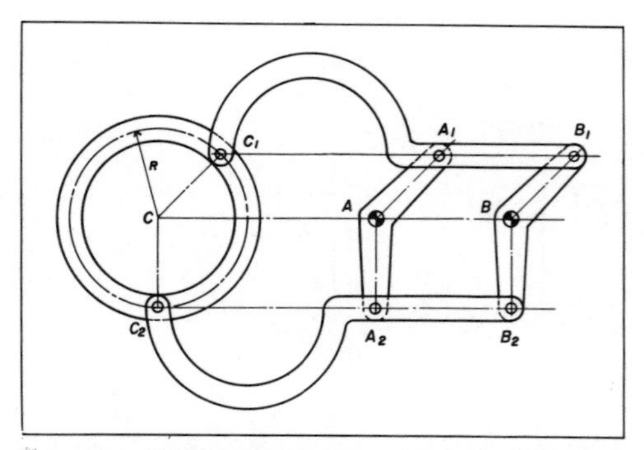

FIG. 2. Constant pivot linkage system rotates circular form C_1C_2 around C when either arm A_1A_2 or B_1B_2 is moved.

points. The triangle B_1BB_2 is equal and similar to triangle B_4AB_3. Similarly, triangle CC_1C_2 is equal and similar to triangle AC_4C_3; also B_1B_4 equals B_2B_3; and finally C_1C_4 equals C_2C_3. When the link CC_1C_2 is moved, the ring-formed piece will rotate around A through 180 degrees of arc.

Converting Rotary Motion from Continuous to Oscillating

A device that transforms a continuous rotary motion into an oscillating rotary motion is shown in Fig. 4. The arrangement is compact and the axis of the rotation of the output is perpendicular to that of the input.

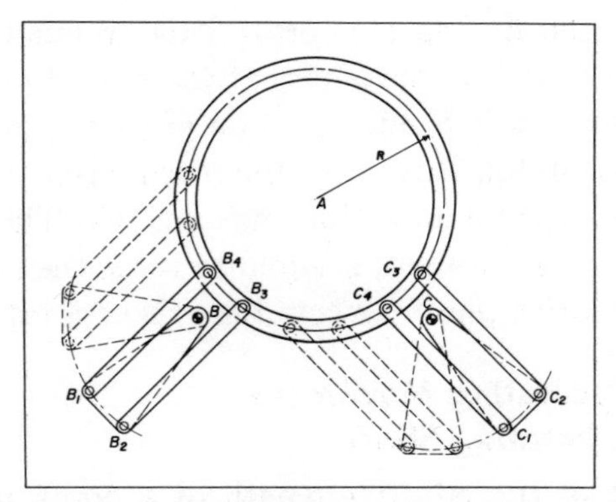

FIG. 3. This pivot system has more levers than that in Fig. 2 but is more compact.

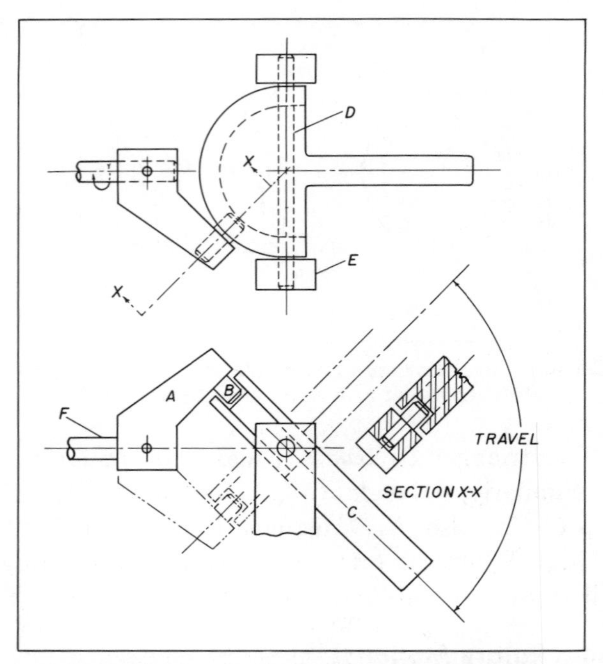

FIG. 4. Arrangement that produces an oscillating rotary output from a continuous rotary input. The mechanism is shown at a different position in each view.

Driving arm *A* has a drive-pin *B* which engages in a semicircular groove in a driven member *C*. This driven member is permitted to pivot freely on pin *D* which is retained at each end in a support block *E*. Pin *D* is press fitted in one block and slip fits in the other. It is important that the axis of pin *D* intersect the axis of drive shaft *F* and the axis of drive-pin *B* at a common point. Oscillating output of the mechanism is derived from the integral arm extending from member *C*. The total movement of this arm, in degrees, is equal to twice the angle at which the axis of the drive-pin intersects the axis of rotation of arm *A*.

Adjustable Oscillating Movement Derived from Rotating Shaft

Adjustment of the effective length of a crank may be made with the device in operation by means of the arrangement shown in Fig. 5. Cranks A' and E', together with their respective shafts, are bored through to receive cylindrical rack F with a sliding fit. Pinion gear G, shown at Y in the illustration, meshes with the cylindrical rack. Gear G is pinned to threaded shaft H. Square nut J, which is threaded on shaft H, rides within a guide-slot in the crank body. This nut is secured to the end of coupler rod C'. The same arrangement is provided in the body of crank E'.

Cylindrical rack F is operated by the mechanism illustrated at Z. Gear wheel K, mounted on shaft L, meshes with the cylindrical rack. Also mounted on shaft L is a worm-gear M that is driven by a worm N. An operating handle and a calibrated dial, both located outside of the crank housing, operate in conjunction with the worm.

When the operating handle is rotated, gear wheel K is moved through the action of the worm and worm-wheel. This rotation of gear wheel K imparts a linear movement to cylindrical rack F.

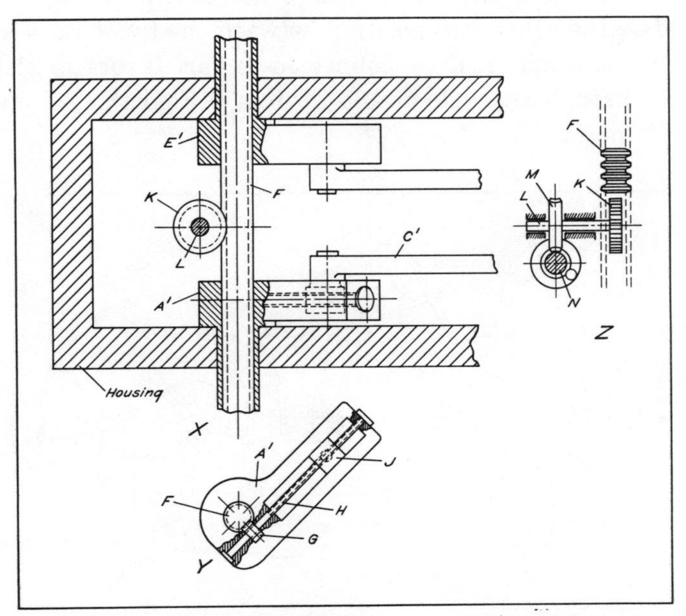

FIG. 5. Effective length of the crank may be altered during operation by means of a gear train.

Pinion gear G and shaft H are then rotated, which, in turn, alters the position of coupler rod C'.

Eccentric Gears Feature of Bag-Folding Device

One functional requirement of folder blades on a machine for making foil bags is that the first fold should not interfere with the second fold. This is necessary to prevent wrinkling and turning of the edge of the first fold. A pair of eccentric gears was arranged as shown in Fig. 6 to accomplish the proper folding action.

Eccentric gear A is secured to the shaft integral with part B which carries folder blade C. Similarly, eccentric gear D and folder blade E are attached to part F. The pitch radii on the gears are indicated by R_A and R_D in the illustration. Bearings in housings integral with the machine frame G supports parts B and F. Eight bag-forms H are mounted on a turret (not shown) having a Geneva motion which intermittently indexes one after the other into position between the two blades for the folding operation. A drive connected to part B rotates blades C and E through an angle of 180 degrees to fold the foil on each form in its turn.

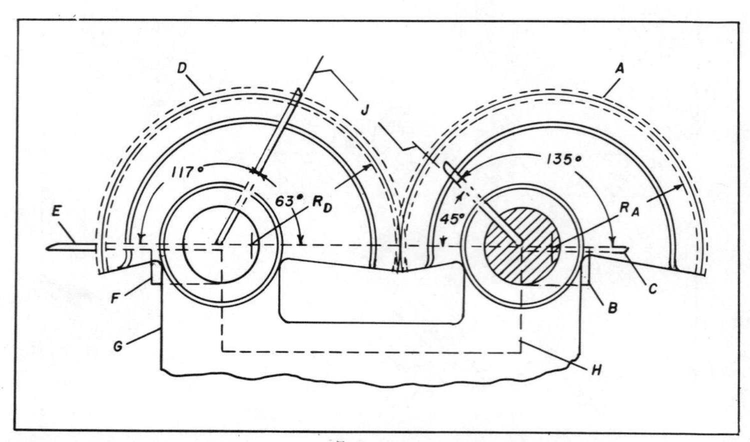

FIG. 6. Eccentric gears arranged to facilitate folding of foil on a form in a bagmaking machine. Interference between the folds is prevented.

In operation, as folder blade C rotates 135 degrees, blade E travels through an angle of 117 degrees. At this instant blade C lags blade E by 18 degrees, and this lag permits the portion of the foil J carried by blade C to clear the portion carried by blade E. During the remaining 45-degree rotation of blade C, blade E gains back the 18 degrees, and both blades complete the 180-degree working cycle simultaneously.

If standard gears were to be used, one gear would have to be retarded slightly to prevent interference between the folds. In this case, both blades would not hold the foil flat against the form at the end of the folding operation.

Simple Linkage Replaces Three Gears

On automatic machines it is sometimes desired to rotate two shafts at a uniform velocity ratio over a limited range of time. In a mechanical calculator being designed this velocity ratio was 2 to 1 and it was required that both shafts should rotate in the same direction. Rotation of the input shaft was through about 180 degrees. Normally this motion would have required three gears because an idler gear would have been necessary in order to operate the two shafts in the same direction. Also, the distance between the two shafts would have been rather large.

A mechanism designed to fulfill all requirements is shown in Fig. 7. As input shaft A turns on its axis, it swings crank C, on the end of which is a roller D. This roller slides in a slot of link E. This link is fastened to output shaft B. Because the radius of crank C equals the center distance between shafts A and B, it can be seen from geometrical considerations that when the shaft A moves through angle Y the output shaft B will move through angle $2Y$.

The total output angle of shaft B is about 120 degrees.

Drilling Parallel Rows of Holes

When drilling a row of holes with a radial drill, a multiple-spindle head is of great advantage, since hole alignment is automatic and a number of holes in the row can be drilled simultaneously.

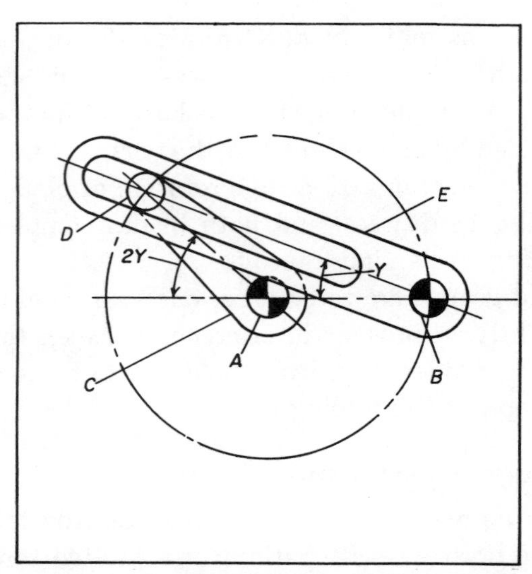

FIG. 7. Rotation of shaft A through $2y$ degrees causes B to rotate y degrees.

If several rows of holes are to be drilled, as in the case of heat exchanger tube sheets, maintaining parallelism between the rows can be something of a problem. A fixed angular relationship between the multiple-spindle head and the workpiece, regardless of the angular position of the radial drill arm, is needed.

The accessory, shown in Fig. 8, solves the problem. It consists of two interconnected double-acting hydraulic cylinders. The piston rod of one of the cylinders is fixed to the outer rotating column of the machine. As the cylinder and piston rod pivot with the external column, a rack integral with the cylinder is driven by a mating gear segment mounted on the stationary inner machine column. This cylinder is thus translated to the left or right, with respect to the piston, as the radial drill arm pivots with the outer column.

Movement of the cylinder varies the volume of hydraulic fluid in the two opposed chambers. The fluid in the cylinder chambers is forced to enter and leave through two ports in the piston rod (one at each side of the piston) passing through the rod and hose connections at either end. If the radial arm is pivoted coun-

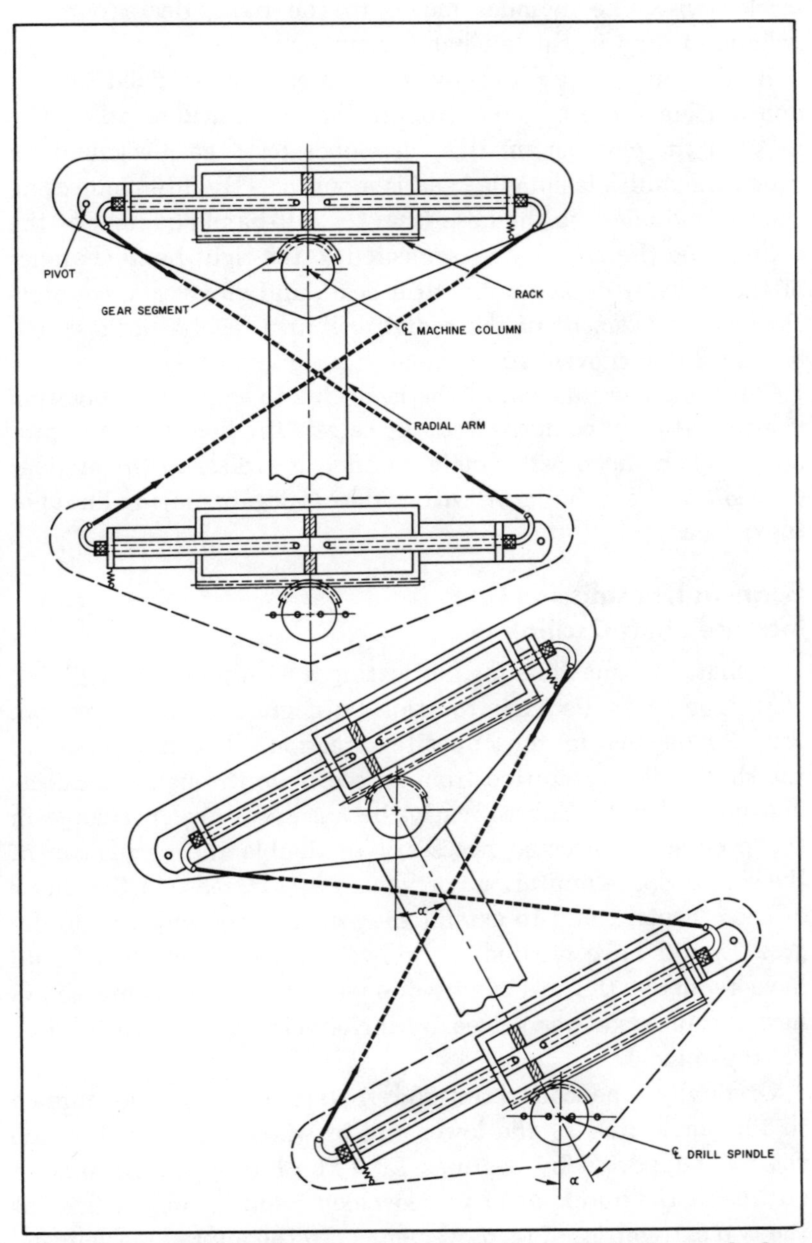

Fig. 8. Spindle located on arm is prevented from rotating by two hydraulic cylinders.

terclockwise, the cylinder moves to the right, decreasing the volume of fluid in the left-hand chamber.

A similar cylinder-and-piston arrangement is fixed to the nonrotating sleeve at the bottom of the radial drill spindle. The rack on the cylinder, in this case, operates a gear segment on which the multiple-spindle head is mounted. By interconnecting the two cylinders so the hose from the left-hand chamber of the cylinder on the machine is connected to the right-hand chamber of the cylinder on the radial drill head, and vice versa, counterclockwise movement of the radial drill arm results in an equal, but clockwise, movement of the multiple-spindle head.

Thus angular rotation of the head due to radial arm pivoting is automatically compensated. Once set, the hole patterns produced by the head will remain parallel regardless of the angular position of the radial arm, within the 90-deg. operating limits of the device.

Gears in Transmission Line Increase Shaft Oscillation

A shaft in a machine for fabricating a wire product oscillated continuously on its axis: rotating 90 degrees in one direction, then 90 degrees in the opposite direction. The movement of the shaft was transmitted from an eccentric through a connecting-rod and link. Subsequently, because of a design change in the product, it became necessary to double the oscillation of the shaft. Space limitations prevented any increase in the throw of the eccentric, and to extend the swing of the link to 180 degrees would have created a dead-center condition that would have rendered the mechanism inoperable. The drawing shows how the problem was solved by introducing gears into the line of transmission.

Originally, the connecting-rod *A* (see Fig. 9), was pinned to the single link *B*, the lower end of which was keyed to the shaft *C* supported by bearings *D*. (The link is shown in solid outline at the midpoint of its movement, and in broken lines at the two extremities of its movement.) In the altered mechanism, two links *B* and *B′* swing freely on the shaft, and straddle a

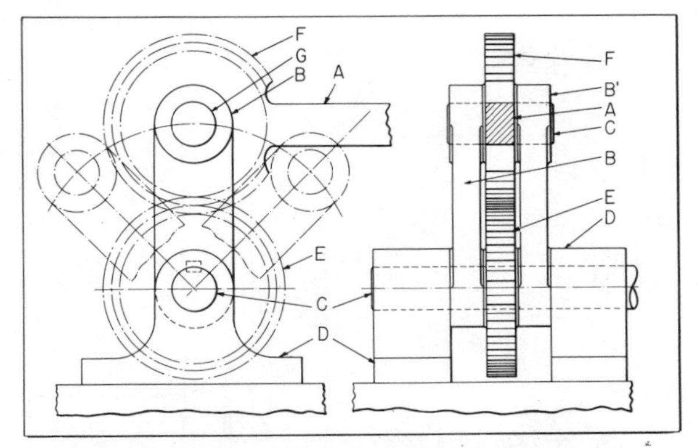

FIG. 9. Driving shaft C through two gears doubles the degree of oscillation.

gear E keyed to the shaft. Welded to the end of the connecting-rod is a second gear F, meshing with gear E and free on stud G.

In operation, the connecting-rod transmits movement to the links, running 90 degrees, as in the original mechanism. However, the action of gear F in rotating around gear E causes shaft C to oscillate the increased amount that is desired. The reason for the increase is that gear F rotates with respect to gear E creating an additional movement of gear E. Both gears are identical in size. Actually, since gear F does not rotate completely around gear E, gear segments instead of complete gears would serve the purpose.

Increasing the Movement of an Oscillating Shaft

In fabricating a wire product, it became necessary to increase the angular movement of an oscillating shaft of a machine tool. Because of space limitations, it was impossible to increase the throw of the eccentric controlling the shaft, and some other means of obtaining the additional movement had to be devised, as shown in the accompanying illustration.

In the original design (see Fig. 10), the eccentric-operated rod A connected to an arm B was keyed to the shaft C supported in a bearing D. In the present design, the arm is free to rotate on

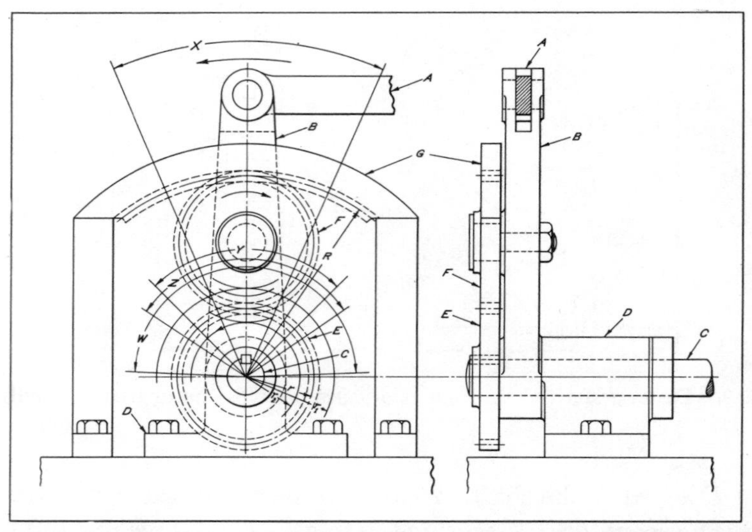

FIG. 10. By introducing an idler gear *F* and a gear segment *G* the oscillation of arm *B* and shaft *C* can be increased without changing the throw of an eccentric-operated rod *A*.

the shaft and a gear *E* is keyed to the shaft. Another gear *F* rotates freely on a stud carried on the arm and meshes with gear *E*. Gear *F* also meshes with a segment of an internal gear *G* fixed to the bed of the machine. (In the right-hand view the supports for the segment have been omitted.)

The arm is shown at its central position moving in the direction indicated by the arrow. Gear *F* moves with the arm, and since this gear meshes with the segment, it is caused to rotate on its stud. The rotation of gear *F* is transmitted to gear *E* and thus to the shaft. In the illustrated application, gears *F* and *E* are of the same pitch diameter, gear *F* being an idler which has the effect of imparting movement to the shaft in the same direction as that of the arm.

Angle *X* indicates the magnitude of the oscillation of the arm. Actually, the gear *F* serves as a lever, with its fulcrum at the pitch line of the segment. In this manner the action of gear *F* causes an increase in the angular movement of gear *E* as compared with the movement of the arm.

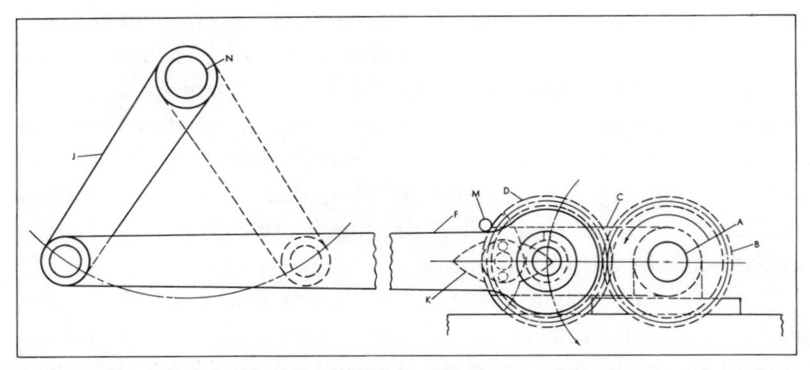

FIG. 11. Front view of intermittent speed-change linkage mechanism in which N is the reciprocating driven shaft and A is the drive. Gear B turns constantly, but at points in a cycle gear is locked by a clutch, causing the whole mechanism to rotate around shaft A.

Sun and Planet Gears Produce Intermittent Speed Change

Figures 11 and 12 illustrate the construction of a mechanism which provides oscillating angular motion to a shaft, with the alternate cycles at different speed relations to the rotation of the driving shaft. This application is used on a machine producing wire textile screening, the object being to produce two different spacings of the wires. Figure 11 is a front elevation and Fig. 12 is a plan view of the intermittent speed-change mechanism.

The driving shaft A, rotating in the direction of the arrow, is keyed to gear B. Gear B meshes with gear D, carried on a pin

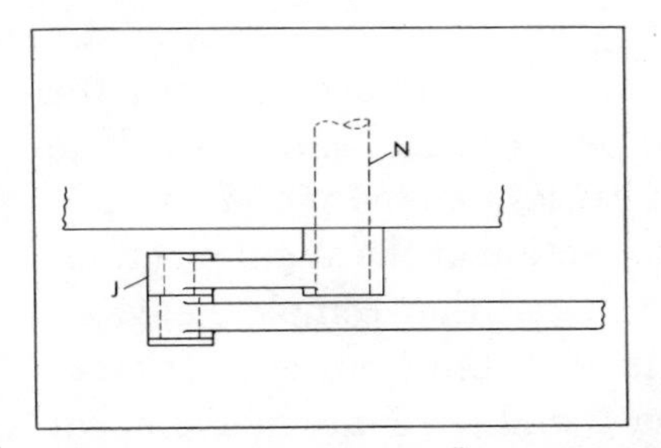

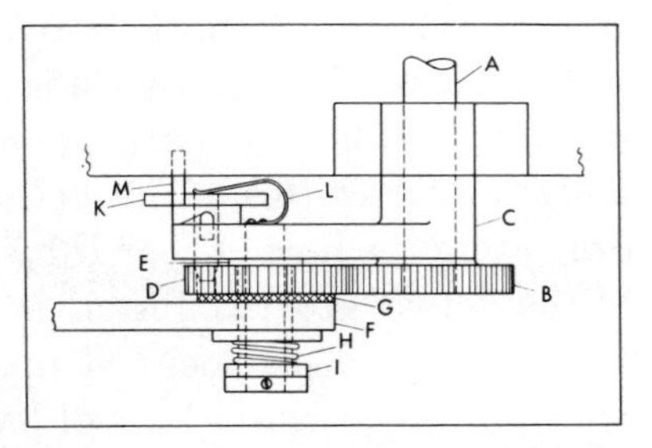

FIG. 12. Details of the clutching device appear in this plan view, with the clutch pin retained by the free end of spring L. The friction material is shown at G.

pressed into the arm of lever *C*, which is carried free on shaft *A*. Gear *D* is pressed onto a flanged sleeve *E*, which is free to turn on the same pin. The connecting-rod *F*, which transmits motion to lever *J* (keyed to shaft *N*), is rotatably supported by sleeve *E*. A disc of friction material *G* is carried between gear *D* and rod *F*. A spring *H*, adjusted by threaded collar *I*, maintains pressure of gear *D* and rod *F* on the friction disc *G*. A collar on the outside end of the pin retains this assembly.

Another pin with a tapered end carries star-wheel *K*, and passes freely through lever *C*. The tapered end of the pin engages a matching hole in gear *D* on alternate cycles. The hub of star-wheel *K* is provided with two opposite detent grooves, shaped as shown, which engage two pins in lever *C*, as controlled by the position of star-wheel *K*. A flat spring *L* insures engagement of the pin with the hole in gear *D*. Another pin *M* is located in a stationary part of the machine, and imparts rotative motion to star-wheel *K* with each rotation of lever *C*.

In the position shown, the mechanism is at the beginning of its cycle, with the lever *J* at its extreme left-hand position. At this point, the tapered pin carrying star-wheel *K* is engaged in the hole in gear *D*, so that gear *D* and lever *C* are locked together. Because gear *D* cannot rotate on its pin, lever *C* becomes a simple crank revolving on the center of shaft *A*, thereby transmitting oscillating motion to shaft *N* through the linkage of rod *F* and lever *J*. At this stage of the cycle, rod *F* is slipping on the friction disc *G*. Thus the entire assembly delivers a conventional crank motion, imparting an oscillating motion to shaft *N* with each rotation of shaft *A*.

The completion of this phase of the cycle occurs when the movement of lever *C* past tapered pin *M* causes star-wheel *K* to make a quarter-turn. This action causes tapered pin *M* to withdraw from the hole in gear *D* by the action of the angular surface of the detent groove. Pin *M* is so placed that complete engagement and disengagement of the tapered pin from gear *D* takes place when the center lines of lever *C* and rod *F* are in alignment as in Fig. 11, so that the speed change will take place at the end of the cycle. Although the tapered pin has now been disengaged from gear *D*, the latter is not free to rotate because it is frictionally locked to rod *F*. In this phase gear *D* now acts as a planet gear revolving about gear *B* as a sun gear. With this type of reducing gear application, the relative revolution of the planet gear to the rotation of the sun gear will be equal to the ratio of the gear tooth count plus 1. As gears *B* and *D* are of the same tooth count, gear *B* must perform 1/1 + 1 revolutions, or 2 revolutions, to produce one revolution of the crank whenever the tapered pin is disengaged from gear *D*.

Producing an Oscillating Movement of Uniform Angular Velocity

The mechanism shown in Fig. 13 was designed to produce an oscillating motion having uniform angular velocity, using a crank *A* rotating at uniform speed as a driver. With the arrangement shown, an oscillating movement is imparted to lever *C* by connecting-rod *B*, but the angular velocity of this movement is not uniform. Although approximately uniform angular velocity could be obtained by arranging a suitable mechanism in front of

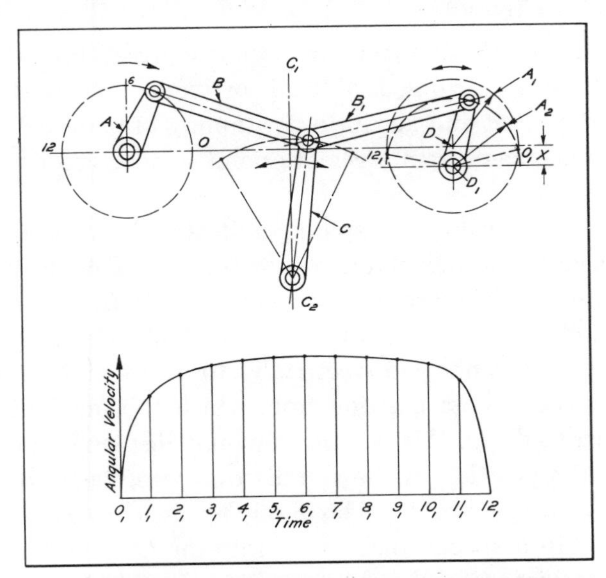

FIG. 13. Diagram showing method of obtaining an oscillating movement of approximately uniform angular velocity from a rotating crank.

the rotary crank, such devices are usually complicated and were not considered applicable in this case.

To achieve the desired results, the center position C_1 C_2 of lever C was taken as the axis of symmetry, and the mechanism on the left-hand side (crank A and connecting-rod B) was exactly duplicated on the right-hand side (crank A_1 and connecting-rod B_1). If crank A_1 were located on the same horizontal center line as crank A the angular velocity of crank A_1 would be uniform, but its oscillating movement would be through an angle of 180 degrees, which is not practical. However, by lowering the center of crank A a distance X (from position D to D_1) keeping radius D_1A_2 equal to DA_1 and leaving the mechanism on the left of the axis of symmetry in its original position, oscillation angles of 120 to 150 degrees, which are permissible, are obtained. Graphical analysis of the angular velocity of crank A_1 will produce a curve like that shown at the bottom of the illustration.

Remote Control Presets Spindle-Quill Travel

An arrangement that can be used to preset the length of travel of a spindle quill, tool-head, or work-table is shown in Fig. 14. Repeated and accurate positioning of one of these machine tool members from a remote point is achieved by the use of synchros.

Linear displacement of a spindle quill A is converted to rotation by means of a rack B and a gear train C. The output shaft of gear train C drives a transmitting synchro D. Components A, B, C, and D are built into the machine tool.

Synchro D is connected electrically to a receiving synchro E located at the remote position from which the machine tool is to be controlled. In this manner, angular displacements which represent the position of the spindle quill are transferred from synchro D to synchro E. The receiving synchro is connected by shaft F to a special dial type indicator G. A pointer H is attached to shaft F, and a second pointer I, which pivots on the same axis and is concentric with dial plate J, is driven through a

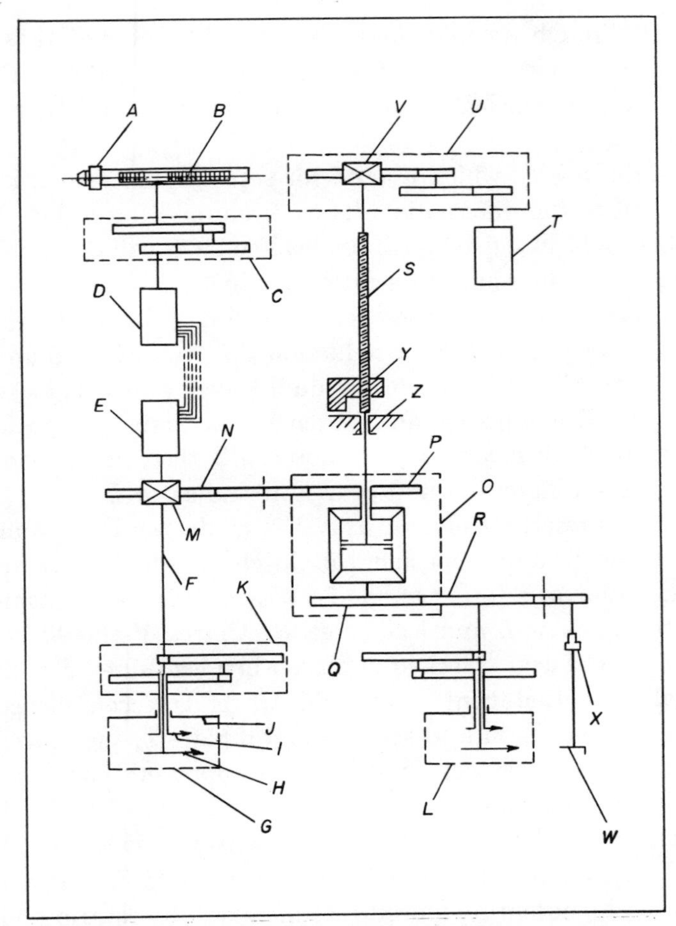

FIG. 14. Arrangement that permits remote presetting of the travel of a spindle quill, tool-head, or work-table.

reduction gear train *K*. Also driven by shaft *F*, gear train *K* reduces the rotation in the ratio of 100 to 1. Since rack *B* and gear train *C* are in a ratio such that one turn of shaft *F* and pointer *H* represents a 0.1-inch displacement of the spindle quill, one revolution of pointer *I* will, therefore, represent 10 inches of quill travel.

An electromagnetic clutch *M*, when actuated, allows shaft *F* to engage a gear *N*. This gear, in turn, drives a differential gear train O through an idler and gear P. The differential is connected through gears Q and R to a gear train and a dial indicator L. These are identical in design to members K and G, respectively.

A lead-screw S, which is used for setting quill travel, is also connected to the differential. Thus, if clutch M is engaged and gear Q is held in a fixed position, lead-screw S will drive shaft F through the differential in a ratio of 1 to 1.

An electric motor T, used for presetting quill travel, is connected to lead-screw S by a reduction gear train U and an electromagnetic clutch V. To preset quill travel, gear P is locked in place, gear R is released, and clutch V is actuated. Rotation of lead-screw S by means of the motor will then be transmitted through the differential and gear R to indicator L.

Manual presetting of quill travel is made possible by means of a handle W and a mechanical clutch X, which is geared to gear R. Manual rotation of handle W is in this way transmitted both to indicator L and lead-screw S. Clutch V should be disconnected and gear P locked in place when lead-screw S is preset manually. Actual control of quill travel is accomplished by means of a nut Y on lead-screw S and a plate Z, the movement of which opens a switch in the circuit supplying power to the quill-feed mechanism.

Operation of the remote quill-travel control is as follows: With gear P locked and a nut Y in contact with plate Z, lead-screw S is rotated by actuating motor T or manually by rotating handle W until the desired length of travel is read on the dial of indicator L. This moves nut Y on screw S away from plate Z a distance proportional to the required quill travel. Movement of the quill for the desired travel with gear R locked and gear P released will then cause nut Y to return the same proportional distance in the opposite direction and contact plate Z, thus stopping the forward motion of the quill. During machining, indicator G shows the position of the quill.

Torque Filter Eliminates Backlash

Design for aerospace programs has resulted in some useful solutions to common problems. The mechanism shown in Fig. 15

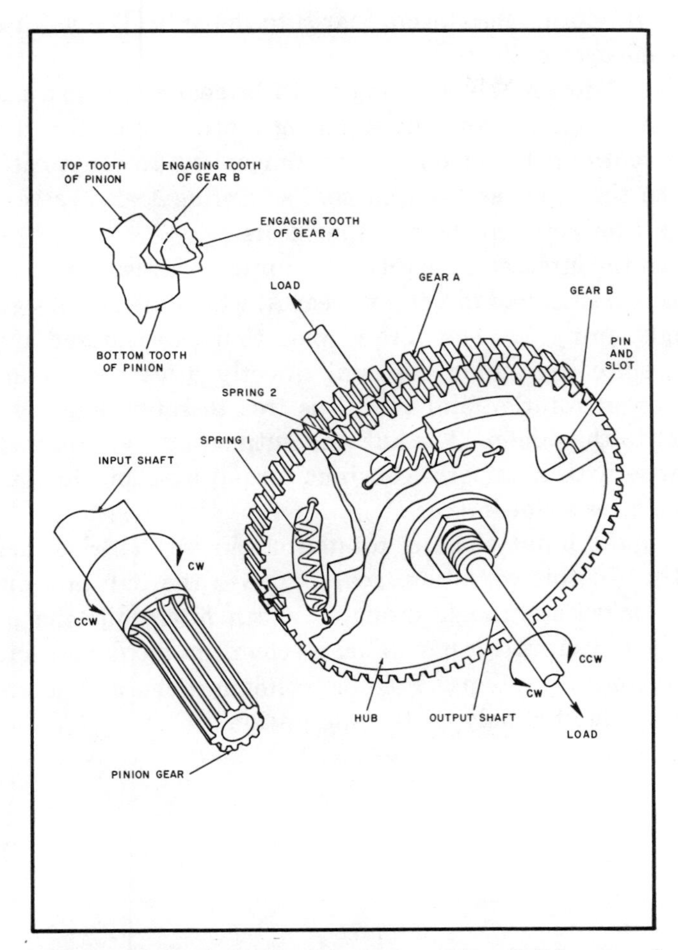

FIG. 15. Smooth output obtained from gears *A* and *B* by use of springs 1 and 2.

enables a constant output torque, free of backlash components, to be maintained from a pulsating input torque.

Two elastic components (springs) connecting a hub and two spur gears absorb torque differentials and provide the desired antibacklash characteristic between input and output shafts. The system performs equally well in either direction of rotation.

The hub is securely attached to the output shaft and two spur gears turn freely on the shaft. Spring 1 connects gear *A* to gear B. Spring 2 connects gear B to the hub. The input shaft pinion engages both spur gears.

Spring 1 forces the engaging tooth of gear A against a lower tooth of the pinion and the engaging tooth of gear B against a higher tooth of the pinion. This arrangement prevents backlash between the input and output shafts. Spring 1 also serves as a torque filter between the two spur gears.

When the input shaft is rotated counterclockwise, its pulsating torque is transferred directly to gear A, which then drives gear B through spring 1. The torque pulsations are filtered by the spring since the pinion does not directly drive gear B in this direction of rotation and gear B is free to rotate slightly with respect to the pinion. The hub and output shaft are then driven clockwise with a smooth, constant torque by a pin in the hub that engages a slot in gear B.

When the input shaft is rotated clockwise, it drives gear B directly. In this condition, gear B drives the hub and output shaft clockwise through spring 2, which filters out the pulsations in the input torque. This mechanism could be useful in precise control systems. Possible configurations and filter materials are limited only by the application.

CHAPTER 11

Mechanisms Providing Combined Rotary and Linear Motions

Mechanisms which provide combined rotary and linear motions are described in this chapter. Similar mechanisms are described in Chapter 11, Volume III of "Ingenious Mechanisms for Designers and Inventors."

High-Speed Spiral Scanner

In radar, television, and other systems involving the reception of electromagnetic waves, the need frequently arises for a high-speed spiral scanner. In this case, the scanner consists of a mirror held in an adapter actuated by a power-operated mechanism. The mirror is tilted at a constantly changing angle with respect to an axis passing vertically through its center. At the same time, the direction of the tilt continuously changes through a complete circle of 360 degrees. Thus, the motion of the mirror is such that it can "see" any object within the angle of its cone of motion.

The mechanism built to provide the required scanning motion, which has been performing satisfactorily, is shown in Fig. 1. The requirements for this particular instrument were these: operational speed had to top 10,000 revolutions per minute; the mirror or antenna adapter *K* (see diagram Fig. 2) had to scan a cone opening angle *O* of 12 degrees, the axis y-y of axle *J* pivoting about point *H* while point *S* at its lower end described a spiral of twenty turns between 0 and 12 degrees — that is, the scanner had to cover its field ten times per second when drive-shaft *B*, Fig. 1, was operating at 12,000 revolutions per minute; and the center of the conic motion had to be at point *H* on the upper surface of the adapter *K*. Finally, it was necessary that the

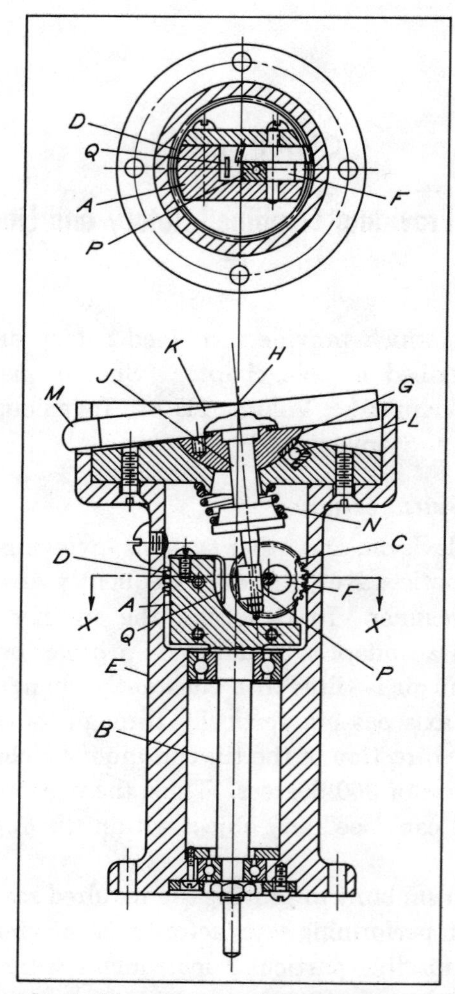

FIG. 1. High-speed spiral scanning mechanism in which the conic motion of the scanner K is set up by gear C turning on its own axis while its carrier A rotates at high speed around the axis of main shaft B.

instrument be simple and easy to construct. Thus, the axle B of the main shaft was designed to (1) revolve at a speed of 12,000 rpm; and (2) move the lower end of axle J outward from its vertical, or zero, position toward the widest cone angle scanning position, and return to the zero position once every forty

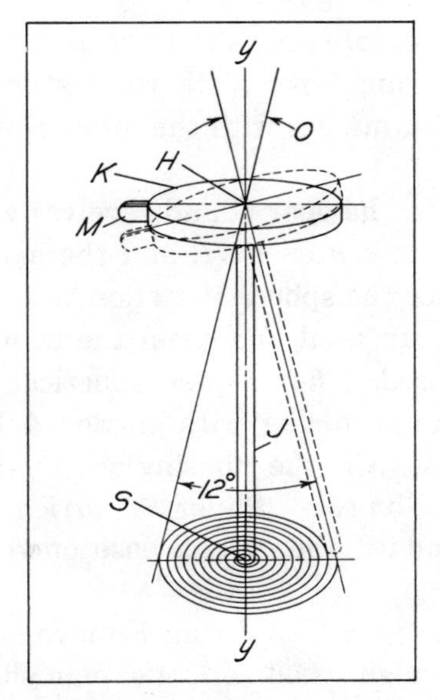

FIG. 2. Diagram illustrating the basic operating principle of the spiral scanning mechanism, Fig. 1.

revolutions of Axle *B*. In Fig. 2, the axle is shown in vertical position (solid lines) and in widest scanning position (broken lines). The relative positions of the moving parts during this operating cycle are shown by the diagrams in Fig. 3.

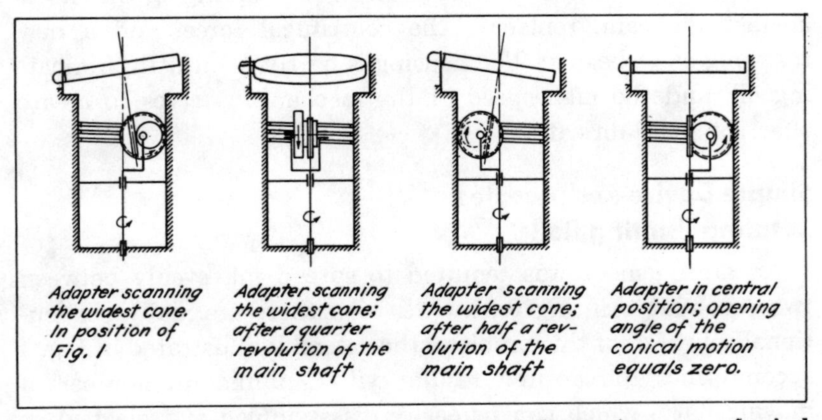

FIG. 3. Diagrams showing relative positions of the moving parts of spiral scanner at different periods of the operation cycle.

Carrier *A*, Fig. 1, is integral with main shaft *B* and is balanced. Helical gear *C*, having forty teeth rotates on a pivot integral with the carrier, and meshes with the internal worm thread *D* in housing *E*.

Cam *F* is fixed to the gear *C* and revolves with it. Spherical section *G*, having its center at *H*, and the axle *J* are attached to adapter *K*. Since the spherical section rests on three equally spaced balls *L*, it can easily be given the continuous tilting or conic motion required. The adapter, spherical section, and axle are prevented from revolving with carrier *A* by projection *M* which engages a slot in the housing *E*. Spring *N* holds the spherical section in its seat. Slider *P*, carried on the axle *J*, is free to rotate around it. The slider is positioned between the gear and the carrier.

In operation, the gear *C* and cam *F* move as a unit with the carrier, rotating at high speed with the main shaft *B* so that the axle *J*, spherical section, and adapter perform the required conic motion. At the same time, the worm thread *D* causes the gear *C* to revolve around its own axis while centrifugal force presses the slider *P* against the cam *F*. Since the gear and cam are also revolving around the gear axis, the opening angle of the cone varies continuously as dictated by the shape of the cam, thus producing the required spiral scanning motion.

When the slider reaches the center, flat spring *Q* thrusts it against the cam, replacing the centrifugal force, and a new scanning cycle begins. The housing is partly filled with lubricating oil, and the high speed of the mechanism serves to create effective mist lubrication.

Simple Device Reciprocates Rotating Printing Roll

An arrangement was required to spread ink evenly between two revolving printing rolls. Although there are many conventional solutions of this problem, the assembly illustrated in Fig. 4 accomplishes the desired results with a minimum number of moving parts which can be easily disassembled and cleaned as needed.

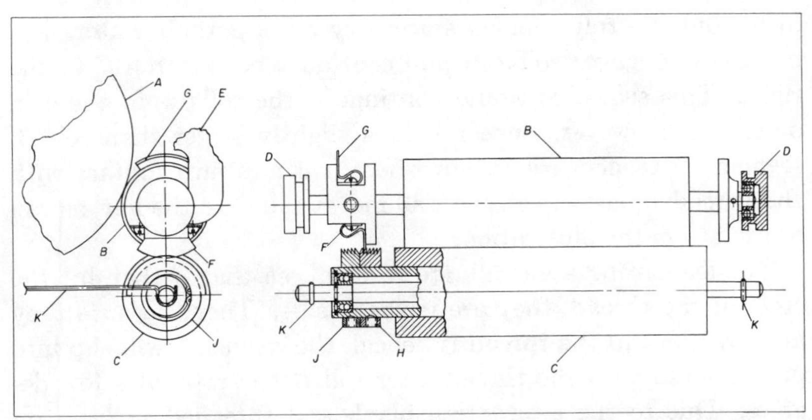

FIG. 4. To spread the ink evenly, roll C is reciprocated as it rotates in contact with roll B. Each stroke is accomplished in short intermittent steps.

Equal distribution of the ink is best obtained by allowing one of the rollers to reciprocate continually in an axial direction during operation. In addition, one roll should be made slightly larger than the other so that the same areas of contact are not repeated on each revolution of the rolls.

The reciprocating mechanism is shown in detail in Fig. 4. In the setup, a drive wheel A rotates printing roll B which, in turn, drives the second printing roll C. Roll A rotates in antifriction bearings which are held in housings D supported freely in slots in machine frame E. Members F and G are leaf-spring segments mounted in diametrically opposed positions on a collar on the shaft of roll B. Two collars H and J are each threaded on one-half of their circumference and relieved on the other half to a diameter less than the minor diameter of the thread. Collars H and J are threaded right-handed and left-handed, respectively, and are mounted side by side on the shaft support roll C. The width of each collar is equal to the length of the total axial stroke to be given to roll C.

As roll B revolves, segment F enters the thread on collar H, moving roll C to the right during the period they remain in mesh. Then, as segment J is rotated into the meshing position, the unthreaded half of the collar is presented to the segment and the roll remains stationary for a period. Later, segment *F* re-engages collar *H* and continues moving roll *C* to the right. This sequence would continue if the rolls were equal in diameter. However, since roll *C* is slightly larger than roll *B*, segment *G* reaches a position where it comes into contact with the threaded part of collar *J* and moves roll *C* to the left, as can be seen from the illustration.

The leaf-spring segments are resilient, so that if they hit the crest of the thread, they are pushed aside. Then, because they are straight and the thread is helical, the segments will slip into proper mesh with the thread after roll *B* has rotated a few degrees. Due to the alternating blank and threaded surfaces on collars *H* and *J*, roll *C* is displaced axially in short intermittent strokes. This type of motion is helpful in spreading the ink.

In the particular arrangement shown, each stroke lasts fourteen revolutions of roll *B*. The ratio of the roll diameters is 15 to 16, and the length of the axial stroke roll C is ⅜ inch or two and one-half times as large as the largest letter to be printed. Each of the collars has an eight-entry thread, the lead of each thread entry measuring ½ inch and the pitch, therefore, being 1/16 inch.

The segments displace roll *C* axially about 1/10 ipr (which is a little more than the pitch) in order to assure the entry of the segment into the next thread. Consequently, each segment occupies 70 degrees of arc. The effective axial displacement, therefore, measures 1/16 inch, and six turns of roll *B* are required to move roll *C* ⅜ inch. During each seventh and eighth turn of roll *B*, roll *C* is idle. In operation, the rotational speed of roll *B* is 230 rpm.

Members *K* are two identical hook-shaped springs that hold roll *C* against roll *B* and thus cause roll *B* to press firmly against driving wheel *A*. With this arrangement, the whole assembly can be taken apart simply by removing hooks *K*. It is unnecessary to adjust the mechanism during assembly, since the arrangement will automatically start the correct cycle after it has been in operation a few revolutions.

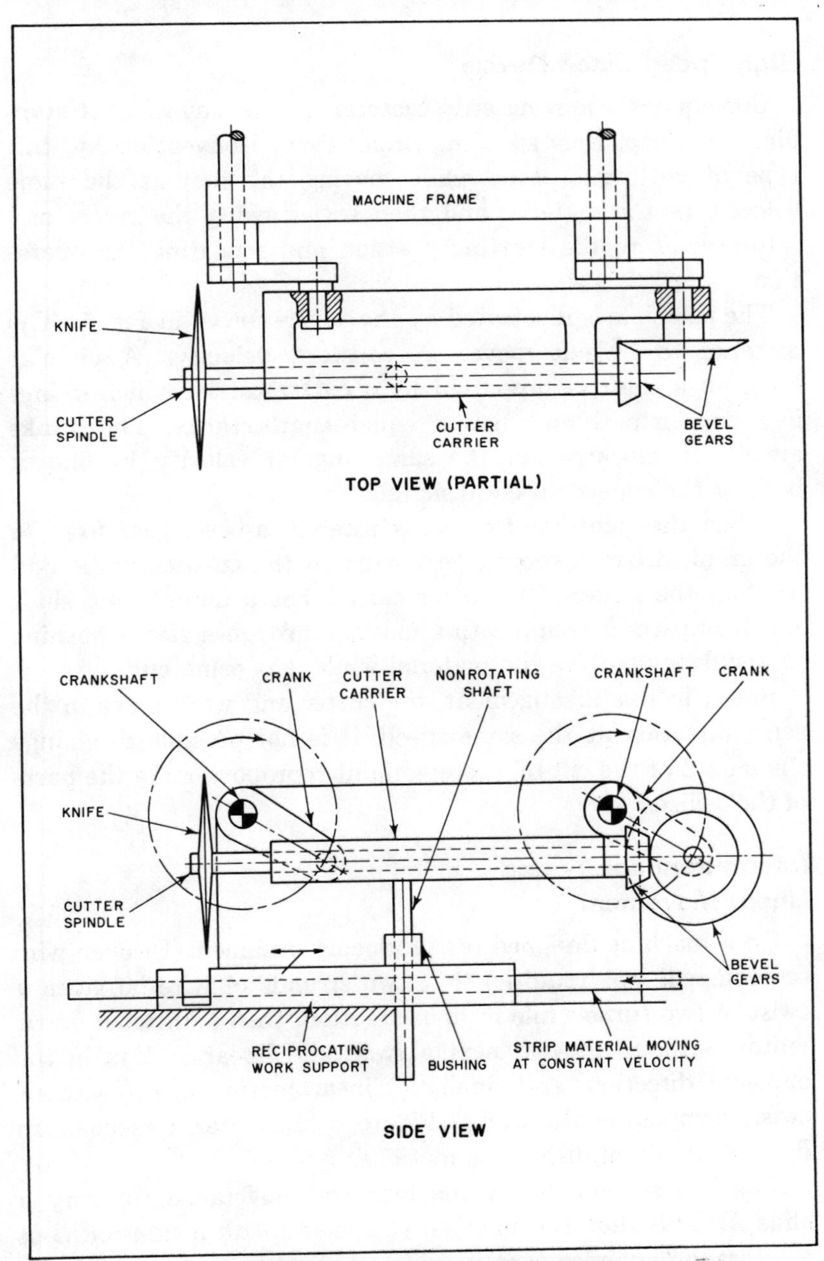

FIG. 5. Stock cut on the fly by knife rotating in a circle.

High-Speed Cutoff Device

Continuously moving strip material can be cut without stopping the strip, thus speeding production. Conventionally, this type of cutting is done while moving the tool at the same velocity as the material and then withdrawing the cutter and returning it to the starting position and repeating the operation.

The principle is illustrated by the device shown in Fig. 5. The material to be cut moves at constant velocity. A circular knife on a shaft is supported by a cutter carrier which swings in a circular path on a pair of equal-length cranks. The cranks are driven clockwise at the same angular velocity by one or both of the connected crankshafts.

When the right-hand crank is rotated, a bevel gear fixed to the crank drives a second gear fixed to the cutter spindle and revolves the knife. The cutter carrier has a nonrotating shaft which imparts a reciprocating motion through a sleeve bushing to a slide supporting the material while it is being cut.

Since, in this arrangement, the cutter and work move in the same direction at the same speed, it is not possible to change the length of the cut-off piece without reproportioning the parts of the device.

Intermittent Rotary and Linear Movement

On a machine designed for producing ornamental woven-wire screening, it was required that two strands of wire be given a twist of two turns while in linear motion, then additional linear moton without a twist, next a twist and linear motion in the opposite direction, and finally a linear motion again without twist, completing the cycle. Figure 6 illustrates a mechanism devised to accomplish these motions.

The bed section of the machine was dovetailed to carry a slide *A*. This slide is connected at one end with piston-rod *B* of a hydraulic cylinder that provides the operating power for the mechanism. Slide *A* carries bearing bracket *C*, which supports

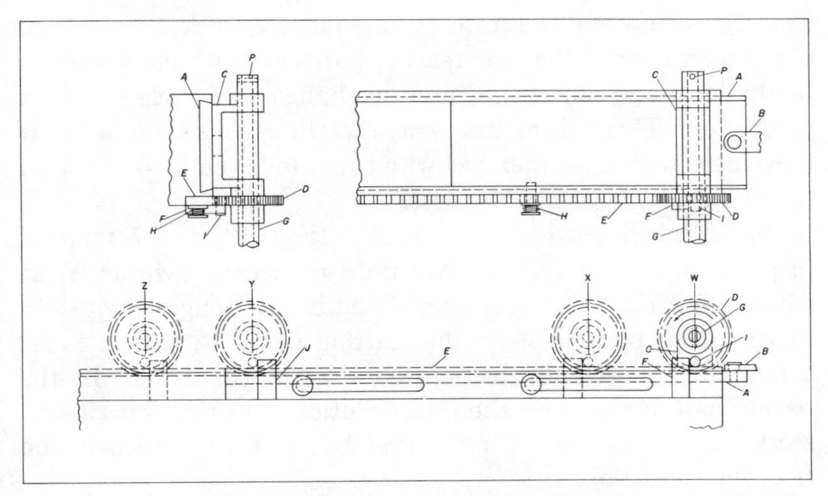

FIG. 6. Mechanism devised for actuating intermittent rotary and linear movements.

a tubular shaft *G*. The far end of shaft *G* carries a pin *P* through its radial center to separate the two strands of wire and apply the twisting force, one strand of wire passing through the tube on either side of the pin.

The outer end of shaft *G* carries two spools of wire which revolve on the shaft axis. Shaft *G* also carries gear *D*, keyed to it. The gear is provided with a pin *I*. Gear *D* meshes with rack *E*, which is mounted on the bed section of the machine by two studs that pass through slots in the rack. Each stud is provided with a spring *H* that produces frictional resistance to the movement of the rack. Two stops *F* and *J* are attached to the ends of the rack.

At the beginning of a cycle, gear *D* is at the right-hand end of the assembly in position *W*. As piston-rod *B* moves to the left and pushes the slide assembly with it, rack *E* does not move because of the frictional resistance applied by springs *H*. Gear *D* is, therefore, caused to rotate counterclockwise, beginning the twisting action on the wires. This action continues until gear *D* reaches position *Y*, at which point pin *I* is in contact with the far side of stop *J*. As the movement of slide *A* continues, gear *D* can no longer rotate. Consequently, rack *E* is carried along with it until the gear reaches position *Z*, which is the end of the left-hand movement, and one-half of the cycle has been completed. This half of the cycle, therefore, consists of a linear movement with a counterclockwise twist, followed by a period of linear movement without a twist.

On the return movement of rod *B* to the right, rack *E* remains in position, and gear *D* revolves until it reaches position *X*, at which point pin *I* contacts stop *F* and can no longer rotate. In moving from position *X* to the starting point *W*, rack *E* is returned to its original position, completing the cycle. In the second half of the cycle the linear motion is in the opposite direction to the movement in the first half and the twist is in the clockwise direction.

Rotating and Sliding Mechanism Used in Polishing Rectangular Frames

The mechanism shown in Fig. 7 was designed to provide the required motion for polishing a rectangular metal frame on all four sides, as well as on the faces adjoining those sides. The diagram in the upper right-hand corner indicates the movement of the frame relative to the wheel *W* during the polishing operation. The frame is indicated by dot-and-dash lines.

In operation, the frame moves along line L_1L_2 in the direction indicated by arrow *A* until point C_2 reaches point C_1. This movement of the work past the wheel results in polishing surface *1*. The frame then turns 90 degrees in a clockwise direction, as shown by arrow *B*, and surface *2* passes the fixed point C_1 in the same way that surface *1* did. Surfaces *3* and *4* pass point C_1 in a similar manner. After the entire periphery of the frame has been polished, the mechanism automatically stops for reloading.

The four intersections of the sides of the rectangle — points C_1 to C_4 — are the centers of the 90-degree angle of rotation at the end of each stroke. These points are analogous to the centers C_1 to C_4 shown in section B-B.

Cross-section A-A shows the work in dot-and-dash lines, mounted on a rotating and sliding nest *RN*. Surfaces *1*, *2*, *3*, and

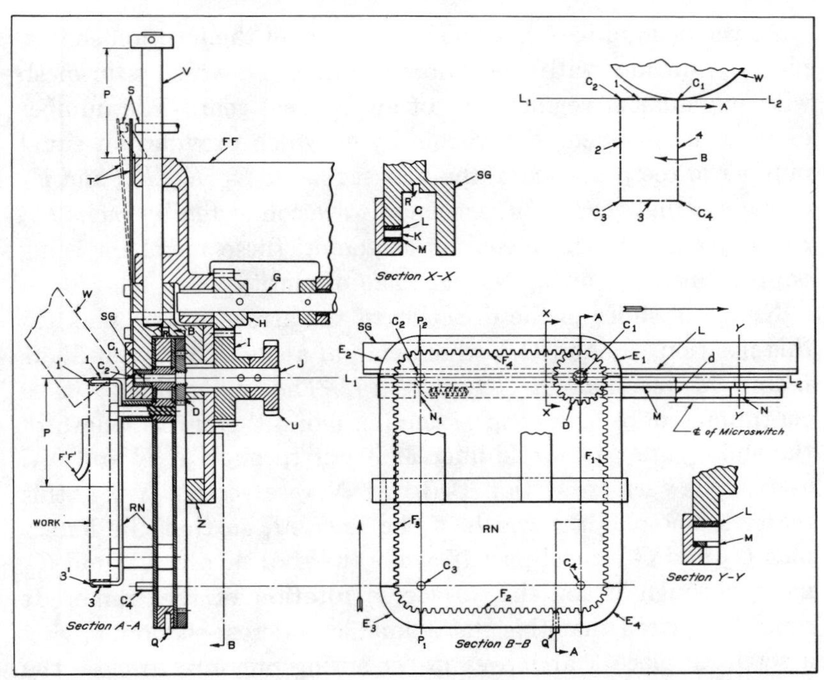

FIG. 7. Rotating and sliding mechanism designed for automatically polishing the periphery and faces of rectangular frames.

4, of which *1* and *3* can be seen in this view, are to be polished. The adjoining faces *1′*, *2′*, *3′*, and *4′* are also to be polished, and for this reason, the wheel *W* is mounted at an angle, as indicated.

Essentially, the mechanism consists of a stationary frame *FF* and *F′F′* in which are bearings for the drive-shaft *G* to which is fastened a gear *H*. Rotating slowly and at constant speed, gear *H* meshes with gear *I* to drive a shaft *J* on which is mounted a pinion *D*. Other members of this assembly include a sliding guide *SG* and the rotating and sliding nest *RN* previously mentioned. The thrust of the polishing wheel is absorbed by a support *Z*. The nest slides (and rotates at the end of each stroke) within the space *C* section X-X. Four pins C_1, C_2, C_3, and C_4 are tightly pressed in nest *RN*, pins C_1 and C_3 being shorter than C_2 and C_4.

At the beginning of the polishing cycle, in the position shown, pin C_1 coincides with the center of pinion D, which is in mesh with one-quarter segment E_1 of an internal gear. The number of teeth in this gear is divisible by 4, which provides an equal number of teeth in each of the four segments E_1, E_2, E_3, and E_4 of the internal gear. These segments are connected by racks F_1, F_2, F_3, and F_4, the means of fastening these members being omitted in the drawing for the sake of clarity.

Nest RN slides in the direction of the arrow on two pins (in this position, C_1 and C_2) which ride in a slot K between liners L and M (sections X-X and Y-Y). The upper liner L is a continuous unbroken strip extending along the entire length of the slide guide SG, while liner M is cut through at N and N_1, as may be seen in section B-B. At N (see section Y-Y), this recess is one-half the width of the liner M, so that the longer pins C_2 and C_4 travel past it, while the shorter pins C_1 and C_3 move through it for the 90-degree rotation of the frame. It should be noted that this slot is angular in cross-section, having a separate piece that moves under spring pressure to close the slot after a pin has passed through it. This provides a smooth, unbroken surface for the pin to travel on after entering slot K.

In operation, the engagement of gears H and I rotates pinion D, which, after disengaging gear segment E_1 engages rack F_4. This moves nest RN along line L_1L_2, in the direction indicated by the arrow, until the pinion engages the second gear segment E_2. By that time, the short pin C_1 is at N and moves through the opening, thereby allowing nest RN to turn in a clockwise direction until pin C_3 enters the guide strip through the opening at N_2, which confines the angle of rotation to 90 degrees.

After the rotation of pinion D has rotated segment E_2, it engages rack F_3 and causes the nest to slide along line L_1L_2. A long stroke now takes place, at the end of which C_2 exits from N_3 as pinion D engages E_3 at the end of 90 degrees of rotation. C_4 enters N_1 and movement along L_1L_2 again takes place. This procedure is followed for the remaining side, so that the frame rotates and slides four times in one complete polishing cycle.

At the completion of a cycle, gears H and I are disengaged automatically to stop the movement of the mechanism and permit unloading and reloading of the work. The first step in this automatic action occurs when a pin Q (pressed into the rotating nest at a location farthest from point C_1) contacts a micro-switch (not shown) moving it a distance d. A slot R (section X-X) in the slide guide provides clearance for the pin during its travel along line L_1L_2. The microswitch energizes a solenoid that moves a rod S, disengaging a spring steel latch T which is connected to slide guide SG and nest RN. The slide guide and nest then drop, by reason of their unsupported weight, a distance P, traveling along two guide rods V (only one of which is shown), thereby disengaging gears H and I.

After reloading the nest, the operator raises the nest and guide by means of a knob U, bringing the assembly to the position illustrated. Since gear H rotates very slowly, its reengagement with gear I is a simple matter.

This mechanism can be applied to any polygon. If it is employed for a regular polygon, the short pins C_1 and C_3, as well as slot N, may be eliminated and four (or more) long pins used, which would drop after reaching the end of slot K.

Vibrating Roll Drive for Printing Press Fountains

In the ink-distributing fountain of a printing press, certain rolls have to reciprocate axially as well as rotate. The device shown in Fig. 8 illustrates a simple drive that produces both of these movements in a small-diameter roll A. This roll is in contact with a large-diameter roll B which rotates but does not itself reciprocate. Keyed to the end of each roll shaft is a spur gear. Both gears have pitch diameters that are equal to the outside diameters of the respective rolls to which they are attached.

On each side of the smaller gear C is a shroud D. These shrouds overlap the rim of the larger gear E. Gear E is mounted so that its axis is canted a few degrees to that of roller B.

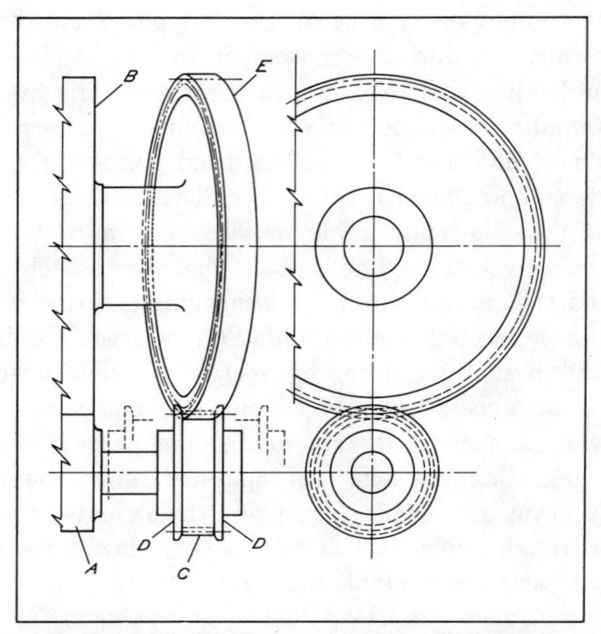

FIG. 8. **Shrouds *D* constrain gear *C* to the helicoidal path of gear *E*, causing roll *A* to vibrate axially as well as rotate.**

In operation, gear *E* drives gear *C*; the shrouds constrain the smaller gear to follow the face of the larger gear. The result is to produce an axial movement of the roll shaft of the smaller gear equal to the throw of the larger gear, as indicated by the dotted lines.

CHAPTER 12

Speed Changing Mechanisms

Providing a fixed or adjustable speed of rotation of a rotating driven member that is different from the speed of rotation of the driving member can be accomplished in many different ways. Mechanisms described in this chapter illustrate the use of gears, ratchets, friction wheels, cams, pulleys and belts in combinations that are noteworthy for some ingenious feature or special function which they perform.

Other speed-changing mechanisms are described in Chapter 11, Volume I; Chapter 10, of Volume II; and Chapter 12, Volume III of "Ingenious Mechanisms for Designers and Inventors."

Ball Bearing Serves as Planetary Reduction Gear

A light-duty mechanism, driven by a 1/10-H.P. motor, was found to be running too fast to function properly. Since no space was available within the mechanism housing to permit the use of a larger driven pulley, another means of speed reduction was sought. This took the form of a conventional single-row, heavy-series type ball bearing, which was used as a planetary reducing mechanism.

As shown in Fig. 1, driving pulley *A* is allowed free rotation on motor shaft *B* by two ball bearings *C*. A spacer *D* is placed between the bearings, and collar *E* retains them. Steel pins *F* are pressed into one side of the driving pulley, their free ends projecting into the rivet space between the cage recesses that contain the balls of the heavy-series type ball bearing *G*.

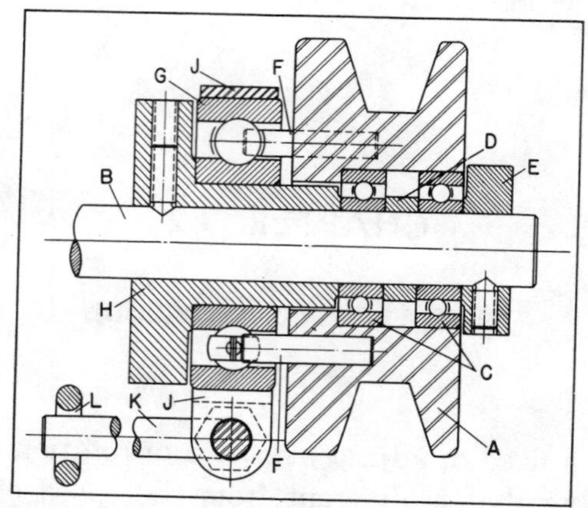

FIG. 1. This drive mechanism makes use of a ball bearing G to serve in the capacity of a reducing planetary gear unit.

The inner race of bearing G is press fitted on sleeve H, which, in turn, is secured to motor shaft B. The outer race of the bearing is held stationary by a split band J. Pressure is applied to the split band by tightening L-shaped bolt K, the end of which passes through eye-bolt L. The eye-bolt is fastened to the frame of the motor.

When the motor shaft rotates, sleeve H and the inner race of ball bearing G move in unison with it. This movement drives the balls and the cage of the bearing, causing the balls to roll along the fixed outer race. Driving pulley A is thus rotated at a reduced speed through the engagement of pins F with the moving cage of bearing G. The slight axial load applied to the balls by pins F tends to increase the power-transmitting capacity of the drive. In practice, no noticeable slip was encountered, even during the instantaneous application of heavy loads to the driving pulley A.

Rotational speed of the inner race of bearing G (also of the motor shaft) is $\left(1 + \frac{D_o}{D_i}\right)$ times the rotational speed of the cage, where D_o is the diameter of the inner-race ball track and D_i is the diameter of the outer-race ball track. With the particular bearing used in the illustrated setup, the speed reduction between the motor shaft and the driving pulley was approximately 2.5 to 1.

Shaft-Mounted Speed Reducer

Shown in Fig. 2 is a geared speed reducer of unusually compact design. The small cylindrical unit mounts directly on the drive-shaft and transmits its torque to the driven member (not shown) by means of a V-belt. Assembled appearance and construction details can be visualized from the partial section in Fig. 2 and the exploded view in Fig. 3.

The center section of the speed reducer consists of a steel sleeve *A*, internal gear *B*, and pinion *C*. Internal gear *B* is pressed into the steel sleeve. Pinion *C*, which is keyed to drive-shaft *D*, meshes with this gear.

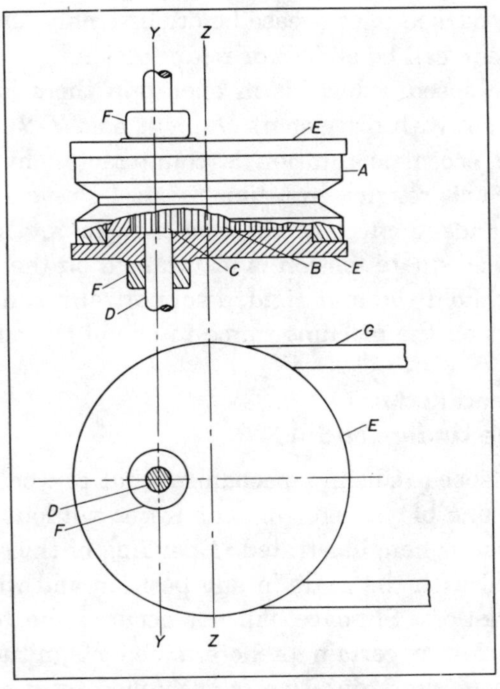

FIG. 2. Compact speed reducer mounts directly on drive-shaft *D*. Driven sleeve *A* is coupled to moving machine member by V-belt *G*.

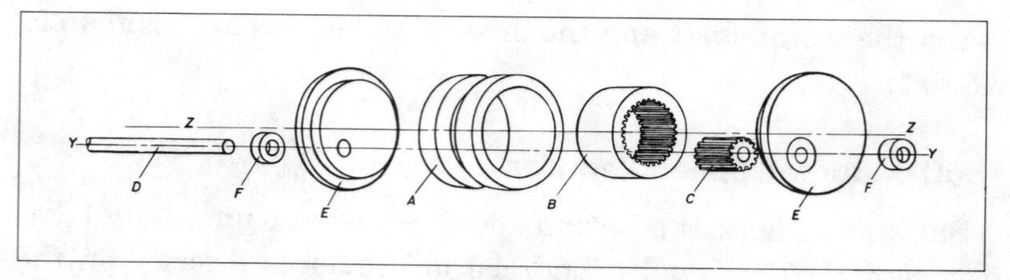

FIG. 3. Exploded view of shaft-mounted speed reducer. For quantity production, internal gear *B* can be eliminated and the teeth cut directly in sleeve *A*.

Two bronze end plates *E* have shoulders that are turned to a running fit with steel sleeve *A*. The bearing holes through which the drive-shaft passes are located off center by the distance necessary to provide proper engagement between the internal gear and the pinion. Collars *F* are locked to the shaft adjacent to the end plates and serve to retain the assembly intact. The unit can be packed with grease before assembly and, if desired, a grease fitting can be added for relubrication.

When the speed reducer is in operation there is a tendency for it to rotate with drive-shaft *D* about axis Y-Y. This proneness toward eccentric rotation is counteracted by the pull of V-belt *G* which restricts rotation to steel sleeve *A* about axis Z-Z. A secondary effect of the tendency to rotate about axis Y-Y is that adequate tension is maintained on the V-belt. If it is desired to hold the unit rigid, a support arm can be provided from a point on the machine frame to one of the end plates.

Geared Speed Reducer Changeable Under Load

A geared speed-reducing mechanism that can be regulated to obtain any one of sixteen different ratios without disengaging the input load is here illustrated. Changing of the speed ratio is accomplished with the gears in any position and while they are idle or in motion. Slippage will not occur if the torque transmitted is below a certain predetermined magnitude, but any over-loading of short duration is cushioned by a spring-loaded shock-absorbing arrangement.

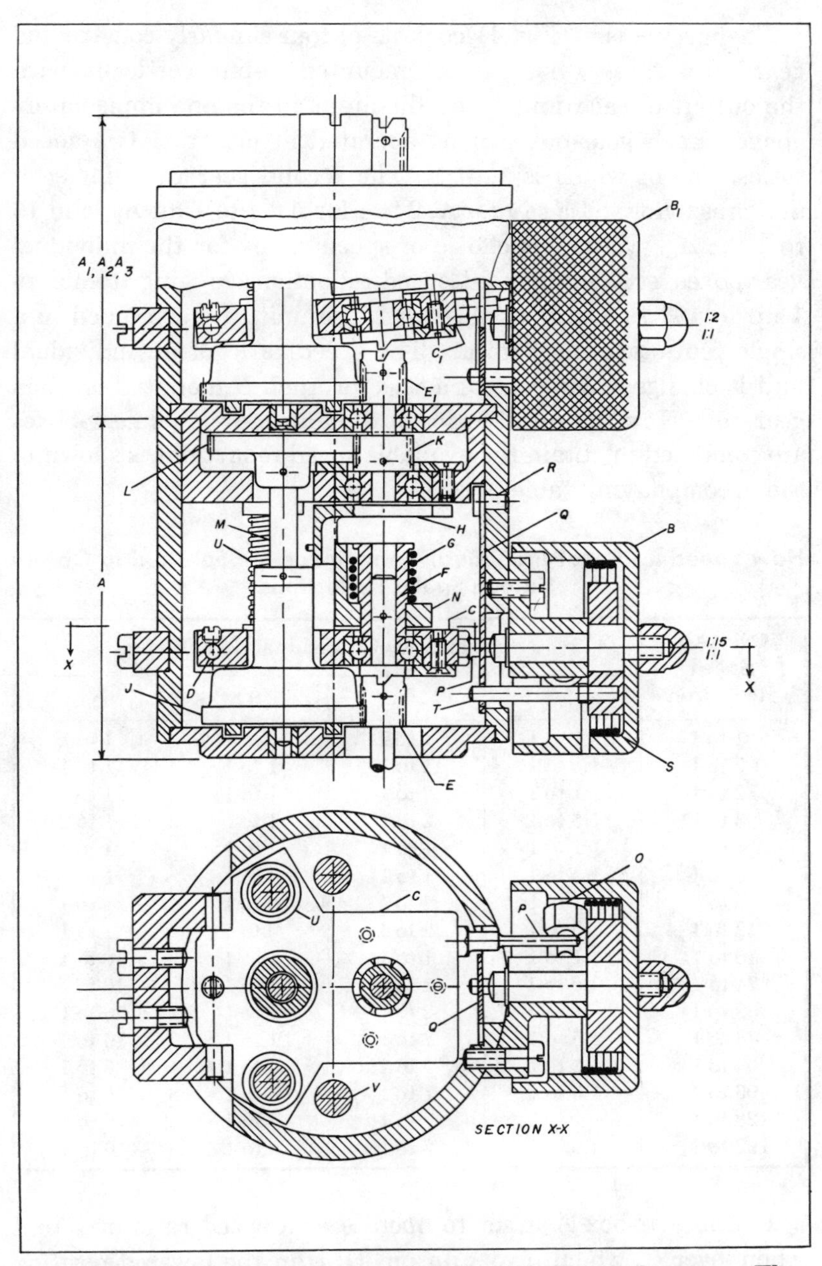

FIG. 4. On the down stroke, cam *A* pulls slide *C* into the press. Then punch *B* pushes the blank through die *E*.

The mechanism (Fig. 4) consists of four similarly constructed gear-boxes A, A_1, A_2, and A_3 mounted in-line vertically with the output of each unit being the input to the one immediately above. Each gear-box can be operated at either of two speed ratios, one of which is 1 to 1. The second gear ratio for each unit is as follows: 1.5 to 1 for A, 2 to 1 for A_1, 4 to 1 for A_2, and 16 to 1 for A_3. With this choice of speed ratios for the individual gear-boxes, sixteen different speed reductions ranging from 1 to 1 up to 192 to 1 are possible when the units are combined in a single four-stage mechanism. The speed ratio of an individual unit is changed by means of a selector knob B mounted on each gear-box. How the speed ratios of the four individual gear-boxes are combined to obtain the available speed reductions is shown in the accompanying table.

How Speed Ratios of Individual Gear-Boxes are Selected to Obtain Sixteen Speed Reductions

Over-All Speed Reduction	Speed Ratio of Individual Gear-Boxes			
	A	A_1	A_2	A_3
1 to 1	1 to 1	1 to 1	1 to 1	1 to 1
1.5 to 1	1.5 to 1	1 to 1	1 to 1	1 to 1
2 to 1	1 to 1	2 to 1	1 to 1	1 to 1
3 to 1	1.5 to 1	2 to 1	1 to 1	1 to 1
4 to 1	1 to 1	1 to 1	4 to 1	1 to 1
6 to 1	1.5 to 1	1 to 1	4 to 1	1 to 1
8 to 1	1 to 1	2 to 1	4 to 1	1 to 1
12 to 1	1.5 to 1	2 to 1	4 to 1	1 to 1
16 to 1	1 to 1	1 to 1	1 to 1	16 to 1
24 to 1	1.5 to 1	1 to 1	1 to 1	16 to 1
32 to 1	1 to 1	2 to 1	1 to 1	16 to 1
48 to 1	1.5 to 1	2 to 1	1 to 1	16 to 1
64 to 1	1 to 1	1 to 1	4 to 1	16 to 1
96 to 1	1.5 to 1	1 to 1	4 to 1	16 to 1
128 to 1	1 to 1	2 to 1	4 to 1	16 to 1
192 to 1	1.5 to 1	2 to 1	4 to 1	16 to 1

Lower gear-box A is set to operate at a speed ratio of 1 to 1 when lever C, which pivots on pin D, is in the lowered position as shown in Fig. 4. In this case, the motion of the input shaft and gear E is transmitted through a positive clutch to shaft F. Spring G further transmits the motion to member H which is a section of a cylindrical cup. Member H is mounted on the same shaft as the input gear E_1 for gear-box A_1. In addition, the motion of gear E is transmitted to gear J, and the motion of member H is transmitted through gears K and L to shaft M. An over-running clutch prevents shaft M from becoming coupled to gear J. This arrangement is possible since the shaft rotates faster than the gear.

The over-running clutch, illustrated in Fig. 5, consists of two rollers that revolve with shaft M inside a bushed hole in gear J. (Rotation of both gear and shaft is always in the direction indicated.) If the shaft rotates faster than the gear, the rollers move

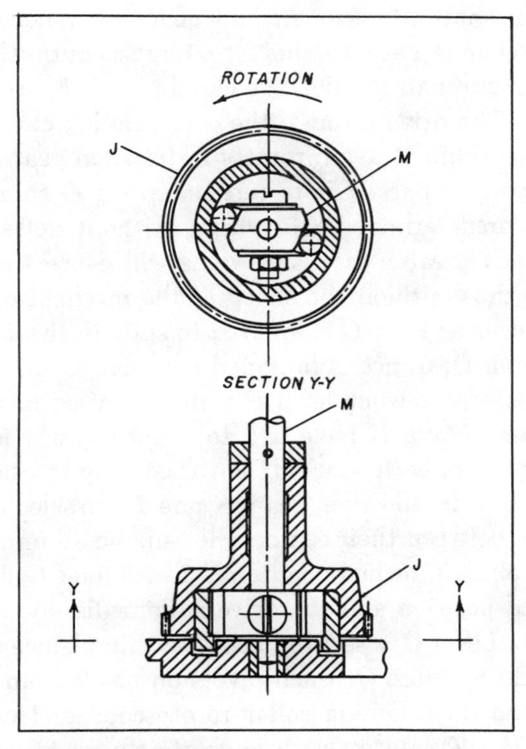

FIG. 5. The over-running clutch that prevents loss of load when speed ratio is changed is seen in detail.

freely with the shaft. Retainers in the form of thin leaf springs are bolted to the shaft to hold the rollers in place. When shaft *F* is disconnected from member *E*, the drive slows in rotation until the relative motion between gear *J* and shaft *M* becomes zero. At that instant, the angular flats on the shaft force the rollers outward until they become wedged against the gear, which will then become the driving member. This is accomplished immediately and practically no relative motion in the opposite direction is obtained.

When the selector knob is turned to obtain speed reduction, the lever *C* is pivoted to the raised position and the positive clutch is disengaged. Lever C_1 is seen in the raised position in Fig. 4. With this arrangement, the motion of gear *J* is transmitted to the output shaft and gear E_1 through the over-running clutch, shaft *M*, gear *L*, and gear *K*. When lever *C* is lowered to change back to the 1 to 1 ratio, output gear E_1 is again driven through the positive clutch, shaft *F*, spring *G*, and member *H*. The drive through the over-running clutch becomes uncoupled as shaft *M* again rotates faster than gear *J*.

The purpose of part *N* is to tension spring *G* so as to transmit only a predetermined safe torque without deflecting. Momentary loads greater than this value will cause the spring to deflect and thus cushion the shock of the mechanism. A shock load may occur as lever *C* is lowered to shift to the 1 to 1 speed ratio. Enough clearance is provided between parts *N* and *H* to prevent interference when lever *C* is in the raised position.

Since gears *J* and *E* have a 4 to 1 speed ratio for all four units, the ratio of each gear-box is varied only by the choice of gears *K* and *L*. In addition, gears *K* and *L* are selected to make the distance between their centers the same in all units.

When knob *B* is turned counterclockwise, bolt *O* lifts lever *C* to the raised position and the drive is immediately shifted to a lower speed. Lever *C* is secured in this position since an integral arm *P* is held by latch *Q* which pivots on headed pin *R*. Spring *S*, pin *T*, and its retaining collar rotate counterclockwise with the knob. Pin *T* rotates latch *Q* so that a protrusion on this member is hooked under arm *P*. Spring *S* also holds the knob in the position in which it was set. Two springs *U* push lever *C* down and provide the necessary pressure to keep arm *P* hooked in place.

Shifting back to a higher speed ratio is accomplished by turning knob *B* clockwise. Bolt *O*, rotating with the knob, pushes latch *Q* aside and releases arm *P*. Springs *U*, in turn, push lever *C* to the lowered position, and shaft *F* becomes coupled to the input gear *E* by means of the positive clutch.

In operation, response of the mechanism is instantaneous when shifting any individual unit to a lower speed. Response to the changing of any unit to a higher speed (that is to the 1 to 1 ratio) is not instantaneous, as a very slight delay is necessary for the positive clutch and spring *G* to become driving members. Use of the overrunning clutch, however, keeps the drive operating under load until it is shifted to these members. The mechanism can be driven in only one direction but it is possible to make the output reversible by adding a special gear-box with operating ratios of 1 to 1 and − 1 to 1.

Double Clutch Permits Reversal of Driven Shaft

In modern plants, it is frequently desirable to have control over the rotational direction of a driven shaft. A simple set-up that provides this control, without altering the direction of rotation of the driving shaft, is shown in the accompanying illustration.

A double electromagnetic clutch (see Fig. 6), consisting of units *A* and *B*, is mounted on driving shaft *C*. Two drive members, gear *D* and sprocket wheel *E*, are mounted on ball bearings and are in contact with clutch units *A* and *B* respectively. Driving gear *D* meshes with driven gear *F*, and driving sprocket wheel *E* is joined to driven sprocket wheel *G* by a link chain, not shown. Both of these driven members are keyed to output shaft *H*.

Shaft *C* rotates in the direction shown, arrow *1*, at all times. When it is desired to have shaft *H* rotate in the same direction, arrow *2*, clutch unit *B* is energized. This results in the transmis-

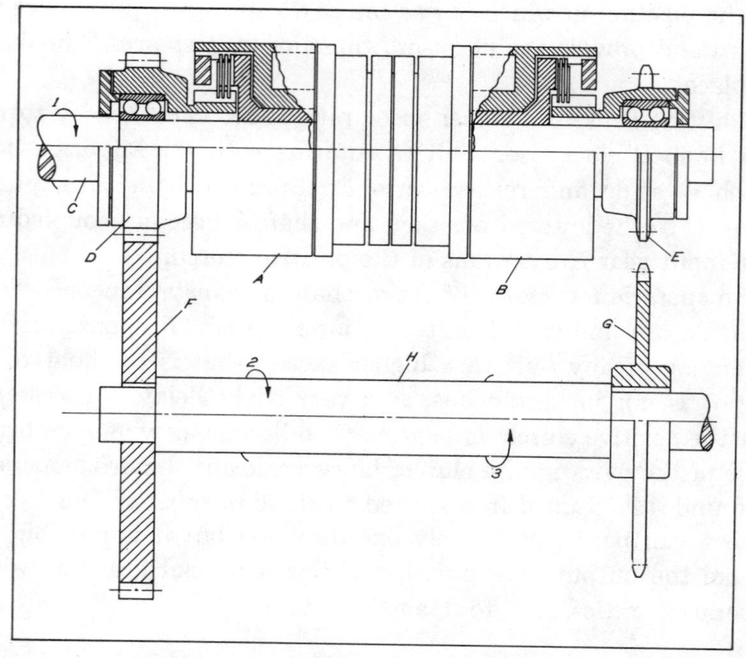

FIG. 6. Rotation of output shaft *H* can be reversed by the action of a double clutch without altering the rotation of driving shaft *C*.

sion of power from sprocket wheel *E* to sprocket wheel *G*, while gear *D* idles. If, on the other hand, it is necessary to rotate the output shaft in the direction shown by arrow *3*, clutch unit *A* is energized. Sprocket wheel *E* then idles, as power is transmitted from gear *D* to gear *F*. If desired, the chain drive can be replaced with a belt drive.

Three-Speed Gear Conversion Unit for Bicycle Coaster Brake

The bicycle three-speed mechanism, as shown in Fig. 7, is assembled on a shaft or axle *K*, which is simply inserted in the regular coaster-brake hub shell *F* in place of the original plain shaft and sprocket. The regular brake plates or disks *H*, the ball bearings at each end of the hub shell, and sleeve *G* remain undisturbed. The brake can be applied when using any of the three speeds by simply back-pedaling in the regular manner.

Essentially, the "Triplspeed" unit consists of a sprocket *A* driven by the bicycle chain; a planet-gear carrier *B* to which the sprocket is attached; four compound or stepped planet gears *C* journaled on the planet-gear carrier studs; a ring gear *D* with which the teeth of planet gears *C* are constantly in mesh (The ring-gear driver *D* is made with a triple-thread extension *E*, which serves as a means of driving the hub shell *F* through the sleeve *G* or applying the brake by exerting pressure on the brake plates *H*.); a sliding sun gear *I* with a larger supplementary sun gear *J*; and an axle *K* to which an axle cone *L* is permanently fixed.

In low gear, the sun gear *I* rotates freely on the two-piece axle sleeve *M*. One set of sun-gear teeth *I* meshes with the teeth at the larger end of planet gears *C* and the other set of sun-gear teeth engages the internal teeth of supplementary sun-gear *J*, whose outer teeth mesh with the teeth of the smaller end of the planet gears. Since it is impossible for the stepped planet gears *C* to revolve about sun gears (*I* and *J*) of unequal diam-

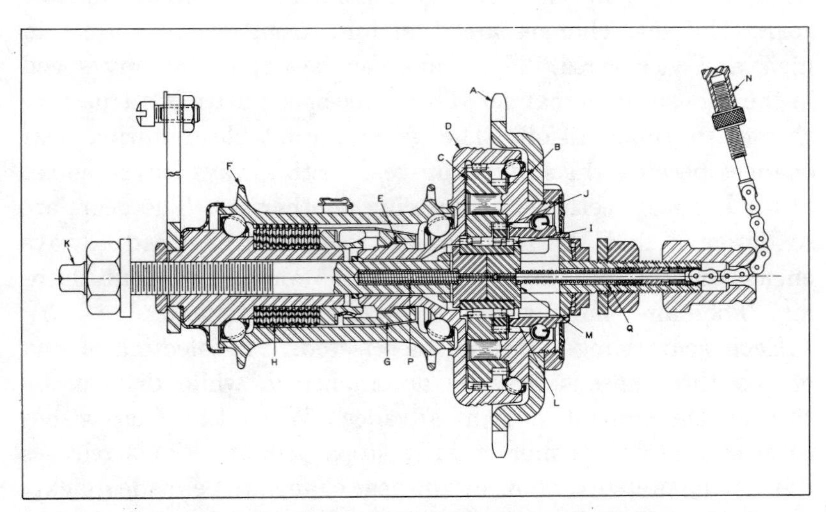

FIG. 7. Cross-section view of "Triplspeed" coaster-brake mechanism.

eters, the mechanism is locked and is driven as a unit about the axle, resulting in direct drive or low gear.

When the sliding sun gear *I* is moved to the right, it disengages the supplementary sun gear *J* and immediately engages the internal teeth of the fixed axle cone *L*. Thus the sun gear *I* becomes stationary, allowing the planet gears *C* to revolve about the sun gear and cause ring gear *D* to revolve. This results in the first over-drive or normal gear.

When the sliding sun gear *I* is moved farther to the right and deeper into the fixed axle cone *L*, the teeth on the left-hand end engage the internal teeth of the supplementary sun gear *J*, allowing the small planet gears *C* to revolve. The teeth at the larger end of planet gears *C*, being made integral with the gear on the smaller end, drive the ring gear *D*. This results in the second overdrive or high gear.

The outward movement of the sliding sun gear *I* is accomplished by pulling the control cable attached to the coupling *N*. The cable is operated by means of the control shifting lever mounted in a convenient position on the handle bar. When the cable tension is released, the sliding gear is allowed to move inward, resulting in successive gear changes to normal and low gear. The gear changes are, therefore, from low to normal to high, and vice versa. The brake can be applied at any speed in the conventional manner. This three-speed drive is a true synchro-mesh transmission. The gears cannot clash during gear changes because the sliding sun-gear teeth always leave one set of mating gear teeth before entering another set. The gears are so designed that the sliding gear, regardless of speed, always engages its mating gear without lost motion and without clashing. Therefore, shifting may be done at any time.

Each gear change may be pre-selected. Pre-selecting of any of the three speeds may be accomplished, while driving, by shifting the control lever in advance. When the rider wishes to change gears, he momentarily stops pedaling. This releases the driving pressure, allowing the gear change to be made quickly and automatically by the actuating spring. Pedaling can then be resumed in the pre-selected speed.

Changing from one speed to another is done as follows: The position of the sliding sun gear I is predetermined by the movement of the two-piece sleeve M on which gear I is mounted. The two sections of sleeve M are backed up by two springs P and Q within axle K. With the sliding sun gear I under torque from pedaling, the shifting lever is moved to the desired gear change position. When the torque is removed from the sun gear by momentarily stopping pedal movement, the axle spring moves the sliding sun gear automatically to the predetermined position. Upon resumption of pedaling, the sliding sun gear has assumed its proper position and the unit is in the desired gear. Similarly, pre-selection may be accomplished from high to low, normal to high, or any other desired combination.

One of the most attractive design features of this drive is that the shifting control-lever assembly is located near the handle-bar grip. Shifting of gears is done without the necessity of removing the hand from the grip. Shifting from low to normal to high is done by pulling up on the shifting lever with the fingers. Shifting from high to normal to low is accomplished by pushing down on the release lever with the thumb.

The calculation of the bicycle "gear number" is as follows: The bicycle "gear" is an indication of the distance traveled by the bicycle per revolution of the pedal crank or front sprocket. The "gear number" multiplied by 3.1416 equals the distance covered in one revolution of the front sprocket. Thus a bicycle having a 69 gear travels 216 inches, or 18 feet, along the road for each crank revolution.

The "gear" of a bicycle is the product of the number of teeth in the front sprocket and the number of inches in diameter of the rear wheel divided by the number of teeth in the rear sprocket. The result is the "gear number" in inches. The trade has dropped the dimensional unit of inches and the gear is known as a number. This calculation may be expressed by the following formula:

$$G = \frac{FW}{R}$$

where $G =$ bicycle gear number;
$F =$ number of teeth in front sprocket:
$W =$ rear wheel diameter, in inches; and
$R =$ number of teeth in rear sprocket.

In this new three-speed drive, the gears are reduced 25 per cent and increased 33⅓ per cent from normal gear, so that the three gears may be computed as follows:

$$\text{Low } G = \frac{FW}{R}$$

$$\text{Normal } G = \frac{4FW}{3R}$$

$$\text{High } G = \frac{16FW}{9R}$$

In a "Triplspeed" equipped bicycle having 26- to 2⅛-inch balloon tires, a 26-tooth front sprocket, and a 13-tooth rear sprocket, low gear would be 53, normal gear 70, and high gear 94.

Therefore, with the new three-speed coaster brake, a bicycle travels about 14 feet in low gear for each revolution of the pedal, about 18 feet in normal gear, and about 24 feet in high gear.

Moving Supports for Long Boring-Bar

When large forged gun barrels are bored, the boring-bar may be as much as 70 feet long. Such a long bar may easily be bent by its own weight. The consequence is a bore that will weave eccentrically in some sections, or be out of round. To eliminate these inaccuracies an effective traveling support system, Fig. 8, was devised.

The boring-bar is supported at three intermediate points on the tailstock by traveling supports X, Y, and Z. The left-hand support W remains stationary. The housing C contains the motor and the necessary gears and controls to feed the bar to the left into the bore at the desired rate. The right-hand end of the boring-bar is attached to this drive. The three intermediate supports move in the direction of the boring-bar feed at

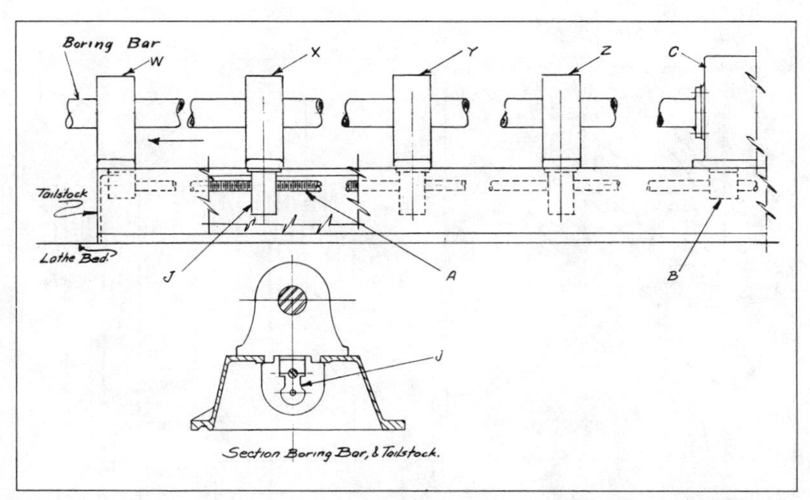

FIG. 8. Movement of boring-bar supports X, Y, and Z toward tailstock W must be proportionate to the total distance traveled by housing C.

speeds arranged so that they are at all times equally spaced between the support W and housing C.

It will be noted that the clear space between supports W and X is one-fourth of the sum of the clear spaces between W and C. Consequently, support X should move one-fourth of the speed of housing C; support Y at one-half the speed of C; and support Z at three-quarters of that speed.

The boring-bar is moved by the lead-screw A which is driven by a gear train in housing C. A nut B carries housing C forward as the screw is rotated. The screw A is provided with a keyway its full length. The bearing at the left end of A is arranged to take axial thrust in both directions. As nut B is rigidly attached to housing C, one revolution of shaft A will move housing C a distance equal to the pitch of the thread. If, however, this nut is rotated in the same direction and speed as the screw, there is no forward motion of housing C.

Figure 9 shows in detail the mechanism used under each support that gives forward motion. The motion consists of a cluster of four gears. Gear D slides on lead-screw A, and is turned by a feather key fastened in the bore of gear D. Gear E,

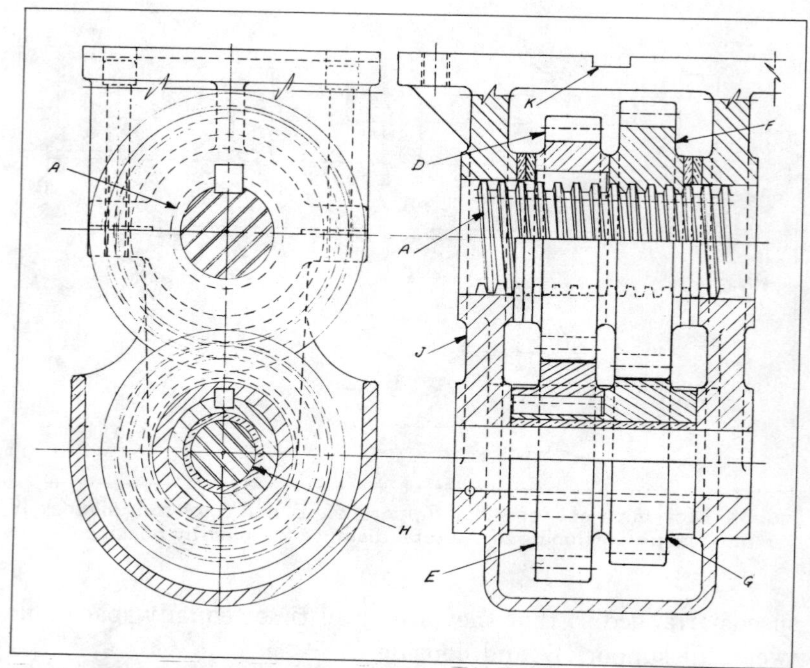

FIG. 9. Each of the boring-bar supports has a reduction gear train driving from the keyway in lead-screw *A*. Gear *D* carries the key but only slides over the lead-screw. Gear *F* drives the nut forward via the reduction gear train from *G* to *E*.

which meshes with *D*, is mounted on the hub of gear *G*. Gear *F* is threaded to fit lead-screw *A*. All four gears are held in a cage *J*, which is attached to the underside of its particular support, and held from axial-movement keyway *K*. By changing the ratios between gears *D, E, G*, and *F*, the supports can be caused to move toward *W* at the desired speeds.

Hydraulic Gear-Shift Control for Speed-Change Mechanisms

For controlling modern speed-change mechanisms that incorporate sliding gears or jaw-clutch couplings, hydraulic systems are being widely used. Generally, the gears of such mechanisms run at a speed that is too high to permit changing the speed under operating conditions. To overcome this difficulty, either slow-motion features are embodied in the design or provision is made to prevent the gears from being shifted prematurely, that is, before the speed has been reduced to a rate suitable for the purpose.

An automatic hydraulic control system that meets the requirements mentioned is shown in Fig. 10. (Note that the views in Fig. 10 are in accordance with the European system of projection.) This system is of conventional type, using an oil pump which serves to lubricate the gearing or the machine driven by it — as for instance, a machine tool. It also supplies oil under the pressure required to effect the control. The system is composed mainly of a small gear type or plunger type auxiliary pump *B* which has its driving shaft *C* coupled to a constant-speed shaft of the gearing; a spring-loaded control valve *D*; and a throttle *E*.

When the gearing is in operation, oil drawn by the auxiliary pump through bore *Q* is carried through bore *S* into cylinder *P*. The oil lifts control valve *D* against the action of spring *J* so that the stream of oil coming from the main pump (not shown) and entering the system at orifice *F* is permitted to pass through annular space *H* and flow freely at orifice *K* to effect the lubrication. The additional amount of oil supplied by pump *B* is also fed to the lubricating pipe through passages *L* and *M*. The oil pressure can be varied by means of set-screw *O* after removing plug *N*.

When the gear mechanism is stopped by disengaging a clutch between the driving motor and the gearing, less oil is supplied by pump *B* and the control valve *D* will then drop because of the pressure exerted by spring *J*. This is accomplished at a rate which depends upon the adjustment of spring *J* and throttle *E*. Oil coming through cylinder *P* is then allowed to escape through passage *M* into the lubricating pipe. In moving downward, control valve *D* opens the pressure pipe *R*. As the main pump continues running, the oil it supplies is fed through a distributor to pistons for moving the gears. Thus, smooth shifting is insured.

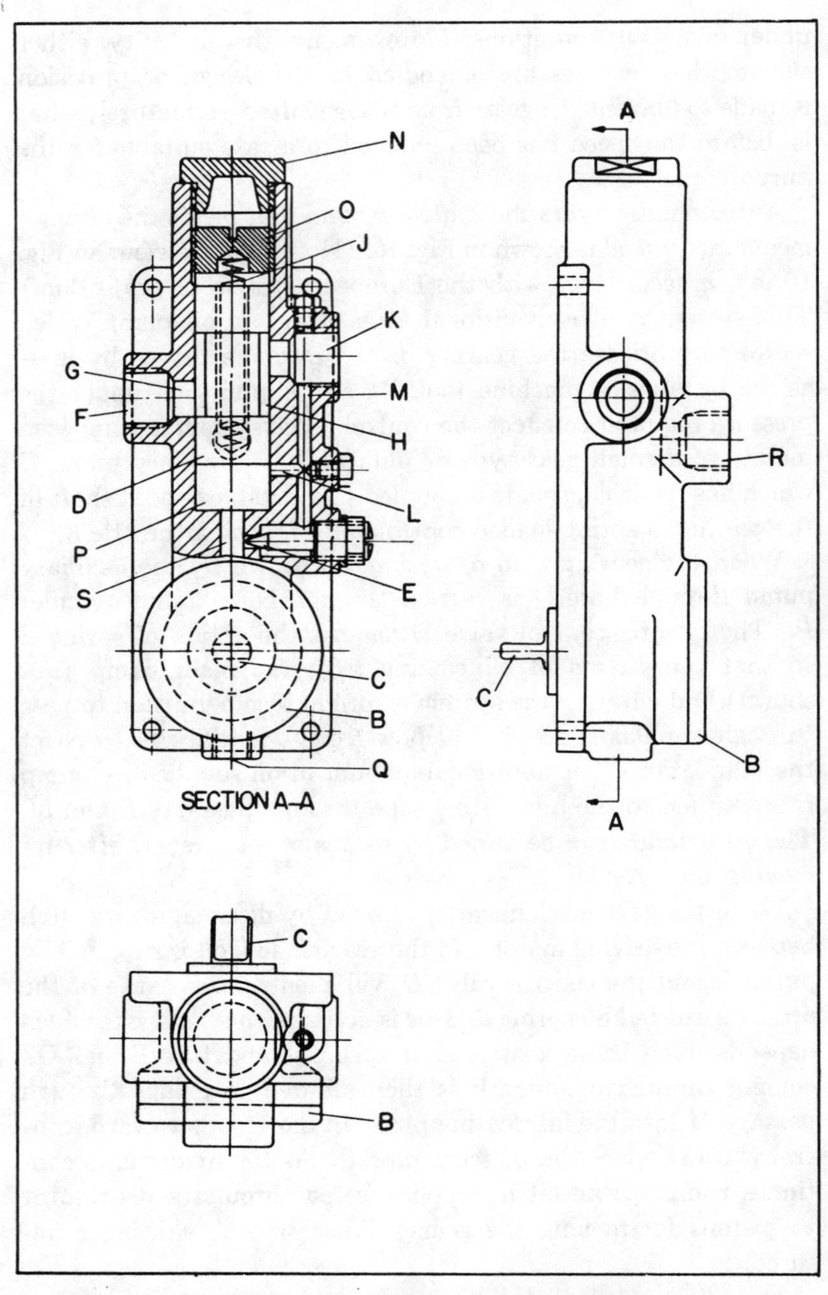

FIG. 10. Hydraulic mechanism assures smooth shifting of gears.

Differential Screw Assembly for a Slide

To enable a slide to travel at a reduced rate during part of its movement, a mechanism consisting of a differential screw assembly was designed. The device, shown in Fig. 11, controls the linear movement of a nut block *A* to which the slide is fastened. The nut block has a 10-pitch internal thread and engages the externally threaded end of a drive-shaft *B*.

At its opposite end, the drive-shaft is square in section, providing a sliding fit for the handwheel *C* which operates the slide. Part of the cylindrical length of the drive-shaft bears in a bushing *D*, which in turn, has a 12-pitch external thread engaging a fixed bracket *E*. Shoulders *F* on the bushing serve to limit its axial movement in the bracket. In addition, the left-hand shoulder is designed as a straight-tooth clutch *G*, the other member of which is developed from the hub end of the handwheel. A spring *H* keeps the two members of the clutch in normal disengagement.

Since another pair of shoulders *J*, fixed on the drive-shaft, prevents its axial movement, turning the handwheel produces a transverse travel of the nut block. When the clutch is disen-

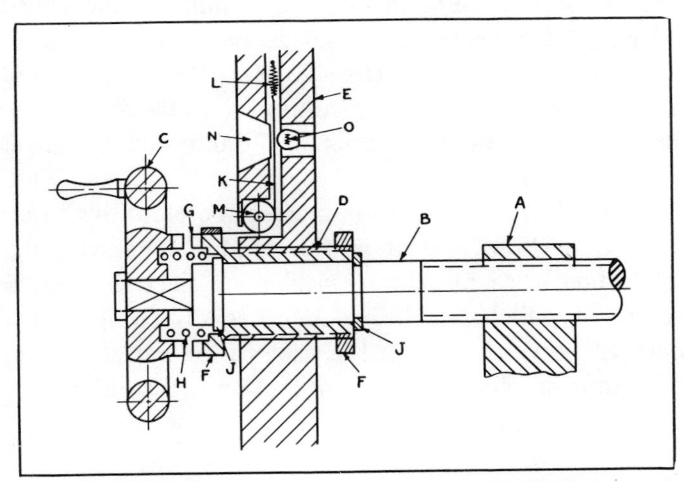

FIG. 11. When clutch *G* is engaged, the rate of travel of the slide is equal to the difference between the pitches of the two threads.

gaged, the bushing remains stationary while the drive-shaft rotates, and the nut block travels $\frac{1}{10}$ inch per revolution.

To produce the speed differential, the clutch is engaged, and the bushing, now rotating in unison with the drive-shaft, also moves axially. This axial movement of the bushing is transmitted to the drive-shaft through the thrust against the shoulders *J*. Movement of the bushing and drive-shaft is opposite to the direction of the nut block. But since the bushing and drive-shaft move only $\frac{1}{12}$ inch per revolution, the net result is to reduce the travel of the nut block to $\frac{1}{60}$ inch ($\frac{1}{10} - \frac{1}{12} = \frac{1}{60}$) per revolution.

Still finer adjustment of the slide is possible if the difference between the thread pitches is still smaller. Thus, with a 9-pitch nut block and a 10-pitch bushing, the travel of the nut block is only $\frac{1}{90}$ inch per revolution. When equipped with a graduated collar, the handwheel will register the amount of slide travel through any small degree of drive-shaft rotation. For example, if the collar has 110 divisions, controlled fine adjustments of 0.0001 inch can be made.

Difficulties may arise if the bushing is rotated continuously in one direction, since eventually, one of the shoulders *F* will jam against the bracket. For this reason, a simple indicating device consisting of a transparent band *K* has been provided. One end of the band is fixed to the shoulder; the other is held by a spring *L*. The band is guided around a roller *M* in front of a window *N* in the bracket which is illuminated by an electric bulb *O*.

If the shoulders of the bushing come too close to the bracket, red-colored portions appear in the window as a warning. Another method is to fit micro switches or electrical contacts to the shoulders, which can close a circuit to a warning light. Either method will serve to expand the use of differential screws, since an important drawback to their operation is thus eliminated.

CHAPTER 13

Speed Regulating Mechanisms

Machines which wind material such as paper, cloth or metal strip on spools or reels or which form or twist wire may require a synchronous rotation of two shifts with or without an occasional momentary acceleration or retardation of one shaft with respect to the other. In other machines the speed of the driven shaft must be maintained within close limits. The mechanisms described in this chapter have been designed to perform such special speed controlling functions. Similar mechanisms are described in Chapter 13, Volume III of "Ingenious Mechanisms for Designers and Inventors."

Instant Acting Centrifugal Governor

A steam turbine in a chemical plant was required to operate below a certain rotational speed. For the particular application, the use of a conventional centrifugal governor would not be satisfactory, since it would reduce steam gradually by starting the decrease before the turbine reached the maximum allowable speed. This would cause a hunting effect when the speed is adjusted by means of the manual valve and would require the use of an especially complicated automatic setup. The desired effect could be achieved by adding a friction brake to a conventional governor. Such an arrangement, however, would result in hysteresis, since the governor would then stop the steam supply at a turbine speed much higher than that at which the supply is renewed. This, again, would be undesirable. The device shown in Fig. 1 provided the necessary speed control.

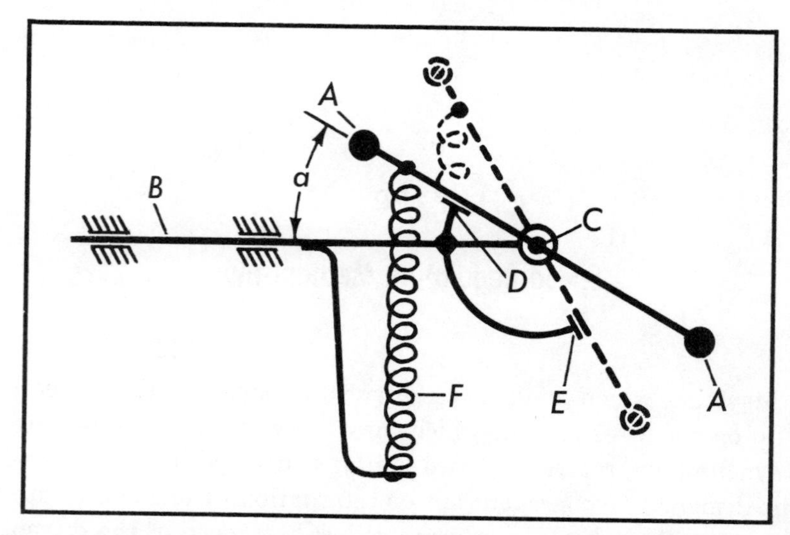

FIG. 1. Schematic diagram illustrating the principal design features of a quick-acting centrifugal governor.

In the schematic diagram of the mechanism (Fig. 1), a rigid bar is shown with weights *A* attached to each end. The bar is free to turn and is mounted on the main shaft *B* by means of a short shaft *C*. This short shaft is carried perpendicular to shaft *B*. Since a full revolution of the bar about shaft *C* is unnecessary, stops *D* and *E* are employed. A spring *F* is attached at one end to the bar as shown, and the other end is secured to a rigid support fixed on shaft *B*.

The centrifugal force f that acts on the mass m of weight *A* as shaft *B* rotates at a speed of n rpm is given by the following equation:

$$f = m\ (r \sin a) \left(2\pi \frac{n}{60}\right)^2$$

$$f = mr \sin a \left(\frac{\pi n}{30}\right)^2$$

where

r is the radial distance between the center of rotation of the bar and the center of gravity of the weight, and
a is the included angle between the bar and shaft *B*.

This force produces a torque on the bar, the lever arm of which is $r \cos a$. Since there are two weights A, the total torque T_A produced is

$$T_A = 2f\ (r \cos a)$$
$$= 2mr^2 \sin a \cos a \left(\frac{\pi n}{30}\right)^2$$
$$= mr^2 \left(\frac{\pi}{30}\right)^2 n^2 \sin 2a$$

In an actual mechanism, the weights and the bar consist of an infinite number of elemental masses, and the total torque T_A will be the integral of the elemental torques these masses produce, or

$$T_A = K_A n^2 \sin 2a$$

where

K_A is a constant depending on the weight and shape of weights A and the bar.

The opposing torque produced on the bar by spring F is obtained in similar fashion. This spring is constructed in such a way that it is not under tension when angle $a = 0$. In Fig. 1, the extension of the spring is

$$r_f \sin a$$

where

r_f is the radial distance between the center of rotation of the bar and the point at which the spring is secured to the bar.

The force exerted by the spring is $cr_f \sin a$ where c is the spring constant of member F. Therefore, the torque T_f applied by the spring on the bar is given by the equation:

$$T_f = cr_f^2 \sin a \cos a$$

In the actual governor (shown in Fig. 2) the spring exerts pressure on two bars. Thus, when each rotates through an angle a, the expansion of the spring will be

$$2r_f \sin a$$

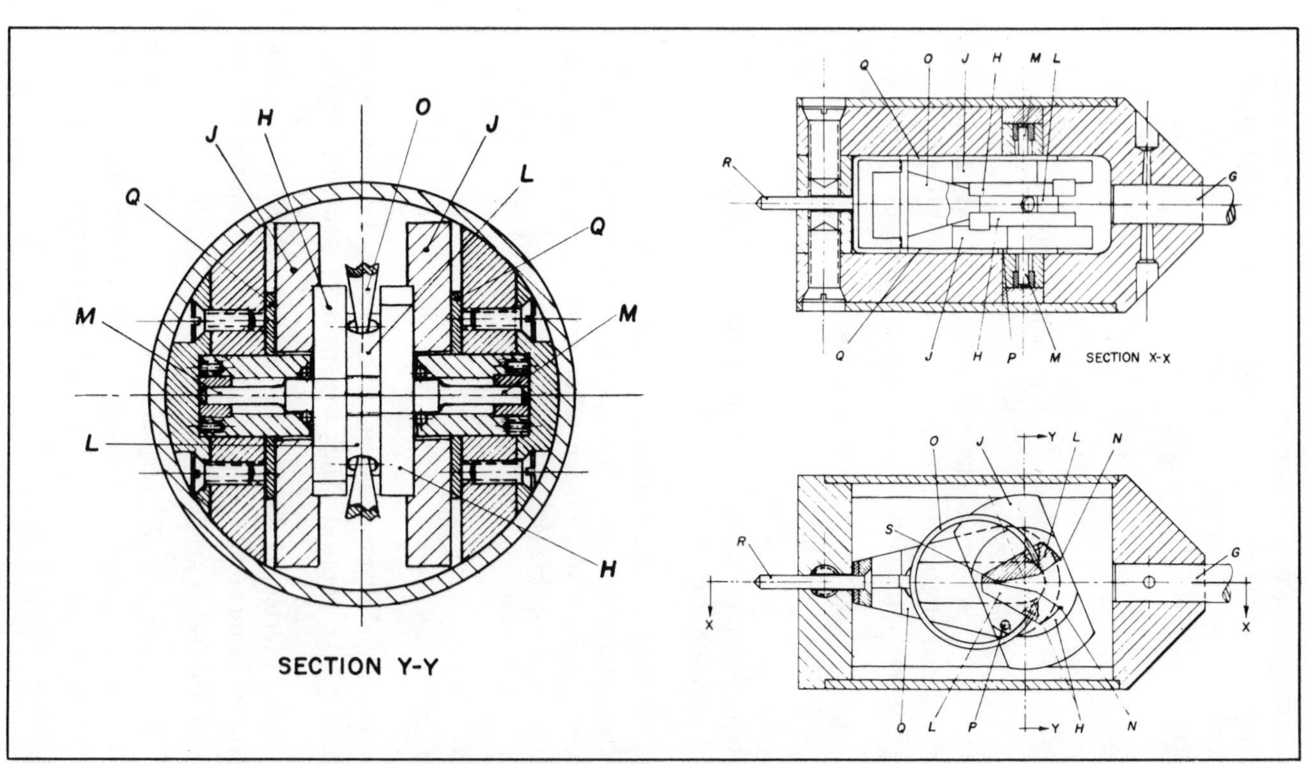

FIG. 2. This governor represents a practical application of the design shown in Fig. 1. In operation, the entire device rotates on the turbine shaft, and pin R provides the output.

and the torque on the bars will be

$$2cr_f^2 \sin a \cos a, \text{ or}$$

$$cr_f^2 \sin 2a$$

since the term cr_f^2 is a constant, it can be expressed as K_f and thus

$$T_f = K_f \sin 2a$$

In this manner the device is arranged so that the torque produced by the spring depends on angle a in the same way as the torque produced by the centrifugal force acting on the weights. Therefore, as long as

$$K_f > K_A n^2$$

the bar is held by the spring against stop D (angle a is at minimum value), but when the rotational speed n increases, $K_A n^2$ increases as the square of the speed until

$$K_f < K_A n^2$$

Then, regardless of the angle, the torque produced by the centrifugal force on weight A will exceed the torque produced by the spring, causing the weighted bar to almost instantly rotate from D to E (maximum value of angle a). When $K_A n^2$ becomes smaller than K_f, the bar will just as rapidly move back to stop D.

The rapid movements are due to the fact that n is squared in the $K_A n^2$ term, and therefore every rise and fall in the speed of the turbine causes a fast increase or decrease of $K_A n^2$ as compared with K_f. The critical speed at which the bar changes position is determined by the shape and weight of the bar and weights A and by the spring constant c. The energy for the movement toward stop E is supplied by the kinetic energy of the turbine, while the return movement is energized by spring F.

When translating the mechanism shown in Fig. 1 into a practical application, three conditions must be met:

1. The weights in the arrangement (Fig. 1) must have opposite counterparts to make it dynamically balanced and

thus prevent the centrifugal force from tending to bend shaft *B* and causing vibrations.

2. The support for the spring must move parallel to shaft *B* in order to permit proper application of spring tension to the bar as angle a changes.
3. A means of obtaining the output must be provided.

The actual governor, shown in the actuated position in Fig. 2, is designed so that the entire mechanism rotates on the main shaft *G*. To achieve dynamic balance, two crossed, weighted bars are employed, each of which is formed of two parts: *H* and *J*. Part *H* is made of steel and has a protrusion *L*. Member *J* is made of brass and serves as the weight. These weights revolve opposite each other and are placed in such a way that after a short rotation about shafts *M* in either direction, the protrusions *L* contact each other at points *N* (or *S*) and do not allow any further rotation in that direction. Parts *H*, *L*, and *M* are machined from one piece of steel.

One end of a spring *O* is set into a depression in each member *L*. An imaginary line extending between these depressions will always be perpendicular to shaft *G*, thus satisfying the second condition.

Spring *O* is triangular when viewed as shown in section X-X. This shape insures an equal distribution of stress through the circumference of the spring. In this way a spring which has linear characteristics and can withstand large deformations in relation to size is obtained.

To translate the rotation of the weights into a linear movement, a pin *P* is fixed on each of the weights *J*. Each pin has a matching groove in a sheet-metal part *Q*. When the critical speed is exceeded, members *H* rotate against the torque provided by spring *O*, moving parts *Q* outward. Members *Q*, in turn, move an actuating pin *R*, welded to them, outward.

Similarly, when the speed of the turbine slows, spring *O* returns parts *H*, which then contact each other at points *S*, and pins *P* pull parts *Q*, and thereby pin *R*, inward. In section X-X only one of the pins P is seen, as the second is fixed to the underside of the upper weight J. Pins P move parts Q in the same axial direction, whereas weights J revolve in opposite angular directions.

The inner ends of shafts M are supported on ball bearings to reduce friction. Sleeve bearings are used at the other ends, as the loads there are smaller. Shafts M are also provided with hardened steel covers which absorb the centrifugal pressure in the direction of their axis.

Since it was not required, no provisions were made in the governor shown for adjustment of the critical speed. Critical-speed adjustment can be easily accomplished, however, by arranging weights so that their distance from shaft M can be varied.

Synchronizer that Insures Precise Speed Measurement

Accuracy approaching that of an electronic counting device has been obtained by the synchronizing unit shown in Fig. 3. This unit is used to control a brake-test dynamometer. It will

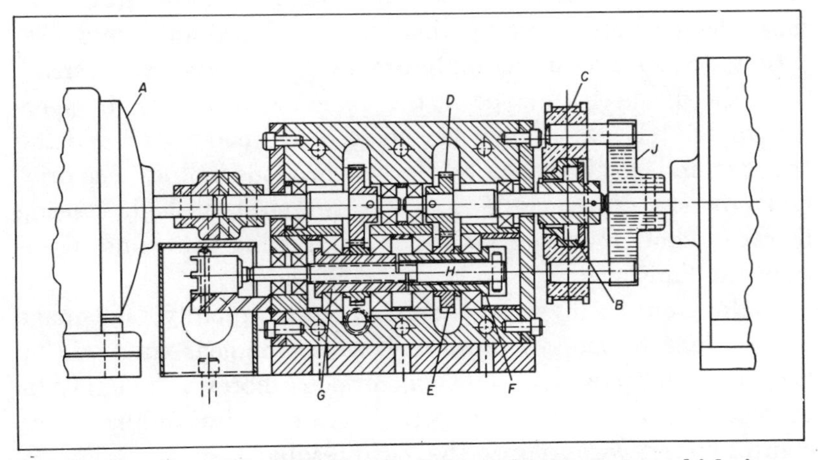

FIG. 3. Speed synchronizer for brake-test dynamometer which is extremely sensitive to speed changes.

sense a speed differential of 1 rpm in a rated speed of 3000 rpm.

The synchronizer receives power input from the synchronous motor *A*, which runs at a speed of 1800 rpm. Through appropriate gearing the power is transmitted at a speed of 3000 rpm to an inner member *B* of an overrunning clutch. Outer member *C* of this clutch is pulley-driven by belt from the flywheel of the dynamometer.

Both members *B* and *C* of the overrunning clutch rotate in the same direction. When the outer member attains a speed of 3000 rpm through the flywheel drive, both the inner and outer members travel at the same relative speed and no locking action takes place within the clutch. A fractional increase in the speed of the outer member, however, will cause the clutch members to lock and the outer member will then be driving the inner member.

This overdrive will cause gears *D* and *E* to run ahead of the synchronous motor input and cause sleeve *F* to advance a fraction of a revolution ahead of sleeve *G*. This advance will cause threads on shaft *H* to screw into threads in sleeve *G*, displacing shaft *H* from its original axial position. Such displacement will actuate a limit switch to trip the flywheel drive motor circuit and cause the flywheel to coast without power. This limit switch also actuates other circuits to apply brakes, timers, and recorders.

As the flywheel loses speed, the inner member *B* of the overrunning clutch is no longer locked to outer member *C*, and the entire train is again driven by the synchronous motor. The original position of sleeves *F* and *G* are re-established, causing threaded shaft *H* to screw back to its starting point and disengage the limit switch.

In the event the flywheel is accelerated too rapidly, no damage can be done to the synchronizer because the gear train will be driven by the flywheel. The synchronous motor will simply be overspeeded for a very short period. An electrical failure of the control circuit will produce the same result.

Sleeves *F* and *G* are so designed that their maximum displacement can never exceed 270 degrees. With 16 threads per inch on the threaded shaft, the axial displacement of the shaft can never exceed 0.047 inch.

Coupling *J* is connected to a zero-speed switch unit which can be used to stop recorders, clocks, and other devices when the pulley has stopped. The synchronizer is made up of stock gears and ball bearings, is oil sealed, and lubricated with light oil. The overrunning clutch is also a standard item designed to have almost zero backlash. A synchronous motor was used to obtain an accurate reference signal.

Constant Horizontal Velocity from a Crank

A paper-converting machine required that an operation be performed on the moving web. The web, however, had to be motionless at the time. Since it was impractical to stop the entire web, the device shown in Fig. 4 was designed for the purpose of stopping a portion of the web.

As illustrated, the web enters from the left, passes under and around roll *A* and over roll *B*. It then passes over table *C*, around and under roll *D*, and returns to the left, to and around roll *E*. At this point, the web leaves the device by moving to the right in the same plane as the web approaching roll *A*. Rolls *A* and *E* are mounted together in a frame *F* which, in turn, is

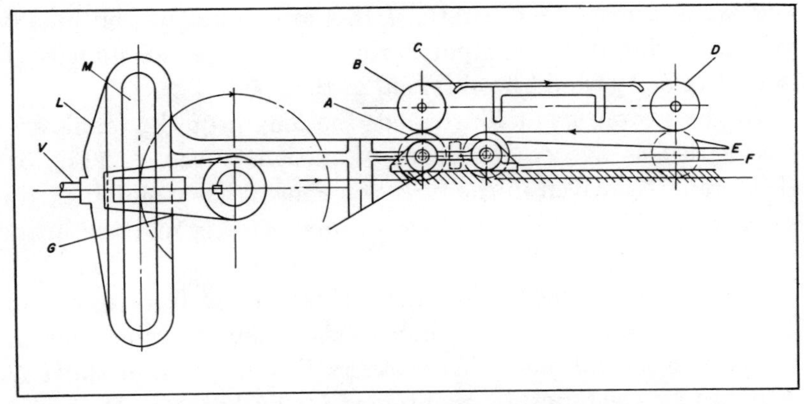

FIG. 4. Device for intermittently stopping a portion of a moving web employs a variable-length crank.

mounted on slides carried in suitable guides in the main frame of the machine.

If frame *F* with rolls *A* and *E* is allowed to move to the right at one-half the speed of the moving web, roll *A* will take up and pay out the oncoming web at one-half web speed. Since roll *E* is paying out the web at one-half the absolute web speed and at the same time is moving away from roll *B* at the same speed (one-half the absolute web speed), the web will remain stationary with respect to rolls *B* and *D* and table *C*. Roll *E* will receive and pay out the web at one-half web speed, due to the relative linear motion of roll *E* with respect to stationary roll *D*. The web, in turn, will be received by the next member of the machine at full web speed, due to the relative motion between that member and roll *E*.

The velocity of frame *F* must correspond to one-half the web speed, for if the speed is less, the web will still move forward over the table; or if the speed is greater, that portion of the web between rolls *B* and *D* will move backward to the left. Since the web must be stationary for an appreciable length of time, the movement of frame *F* to the right must remain at a constant speed during this period. A modified crank mechanism gives the frame constant motion.

The horizontal velocity of a crank movement is normally variable through an entire cycle, due to the constant length of the crank arm. If the length of this arm could be continually varied to suit through a portion of the cycle, a constant horizontal velocity would be obtained in that period.

In the arrangement illustrated, the length of the crank arm was varied as needed by means of stationary cam *U* (Fig. 5). The crank *G*, driven in the direction shown, has a slot *H* at the outer end which carries a slider *J*. Member *J* is retained in the slot by a cover plate *K*.

The horizontal motion of frame *F* is derived from the crank through a yoke *L*. This member has a slot *M* which accommodates a second slider *N*. Washers *P* and *Q* retain slider *N*, and proper clearance is maintained by bushing *R*. Both sliders are mounted on crankpin *S*, which also carries a cam roller follower T. The latter, in turn, engages the groove in stationary cam U.

The profile of the cam groove, Fig. 6, from points 0 to 18 will provide the varying length of the crank arm, whereas the groove

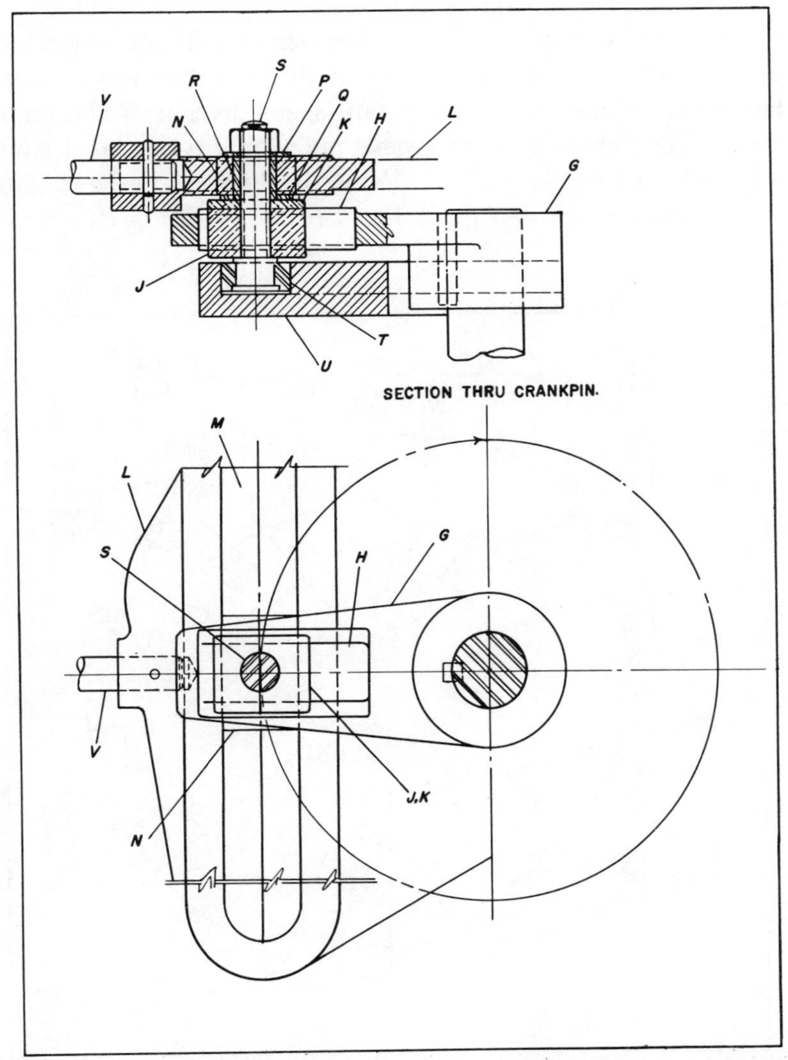

FIG. 5. Enlarged view showing details of mechanism which allows the effective length of the crank arm to be varied.

from points 18 to 0 may be concentric with the center of the cam. The horizontal cam displacements shown above the cam layout, Fig. 6, are variable from points 0 to 4, are equal from points 4 to 14, and vary again from points 14 to 18, but inversely, as from point 4 back to point 0. In this layout, the angular divisions are 10 degrees each.

The radius of the center line of the cam groove from the center of the cam at point 9 is calculated so as to give the required horizontal velocity for the constant-speed portion of the crank stroke. The balance of the groove radii for this portion (points 4 to 14) are then determined. Analytically, the lengths of these various radii would be: $d/\sin 10° = R$; $2d/\sin 20° = R_1$; $3d/\sin$

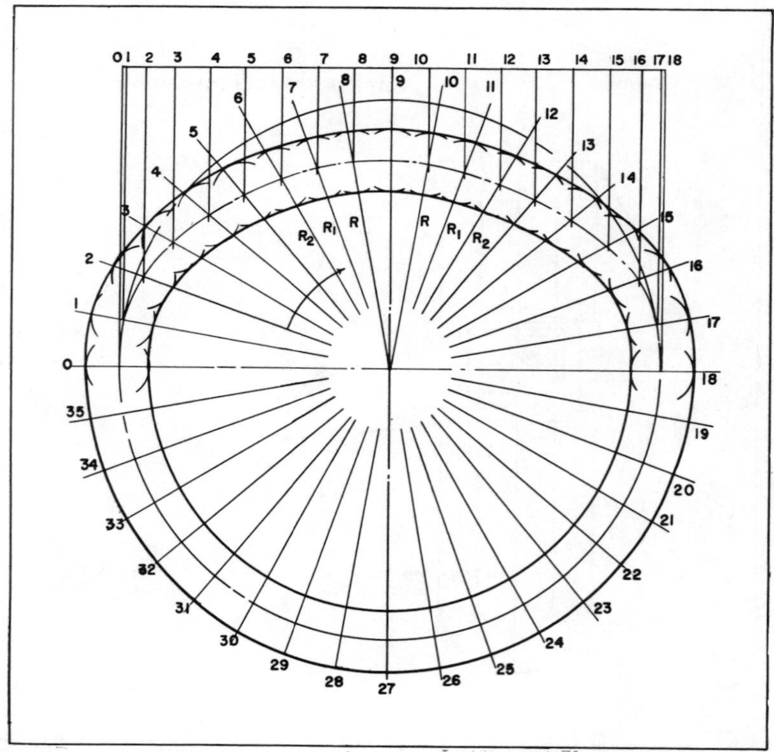

FIG. 6. Layout of the groove in cam U which gives uniform horizontal velocity to follower T, yoke L, and frame F.

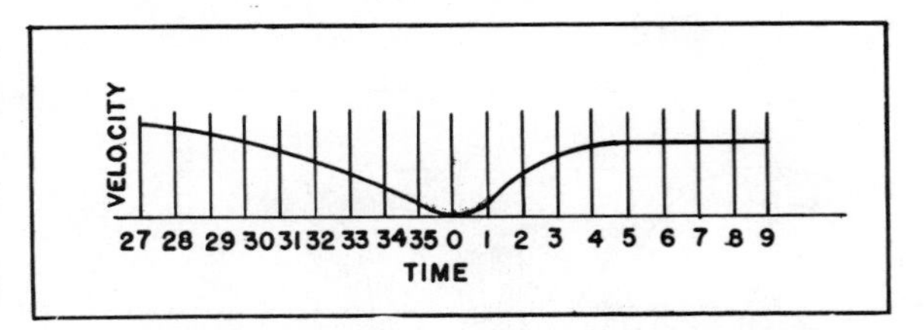

FIG. 7. Time-velocity diagram for one-half of the cam shown in Fig. 6. Constant horizontal velocity of the follower, yoke, and frame is obtained between points 4 and 9, and between points 9 and 14.

$30° = R_2$; etc. where d = total yoke displacement during constant-velocity portion of stroke divided by 10, the number of angular subdivisions in the 100-degree, constant-speed zone of the cam.

As will be seen from the time-velocity diagram, Fig. 7, the horizontal velocity curve from points 27 to 0 is typical of a normal crank motion. From points 0 to 4 the velocity increases until the required value is attained at point 4, and from there to point 14, is constant. Thus, the horizontal velocity of the yoke and frame will be constant through 100 degrees of crank rotation.

Operation of the device is as follows: When the crankpin passes point 27 on the cam, frame *F* will be moving to the left at maximum velocity. At this instant that portion of the web between rolls *B* and *D* will be moving to the right at a velocity much higher than that of the balance of the web. Then, as the crankpin reaches point 0, the frame will have zero velocity, and the entire web will travel at the same velocity.

From point 0 to point 4 the crank radius will be decreasing. Hence the speed of the web between *B* and *D* will be modified until, at point 4, the crankpin will have reached a radius that will give the frame a horizontal velocity equal to one-half of the oncoming web speed. At this point, that portion of the web over table *C* will be traveling at zero velocity and will continue to do so until the crankpin arrives at point 14. During this interval the required operation may be performed on the stationary web which is then supported by table *C*.

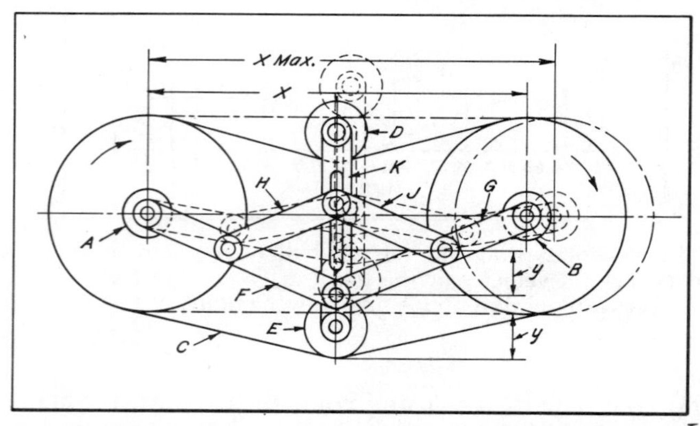

FIG. 8. Idler pulleys D and E, guided by a pantograph linkage mechanism, maintain uniform tension on steel band C when the center distance between shafts A and B is varied.

Transmitting Uniform Speed Between Shafts Having Variable Centers

Driven shafts can be rotated at uniform speeds regardless of variations in the distance from their driving shafts by means of the simple pantograph linkage mechanism seen in Fig. 8. The mechanism maintains uniform tension on a drive belt between shafts having a variable center distance. While a number of plane link mechanisms utilizing as many as eighteen joints and twelve members have been devised for this purpose, the device here described requires only six joints.

Driving shaft A and driven shaft B, having a variable center distance X, are provided with flat-belt pulleys connected by a steel band C. To provide a uniform tension on this band — independent of any changes in the center distance — the drive is equipped with two idler pulleys, D and E, which are guided by a pantograph linkage consisting of levers F, G, H, and J. The long levers F and G are free to pivot about pins pressed into the ends of shafts A and B. Slotted bar K, carrying pulleys E and D, is guided in a direction perpendicular to the common center line of the shafts when the center distance is varied. The relative positions of the mechanism components when shaft B is at its maximum distance from shaft A are shown by broken lines.

CHAPTER 14

Feed Regulating, Shifting, and Stopping Mechanisms

In all machines which perform operations on parts or on material, means must be provided for regulating, shifting and stopping the feed of either a tool or the work. Such mechanisms which provide for this are described here.

Other mechanisms which perform similar functions are described in Chapter 16, Volume I; Chapter 14, Volumes II and III of "Ingenious Mechanisms for Designers and Inventors."

Machine "Stops" Roll Labels Momentarily for High-Speed Die-Cutting

Repetitive operations are sometimes performed on lengths of material that are moving at high speed. Generally, the tool is allowed to move with the work, but the patented mechanism, shown in Fig. 1, momentarily "stops" the moving material long enough for a stationary tool to function. The work, however, passes through the device at a constant, high speed.

The mechanism was designed for die-cutting labels previously printed on rolls of paper stock called "web." Six rows of labels, printed on the web, are cut simultaneously by the die. Enough material is left between the labels within the rows so that the labels can be removed from the machine in rolls. Thin strippings which are left between the rows of labels are separated from them and diverted downward by a stripping mechanism (not shown).

A roll of printed label web *A*, subsequent to being placed on spindle *B* of the machine, is unwound a few turns. The loose

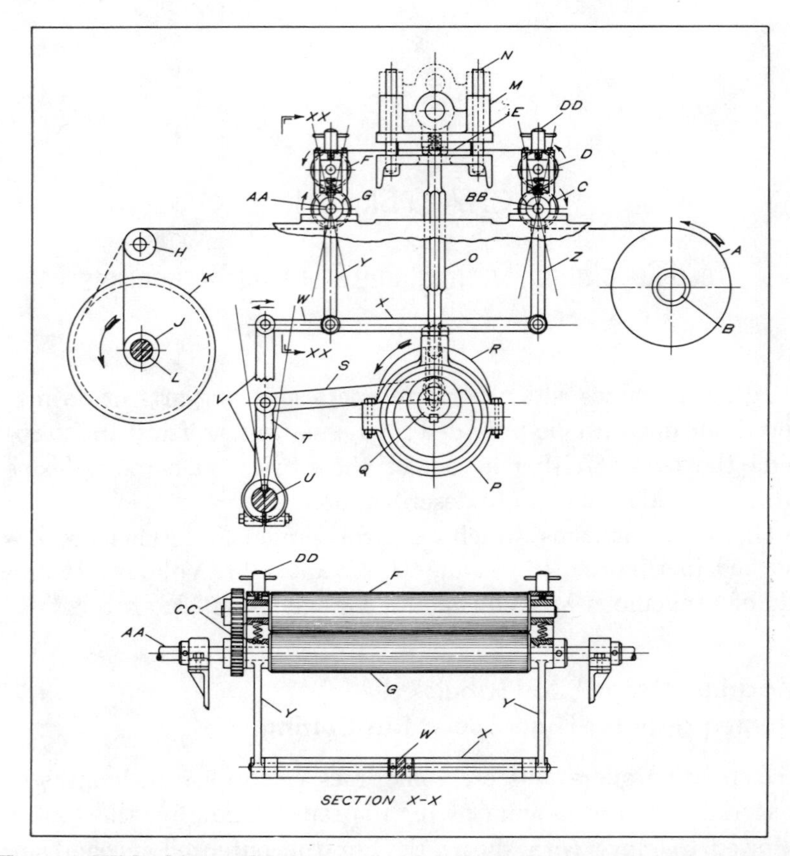

FIG. 1. With this mechanism, printed roll labels are momentarily held stationary and die-cut. The label web enters and leaves the device at a constant high speed.

end of the web is then threaded around rollers *C* and *D*, over die-plate *E*, around rollers *F* and *G*, over roller *H*, and onto cores *J*. Web is built up on the core by a drive through friction discs *K* which are held together under the pressure of light springs (not shown). This allows the core to rotate at a variable speed while taking up the web at a constant rate. Spindle *L* and the friction discs are driven by a separate small motor which runs constantly to keep the labels taut and thus prevent them from being torn by the stripping mechanism.

The upper member *M* of the cutting die reciprocates vertically on guide posts *N*, which are secured in the fixed lower die member. Member *M* is driven by a pair of connecting-rods *O*, eccentrics *P*, and a shaft *Q*. In addition, shaft *Q*, by means of a third eccentric drive and connected linkages, reciprocates rollers *D* and *F* in a short arc.

Connected linkages include driving disc *R*, connecting-rod *S*, rocker shaft *U*, lever *V*, link *W*, frame *X*, and two pairs of parallel levers *Y* and *Z*. Driving disc *R* has a radial T-slot for adjusting the length of stroke of connecting-rod *S* by repositioning its pivot pin. Levers *Y* and *Z* pivot on shafts *AA* and *BB*, respectively, but are always parallel.

Each pair of rollers is driven in opposite directions by means of identical gears *CC* and roller shafts *AA* and *BB*. These shafts are rotated at the same speed in the same direction by a roller chain drive. Pillow blocks mounted on the machine frame support the shafts.

In operation, the web is driven at a continuous speed by the rollers, but the rocking action of rollers *D* and *F* varies the absolute motion of the section of paper stock located between these rollers and under the die. In order to "stop" the work momentarily for the die-cutting operation, the backward motion of the rollers must be made approximately equal to the forward motion of the web relative to the rollers. This is accomplished by varying the rotational speed of the rollers so that they feed the proper length of web forward for the cutting die. To eliminate any strains in the web the stroke of rollers *D* and *F* is then reset by adjusting the stroke of connecting-rod *S*. On the forward stroke, the absolute speed of the web will be accelerated and will average out to the rate at which it is fed over the rollers.

The eccentrics are timed so that the die cuts the label as the web is "stopped" on the die-plate. In the illustration the mechanism is shown in this position. As the labels are made in various lengths, the speed of the web over rollers and the stroke of rollers *D* and *F* must be adjusted for each size.

Table Feed Mechanism Designed to Eliminate Manual Re-Engagement

On a special grinding machine, the work-table was driven by a screw geared to the power source. At the termination of the

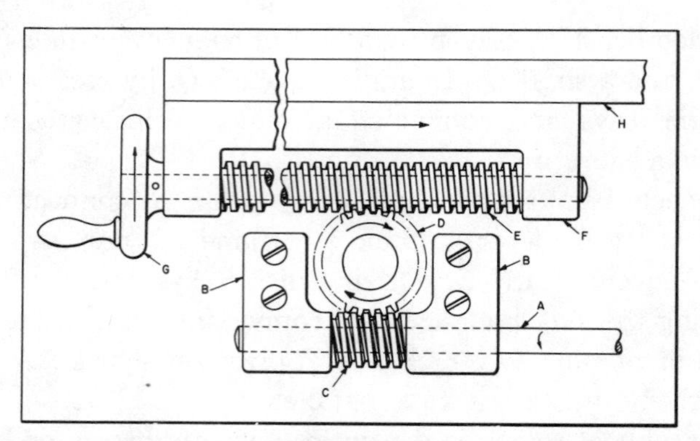

FIG. 2. Table feed mechanism that eliminates manual re-engagement of the clutch and feed-screw.

working cycle, a clutch dog was disengaged to permit the operator to return the table manually by means of a handwheel on the feed-screw. As the table movement was fairly slow, it was frequently impossible for the operator to engage the clutch immediately upon the completion of the loading cycle because the mating teeth were not in position for engagement. A period of several seconds was lost many times an hour in this way, making it advisable to change the feed mechanism to obtain an increase in production.

The design shown in Fig. 2 provided the desired results. A drive-shaft *A*, which rotates in the direction indicated by the arrow and has a worm *C* mounted on it, is supported by two bearings *B* attached to a stationary part of the machine. Worm *C* meshes with a worm-gear *D*, which rotates freely on its supporting stud. The worm-gear also meshes with the screw *E* on the opposite side, screw *E* being supported by member *F* attached to the table *H*.

During the working cycle, shaft *A* transmits rotary motion to the worm-gear *D*, in the direction indicated by the arrows, through the worm *C*. This provides linear motion to table *H*, since screw *E* does not rotate, but acts as a rack. On completion of the working cycle, the screw *E* is rotated by handwheel *G*, in order to move table *H* in the opposite direction. While this is taking place, worm-gear *D* continues to rotate.

While being loaded, the table slowly moves toward the working position, as it is still connected with the drive-shaft through the worm-gearing. Should the loading be completed before the table *H* has reached the working position, the handwheel *G* can be turned in a direction opposite to that in which it was turned previously to accelerate the table movement.

As none of the parts are disengaged at any time, there is no waiting period, resulting in a considerably shortened cycle.

Fine Feed Arrangement for a Surface Grinder

A patented mechanism by which a fine feed can be given to the wheel-head of a vertical spindle surface grinder is shown in the accompanying illustration. Coarse adjustment of the wheel-head slide *A* in relation to the column *B* is effected by rotation of a handwheel attached to the upper end of screw *C*, see Fig. 3.

When the fine feed is to be applied, screw *C* is prevented from rotating by tightening the internally threaded cup-shaped member *D* on the threaded boss that is integral with the column *B*. This is accomplished by means of the attached lever. The

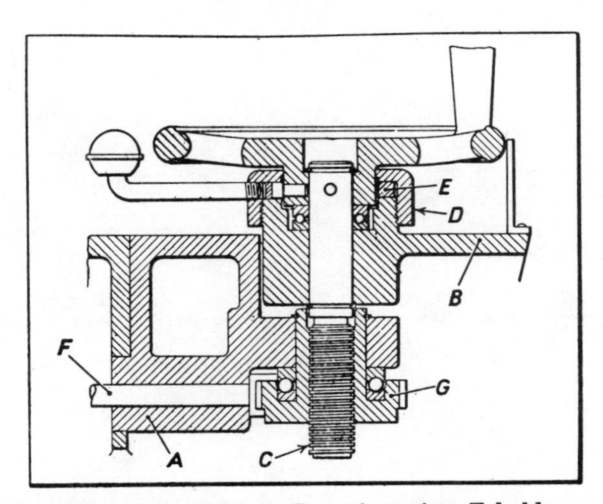

FIG. 3. Tightening threaded cup *D* against ring *E* holds screw *C* in a fixed position. Fine adjustment is then accomplished through the rotation of nut *G* by means of a worm drive on shaft *F*.

action causes ring E, which is keyed to the handwheel, to be clamped between the face of the boss and the bottom of the bore in cup D.

Fine vertical adjustment of the slide A is then effected by manual rotation of shaft F. A worm attached to this shaft drives a worm-wheel integral with nut G which moves along screw C.

An Intermittent Variable-Speed Movement

The device shown in Fig. 4 is used to feed strands of wire at a varying rate of speed through a portion of a machine that produces a woven wire product. A complete feed cycle consisting of a period of movement and an equal period of rest is accomplished by a ratchet and pawl arranged in combination with a pair of levers. The interesting feature of the mechanism is the method of providing the variable-speed motion during the feeding portion of the cycle.

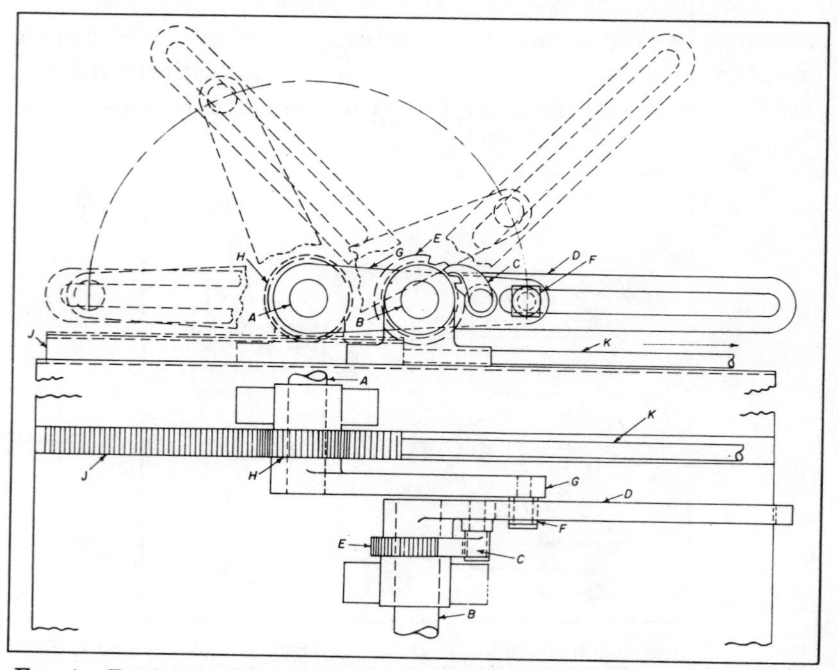

FIG. 4. Device used to convert a reciprocating motion into one that is intermittent and of variable speed.

Shafts *A* and *B* are both free to rotate in bearings attached to the machine. Pawl *C* is mounted on a lever *D* which, in turn, is pivoted on shaft *B*. A spring (not shown) holds the pawl in engagement with a ratchet wheel *E* keyed to shaft *B*. In addition, lever *D* is slotted to receive a slide block *F*. This block, in turn, pivots on a stud secured to the lower end of a lever *G* keyed to shaft *A*. A gear *H*, also keyed to shaft *A*, is constantly in mesh with a rack *J*, which is fitted into a groove in the machine table for guiding during its reciprocating motion.

In operation, rod *K* extending from rack *J* is given a uniform reciprocating motion by another part of the machine. As seen in Fig. 4, the assembly is at the end of the rest period of the cycle and rack *J* is about to be moved to the right. This action causes gear *H* and lever *G* to rotate counterclockwise, and lever *G*, through its slide-block and stud, transmits motion to rotate lever *D* in the same direction. Pawl *C* then engages the ratchet wheel *E* and causes shaft *B* also to rotate in the same counterclockwise direction as levers *D* and *G*.

The levers are shown dotted at three positions in their movement. Since they rotate on different axes, there is a continual change in their relative angular positions. This causes slide-block *F* to move toward the outer end of lever *D*, thus increasing the length of the effective lever arm. The movement of lever *G* is uniform throughout the cycle, and therefore, the slide-block transmits a continuously decelerating movement to shaft *B* until both levers reach the extreme left, where they are in a position of alignment. The rest portion of the cycle is accomplished during the return stroke of rod *K* by action of the ratchet and pawl arrangement.

Piloted Feed Control Mechanism

Auxiliary tooling for a copying lathe may be carried on an automatic overhead slide. A hydraulically operated mechanism for the independent control of the vertical feed of the slide is shown in Figs. 5 and 6.

Three basic units comprise the complete feed mechanism: a high-pressure hydraulic system to provide the necessary thrust

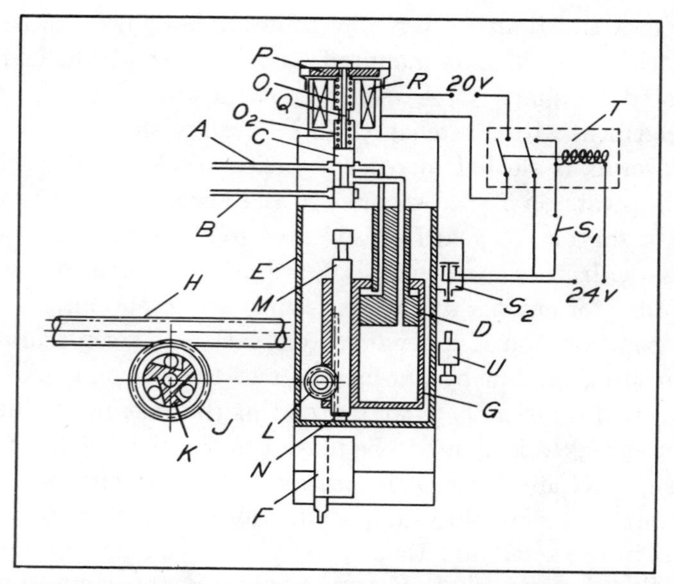

FIG. 5. Mechanism for controlling the feed of a hydraulically operated overhead slide used on a copying lathe.

for rapid approach, working feed, and rapid withdrawal; a lightly loaded mechanical unit to insure precise feed control; and an electrical system to afford positive control over the entire device. Pressure in the hydraulic system is built up by a motor-driven pump that is mounted in a support, forming the hydraulic reservoir, located beneath the lathe headstock. Fluid under pressure enters the slide through line *A*, Fig. 5, and returns to the reservoir through line *B*. Fluid flow from these lines to the slide is controlled by a double-acting control valve *C*.

Differential piston *D* is attached to moving slide *E* so that tool-holder *F* and the piston will move in unison. Cylinder block *G*, in which the piston rides, is mounted to the frame of the slide unit.

Pilot lead-screw *H*, the restrictive component of the feed control unit, is driven from the lathe spindle through a separate gear-box, providing it with a selection of eleven feeds. Located within the slide support, and meshing with the lead-screw, is tangential gear *J*. This gear fits over a roller type clutch-wheel

K. As the lead-screw rotates gear *J* in a clockwise direction, the rollers in the clutch disengage, providing a free-wheeling condition. It should be noted that the gear and clutch assembly is shown outside of the slide unit for clarity.

Pinion *L*, mounted on the same shaft as clutch-wheel *K*, meshes with vertical rack *M*. The rack is situated along the same axis as is control valve *C*. When the rack is pushed downward, pinion *L* rotates. As a result, the clutch-wheel rotates in the same direction as, but faster than, gear *J*. This causes the clutch rollers to engage, thus restricting the speed of the descending rack to the selected speed of lead-screw *H* as long as pressure is maintained on the rack. The lower end of the rack is positioned by stop *N* which is integral with slide *E*.

When the unit is inactive, spring O_1 forces plate *P*, which is free to slide on headed valve-stem *Q*, against the cover of solenoid *R*. This holds the control valve in a raised position so that the feed-back orifice is closed as shown at X in Fig. 6. In this position the hydraulic fluid under pressure is channeled to both the small chamber above the piston and to the large chamber below the piston. Although the pressures in both chambers are equal, a larger piston-face area is exposed in the lower chamber so that the total force pushing upward is approximately twice

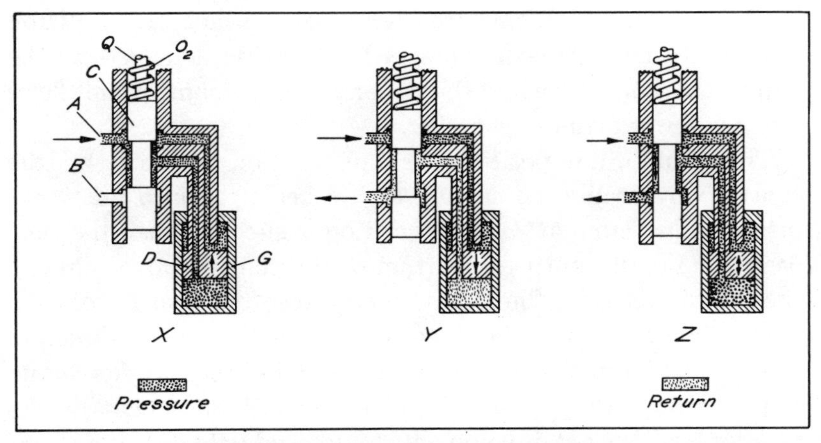

FIG. 6. Positioning of valve *C* results in either of three movements — rapid approach, rapid withdrawal, or working feed.

that pushing downward. Therefore, the piston, together with the slide, is maintained in a raised position.

With the machine in operation, movement of the overhead slide is initiated either by rotation of the template carrier or by the arrival of the lathe saddle at a chosen longitudinal position. Upon the closing of the switch S_1, relay T is closed, and remains so even when the switch reopens. This energizes solenoid R which attracts plate P to it, compressing spring O_1 and releasing the valve-stem.

Under the influence of spring O_2, the control valve is now forced downward to its lowest position as shown at Y, closing the connecting passage between the two cylinder chambers and opening the return line to the reservoir. Oil delivered by the pump is now directed only to the small chamber above the piston. The piston is thus forced downward imparting a rapid approach to the cutting tool as, at the same time, the oil leaves the large chamber below the piston and flows into return line B.

During ascent of the piston, or rapid withdrawal of the cutting tool, the control valve is in the same position as it is when the unit is inactive. This is the raised position that may be seen by referring back to X, in which the connecting passage between the two cylinder chambers is opened, and the feed-back orifice is blocked. The pressure in each being equal, the greater total force exerted against the large bottom face of the piston forces it to rise at maximum speed. Oil being delivered by the pump joins the oil leaving the upper cylinder chamber and flows into the lower cylinder chamber.

The equilibrium position, or the position assumed by the control valve while the cutting tool is being fed into the work-piece, is illustrated at Z. This position is effected when the slide descends rapidly until the bottom of the control valve contacts the top of rack M. The rack is then forced downward, rotating pinion L and causing clutch-wheel K to rotate in the direction shown. The rotative speed of the clutch-wheel results in its engagement with gear J. Rapid downward movement of the rack is thus checked, it being able to descend only as fast as lead-screw H, through gear J, will permit. This speed is selected by the lathe operator.

As the speed of the rack is reduced, the control valve is pushed upward allowing the feed-back orifice to close. The hydraulic fluid, being now diverted to both sides of the piston, forces the slide in the opposite direction. Because spring O_2 constantly tends to push the control valve downward to effect a rapid approach, a series of valve movements occur until the opposing pressures on the piston are stabilized.

In this position, oil enters the upper annular position of the valve housing and passes immediately to the small chamber. The valve is positioned so as to leave a bleed opening in each of the two annular spaces. Oil bleeds through the first opening to the lower cylinder chamber causing the cutting tool to raise slightly. The oil then bleeds through the second opening and returns to the reservoir, causing the cutting tool to lower slightly. All excess oil supplied by the hydraulic pump by-passes the valve and returns to the reservoir.

At the end of the cutting stroke, microswitch S_2 is actuated by adjustable stop U. When the circuit is broken at this point, relay T is opened with the result that solenoid R is de-energized. Control valve C is pulled upward by the action of spring O_1 against plate P, and a rapid withdrawal is effected. As the slide is raised, it causes the rack to travel with it. The unit is held in this raised position until again activated. If necessary, the solenoid cover can be adjusted to limit valve displacement, thereby varying the speed of withdrawal.

Pi-Ratio Universal Rack-Indexing Attachment

Racks of different pitch can be cut on a milling machine equipped with the indexing attachment shown in Fig. 7, without any change of gears being required. One particular two-gear combination and one or more commercially available index-plates can be used to accurately index a milling machine table for cutting racks in all the commonly employed diametral pitches. Although this gear set is for use on machine tables having a feed-screw with a ¼-inch lead, effective gear arrangements may be set up for other leads.

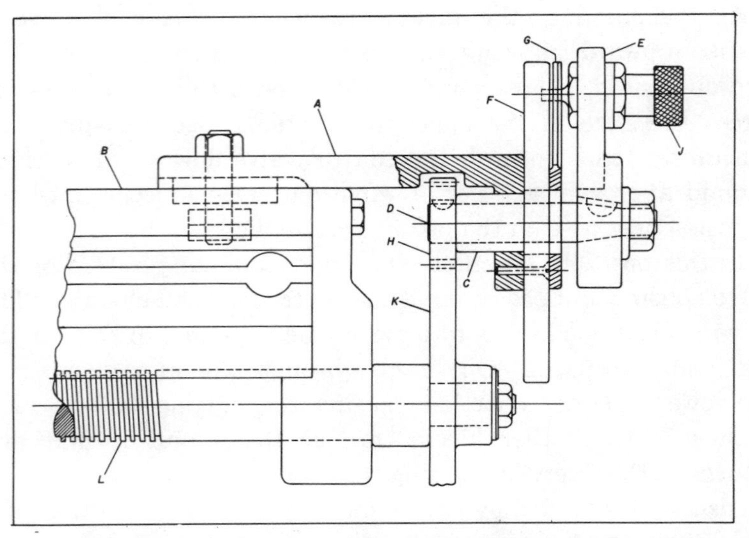

FIG. 7. This attachment for the milling machine facilitates rack-cutting. The same gears can be used to produce racks in all commonly used diametral pitches.

The linear pitch of a rack is equal to π (3.1416 inches) divided by the diametral pitch. Consequently, the number of teeth in 3.1416 inches of rack will be equal to its diametral pitch. A gear set chosen for the indexing attachment must be able to move the milling machine table 3.1416 inches with a number of turns of the crank-handle that can be readily subdivided by the diametral pitch. An ideal gear set is one that can be used with a small selection of index-plates to index the table the exact amount for a rack of any standard diametral pitch.

The ideal condition can be obtained on machines having ¼-inch lead feed-screws with a 71- and 113-tooth gear used in combination. Ideal arrangements or close approximations may be set up for other leads with two- or four-gear combinations.

Construction of the attachment as set up for two-gear operation is illustrated. Bracket *A*, which is keyed and bolted to the machine table *B*, supports bushing *C*, shaft *D*, crank *E*, index-plate *F*, sectors *G*, and gear *H*. A spring-loaded plunger *J* for

Index Settings for the Pi-Ratio Universal Rack-Indexing Attachment Based on a Commercially Available Index-Plate

Diametral Pitch	Number of Complete Index Turns	Fraction of Turn to be Indexed	Number of Index Holes in Circle	Number of Index Holes for Setting	Diametral Pitch	Number of Complete Index Turns	Fraction of Turn to be Indexed	Number of Index Holes in Circle	Number of Index Holes for Setting
128	0	5/32	96	15	8 1/2	2	6/17	68	24
120	0	1/6	54	9	8	2	1/2	66	33
96	0	5/24	72	15	7 1/2	2	2/3	54	36
80	0	1/4	72	18	7	2	6/7	84	72
72	0	5/18	54	15	6 1/2	3	1/13	78	6
64	0	5/16	96	30	6	3	1/3	54	18
56	0	5/14	84	30	5 1/2	3	7/11	66	42
48	0	5/12	72	30	5	4	...	any	0
44	0	5/11	66	30	4 1/2	4	4/9	54	36
40	0	1/2	66	33	4	5	...	any	0
36	0	5/9	54	30	3 1/2	5	5/7	84	60
32	0	5/8	72	45	3	6	2/3	54	36
28	0	5/7	84	60	2 3/4	7	3/11	66	18
24	0	5/6	54	45	2 1/2	8	...	any	0
22	0	10/11	66	60	2 1/4	8	8/9	54	48
20	1	...	any	0	2	10	...	any	0
19 1/2	1	1/39	78	2	1 7/8	10	2/3	54	36
19	1	1/19	76	4	1 3/4	11	3/7	84	36
18	1	1/9	54	6	1 5/8	12	4/13	78	24
17 1/2	1	1/7	84	12	1 1/2	13	1/3	54	18
17	1	3/17	68	12	1 7/16	13	21/23	92	84
16 1/2	1	7/33	66	14	1 3/8	14	6/11	66	36
16	1	1/4	72	18	1 5/16	15	5/21	84	20
15	1	1/3	54	18	1 1/4	16	...	any	0
14 1/2	1	11/29	58	22	1 3/16	16	16/19	76	64
14	1	3/7	84	36	1 1/8	17	7/9	54	42
13 1/2	1	13/27	54	26	1 1/16	18	14/17	68	56
13	1	7/13	78	42	1	20	...	any	0
12 1/2	1	3/5	60	36	15/16	21	1/3	54	18
12	1	2/3	54	36	7/8	22	6/7	84	72
11 1/2	1	17/23	92	68	13/16	24	8/13	78	48
11	1	9/11	66	54	3/4	26	2/3	54	36
10 1/2	1	19/21	84	76	11/16	29	1/11	66	6
10	2	...	any	0	5/8	32	...	any	0
9 1/2	2	2/19	76	8	1/2	40	...	any	0
9	2	2/9	54	12					

indexing is mounted on the crank. Gear K is keyed to the feed-screw L of the milling machine table.

If the milling machine screw has an 0.250-inch lead, then 4 times 3.1416, or 12.5664 turns, will be required to move the table 3.1416 inches. An easily subdivided number of turns of the crank should be used to produce this table movement. Twenty revolutions of the crank are required to move the machine table 3.1416 inches when the 71- and 113-tooth gears are used, and the accompanying table shows how commonly used pitches are indexed. This ideal combination will theoretically move the machine table 3.141593 inches or π inches to six places with 20 turns of the crank, as 20 turns times 71/113 gear ratio times ¼-inch lead of feed-screw equals 3.141593 inches of table movement. The 71-tooth gear should be mounted on the crank-shaft D and the 113-tooth gear on the feed-screw L. It should be emphasized that with this arrangement these gears will not have to be changed to produce racks in any of the commonly used diametral pitches. Furthermore, racks based in design on the metric module system may be indexed using only a 127-hole circle.

CHAPTER 15

Automatic Work Feeding and Transfer Mechanisms

This chapter deals with the automatic delivery of workpieces in the proper position for the operation to be performed on them. Other automatic feeding mechanisms are described in Chapter 16, Volume I; Chapter 14, Volume II; and Chapter 15, Volume III of "Ingenious Mechanisms for Designers and Inventors."

Escapement Mechanism Feeds Rods of Various Diameters

Round bar stock of random diameters can be fed one at a time, regardless of the differences in diameter of adjacent bars, by a battery of identical escapement mechanisms that operate from a common drive-shaft. The design and operation of this device are shown in Fig. 1.

The rods are loaded on a feed-table consisting of parallel steel strips *A*, as can be seen in the plan view at V. A table slope of ½ inch per foot tends to make the rods roll. When the escapement mechanism is in neutral position, as shown in view W, the rods are restrained from rolling by the heel portion of feed-arm *B*. The center of the radius of the curved surface is coincident with that of square drive-shaft *C* on which all of the escapements are mounted.

To initiate the delivery cycle, shaft *C* rotates in a clockwise direction through an arc of 45 degrees, ending up in the position illustrated in view X. This permits the entire stock of rods to roll forward until the first rod strikes the long edge of feed-arm *B*. A short dwell period is provided to allow all the bars to complete their forward travel. Shaft *C* then moves in a

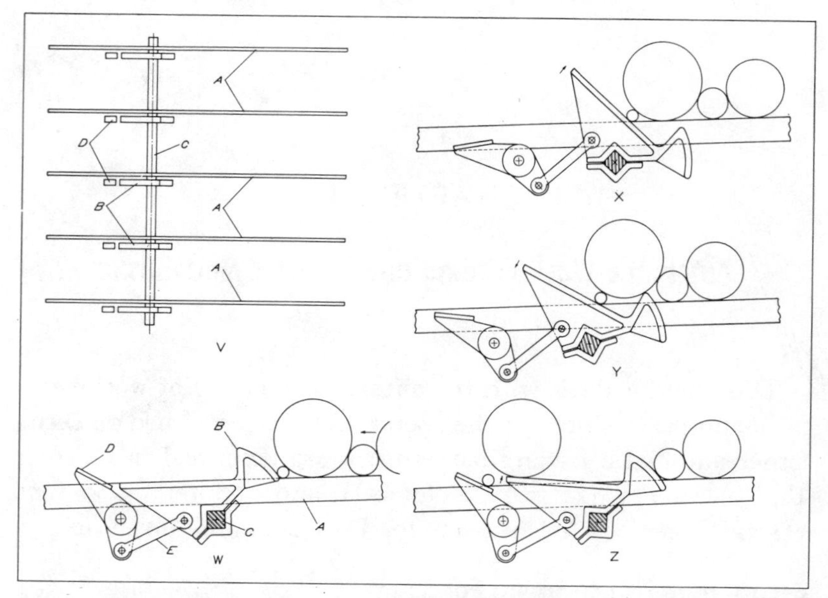

FIG. 1. Escapement mechanism permits feeding of round bar stock of assorted diameters. Arms *B* and *D* function together to allow only one rod at a time to be released, regardless of its diameter.

counterclockwise direction, view Y, until the mechanism has returned to its original position, trapping one or more rods in the space between feed-arm *B* and stop-arm *D*.

During the next clockwise movement of shaft *C*, view Z, stop-arm *D* is lowered, permitting the first rod to roll away. At the same time, feed-arm *B* begins to rise, causing the second bar to roll backward. This is due to the spacing between the two arms which maintains the center of gravity of the second bar to the right of the end of arm *B*. Link *E* that connects arms *B* and *D* should be adjustable to facilitate alignment of all the stop-arms across the width of the feed-table.

Escapement Feeds Cylinders One at a Time Down Ramp

In mass-production plants, cylindrical parts are frequently rolled downhill in chutes from one machine location to the next. An example is the gravity handling of automotive pistons in partly finished condition. Because the force is gravity, the back-up of parts in a chute is a convenient feed magazine. The automatic releasing of one work-piece at a time to feed a machine tool is often a problem.

Figure 2 shows an air-powered escapement device for installation in a gravity feed chute for handling cylindrical parts. Its operation can easily be connected in the electrical system of a machine tool. Compressed air entering cylinder port *A* swings the cage of rollers *E* and *F* in direction *C*. Rollers *E* and *F* are freewheeling. As the roller cage swings, the work cylinder *G* will roll off to ramp *H*. At the same time roller *F* rises to hold cylinder *J* on ramp *K*.

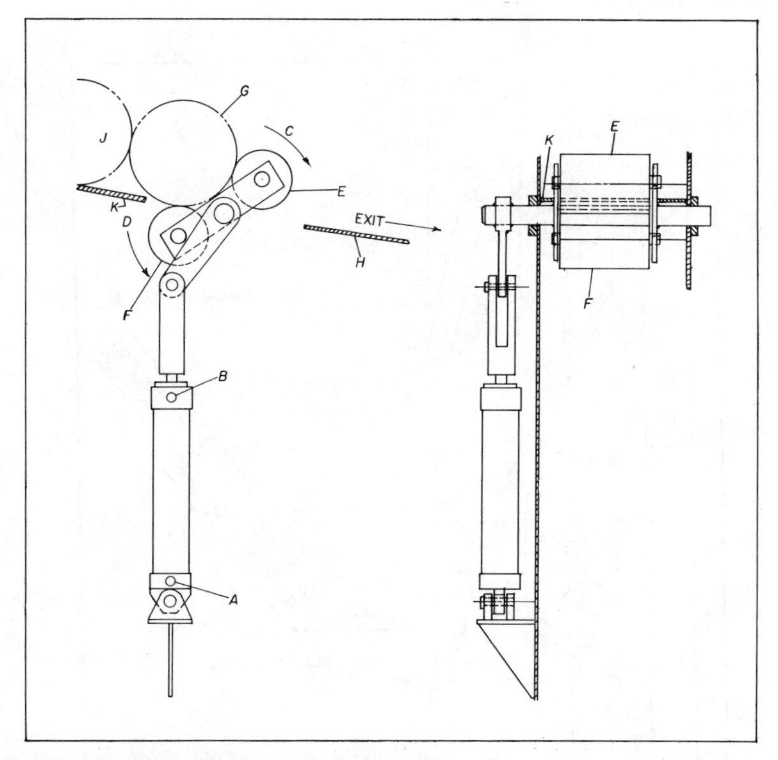

FIG. 2. Cylindrical work parts *G* from chute *K* are fed one at a time down ramp *H* by the rocking of roller cage *E-F*. The air cylinder has a clevis fastening below port *A* permitting it to swivel. Flexible air hoses lead to ports *A* and *B*.

For the return stroke, compressed air enters port *B*, exhausting air from port *A* and swinging the roller cage back to its original position with cylinder *J* set to be ejected down the line.

Semi-Automatic Feeding Device for Small Headed Parts

For a particular application, it was necessary to form a longitudinal knurl on the shanks of small rivets. One half of the knurling die employed was mounted in a fixed position on the frame of a threading machine, while the other half was mounted on the rotary machine table. It was still necessary to provide a device for introducing a single rivet into the die at each revolution of the table.

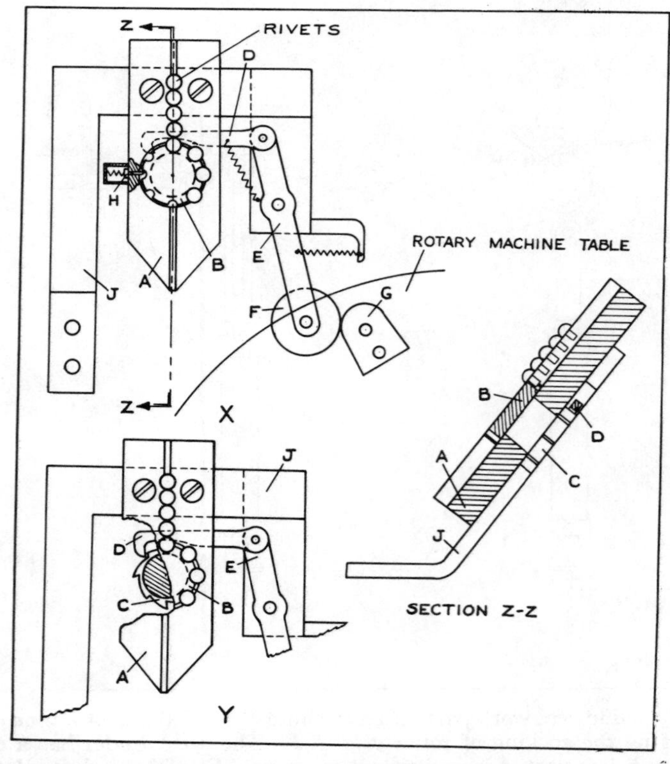

FIG. 3. Feeding device synchronizes rivet flow with rotation of the machine table.

Satisfactory operation was obtained with the feeding device shown in Fig. 3. The main member of the device is central body *A*. A channel, wide enough to accommodate the rivet shanks freely, is milled in the center of the upper surface of the member. With the center of the channel serving as one locating line, a hole is drilled through the body and counterbored to receive transfer wheel *B*, as shown at X in the illustration. Around the periphery of the transfer wheel are machined eight equally spaced slots of a size suitable for carrying the rivet shanks.

A ratchet wheel *C*, which may be seen at Y, is mounted on the underside of the transfer wheel. The remaining parts of the advancing mechanism are ratchet *D*, lever *E*, roller *F*, and actuating finger *G*. Two tension springs are included to insure proper functioning of the lever system. A spring-loaded pawl *H* restricts rotation of the transfer wheel to a clockwise direction. All of these units, with the exception of actuating finger *G*, are mounted on a welded-steel support frame *J*, which is situated at an incline of approximately 35 to 40 degrees from the horizontal. This support frame is bolted directly to the frame of the machine. Actuating finger *G* is screwed to the rotary machine table on which the moving member of the knurling die is mounted.

With the machine functioning, the operator loads the upper portion of the channel in central body *A* with the rivets to be knurled. The rivets are placed with their shanks down as shown in section Z-Z, being supported on the underside of their heads. Normally, four of the eight slots in transfer wheel *B* contain rivets. When actuating finger *G* contacts roller *F*, lever *E* pivots on pin *K*. Ratchet *D* is, in turn, pulled to the right, engaging a tooth on ratchet wheel *C* and rotating the transfer wheel one-eighth of a revolution in a clockwise direction. A rivet is thus aligned with the lower portion of the channel in the central body and, due to the force of gravity, travels downward to the knurling die. As the actuating finger passes by the roller, lever *E* and ratchet *D* are returned to their original position by means of the two tension springs.

Handling Mechanism Turns Strip in Transfer

In processing fiberboard strip for a firelighting device, an interesting materials-transfer mechanism is used. This mechanism picks up the strip as it leaves the saw table, rotates it 90 degrees so that a combustible fluid can be injected into one edge, then rotates it another 90 degrees for ejection.

In Fig. 4, several strips *X* can be seen leaving an extension *A* of the saw table. The strip is advanced manually between raised guides into the fingerlike end of the long leg of a bellcrank *B*.

The bellcrank is keyed to a stud *C* free to revolve on a rectangular slide *D*. Connecting-rod *E*, pivoting at *F*, reciprocates the slide in body casting *G*. The opposite end of the connecting-rod (not shown) is actuated by a conventional eccentric disc. The slide is T-shaped in vertical section, so that it can be retained by keeper plates *H*.

The short leg of the bellcrank forms an angle of 102 degrees with the long leg. At its end, it carries a roller *J* projecting over the front of the body. The roller operates over the upper edge of a guide plate *K* fastened to the front of the body.

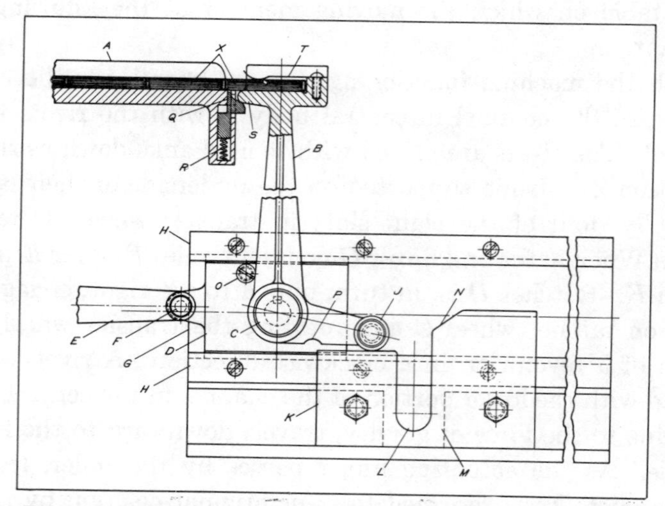

FIG. 4. When bellcrank *B* starts its swing, roller *J* rides over the lower horizontal surface of guide plate *K*.

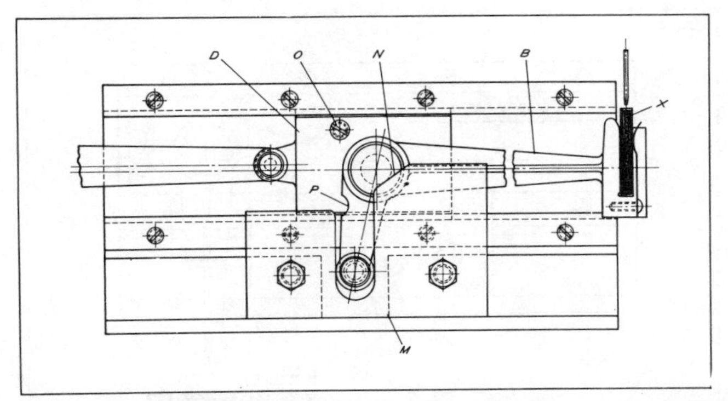

FIG. 5. A momentary dwell of the eccentric disc permits injection of combustible fluid.

When the slide moves to the right, the bellcrank carries along one of the fiber board strips, moving until roller *J* contacts the right-hand wall of slot *L* in the guide plate. The roller then is forced down in the slot, approximately 0.005 inch wider than the roller diameter. Meanwhile the bellcrank, swinging on stud *C*, rotates the strip 90 degrees, Fig. 5. A clearance channel *M* accommodates the short leg of the bellcrank.

Now there is a momentary dwell of the eccentric disc to allow the combustible fluid to be injected. Then, continued movement of the slide in the same direction raises the roller out of the slot, first onto an adjacent 40-degree incline *N*, then onto the higher straight edge of the guide plate. Simultaneously, the strip is rotated downward 90 degrees more, Fig. 6. After the strip is ejected, the slide moves to the left, and the bellcrank returns to its initial position.

Pin *O* (pressed into the slide) offers a positive stop for the bellcrank, bearing against the long leg at the start of the cycle, and against a recess *P* in the short leg when the bellcrank reaches the position shown in Fig. 6. A small, vertical slide *Q*, Fig. 4, prevents the strips remaining on the extension from being pushed off once the long leg is loaded and the bellcrank starts its swing. This slide is in a lip on the extension bottom, and a spring *R* keeps it raised, once the bellcrank swings away, so that the end of the slide slightly intersects the path of the strips.

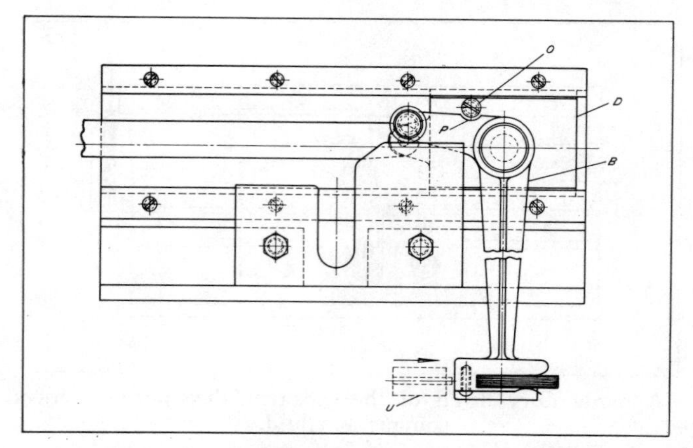

FIG. 6. Recess *P* in the short leg contacts pin *O* in rectangular slide *D*, limiting the movement of bellcrank *B*.

At the start of the cycle, a projecting surface *S* of the long leg depresses the slide, and the foremost strip is advanced into the leg. There, the strip is retained by a leaf spring *T*. To release the strip when it has reached the position shown in Fig 6, a forked ejector plate *U* is actuated by separate mechanical means at the proper instant.

Mechanism Simultaneously Transfers and Reverses Position of Pad

In the processing of paper pads stuffed with excelsior or shredded paper, they have to be transferred a distance of 36 inches between work stations. The pads, which are rectangular in shape, are picked up by a cam-tripped gripper device (not described) along a folded-over bottom edge. During the transfer, the pad has to be reversed so that the leading edge becomes the trailing edge. The diagram, Fig. 7, shows pad positions during the transfer. It will be noted that the locus of the pad center remains along the line A-A.

The transfer mechanism combines simplicity and smooth operation. Drive-shaft *B*, Fig. 8, in frame *C*, rotates at a constant speed of 40 rpm, affecting one transfer per cycle. Lever *D*, keyed to the drive-shaft, contains at its other end free pin *E*.

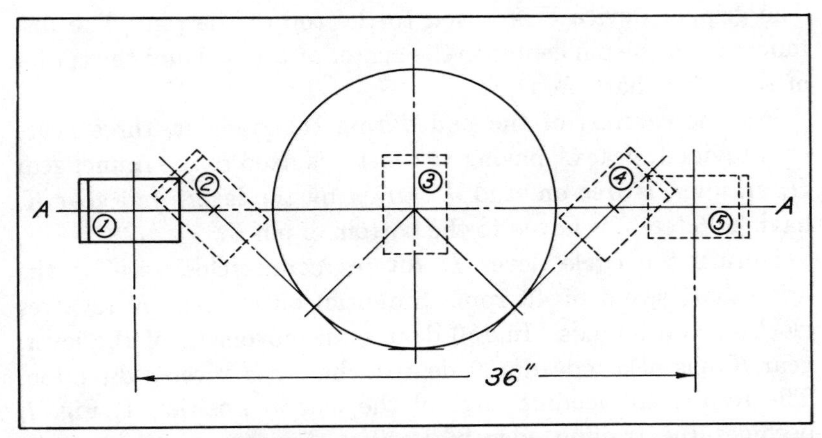

FIG. 7. The transfer mechanism reverses the pad as its center moves along line *A-A*.

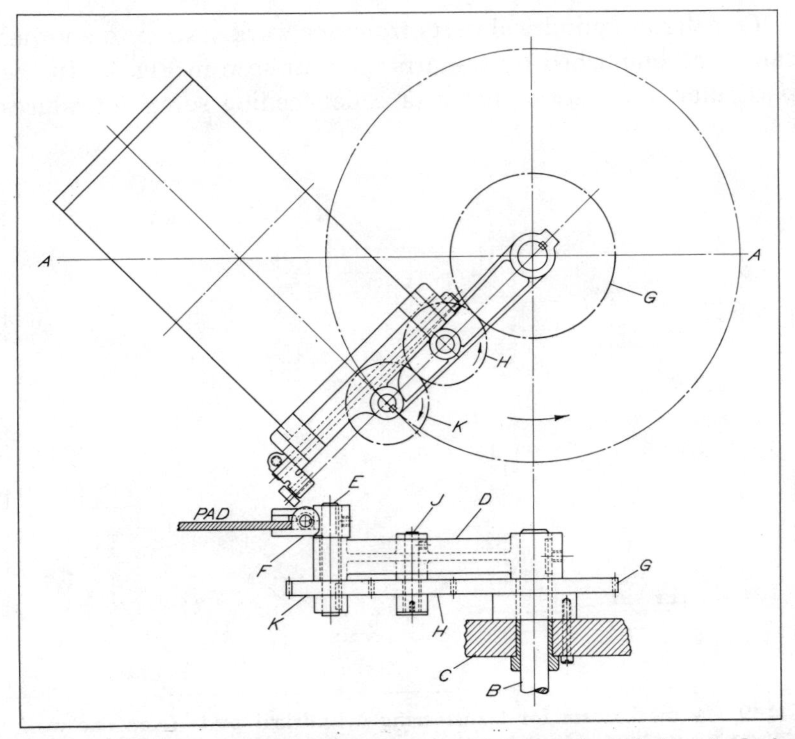

FIG. 8. The counterclockwise movement of lever *D* and the clockwise movement of gear *K* act to reverse the pad in its transfer along line *A-A*.

The gripper device *F* is keyed to the top of the pin. The distances from the pin center to the center of the pad and the center of the drive-shaft are equal.

For the reversal of the pad during the transfer, three gears are provided: gear *G*, having 192 teeth, is fixed to the frame; gear *H*, an idler, is free on stud *J* carried by the lever; and gear *K*, having 96 teeth, is keyed to the bottom of pin *E*.

During the cycle, lever *D* rotates counterclockwise at the drive-shaft speed of 40 rpm. Simultaneously, gear *K* revolves clockwise on its axis. In 180 degrees of movement of the lever, gear *K* has also moved 180 degrees, but in a reverse direction. The result: the leading edge of the pad in Position 1, Fig. 7, becomes the trailing edge in Position 5.

Transfer Device for Cylindrical Parts

Transfer of cylindrical parts from one work station to another can be accomplished by the arrangement seen in Fig. 9. In the particular application shown, a collet-feeding setup introduces

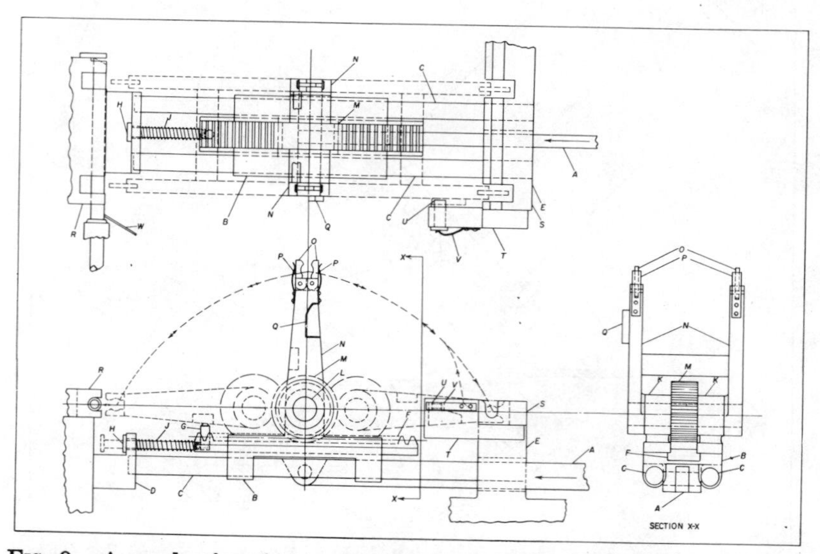

FIG. 9. A mechanism for transferring cylindrical parts from one workstation to another. Drive-bar *A* is given a variable reciprocating motion by a cam.

the work into the first station in the form of tubing. After the tubing is cut to length at the required angle by a slitting saw, the resulting work-piece is picked up by transfer arms which rotate 180 degrees and deposit it at the second station. A cam-operated drive activates the mechanism.

The transfer mechanism is illustrated at the mid-point of its cycle. Drive-bar *A*, which is given a variable reciprocating motion by a cam (not shown), transmits this movement to a slide-block *B* that slides on two rods *C*. These rods, in turn, are supported at the ends by members *D* and *E*. Slide-block *B* is grooved to accommodate a sliding rack *F*, which has a pin *G* and a headed stud *H* at one end. Stud *H* passes through a hole in a projection on member *D* and carries a spring *J*. This spring exerts pressure on the rack, thus tending to hold the head of the stud in contact with member *D*.

Bearings *K*, the bases of which retain the rack in block *B*, support a shaft *L*. Keyed to this shaft is a gear *M* that meshes with the rack. In addition, two transfer arms *N*, connected by a bar, are keyed in alignment on the shaft. Gripping jaws *O*, located on the end of each arm, are held in the closed position by springs *P*. One of the transfer arms carries a small block *Q*.

Member *R* is grooved to guide and back up the entering tubing, and a receiving block *S*, mounted on member *E*, is grooved on the top to accept the severed piece of tubing. Another block *T*, also mounted on member *E*, carries a latch-pin *U*, which is held under tension by a spring *V*.

In operation, drive-bar *A* moves to the left from the center position, carrying slide-block *B*. Since gear *M* is supported on block *B*, and rack *F* cannot move with it due to the resistance of spring *J*, gear *M* is caused to rotate counterclockwise in mesh with rack *F*. Thus, movement continues until the bar connecting the transfer arms *N* contacts pin *G*, as shown in the phantom view of the arms at the left. Pin *G* is so adjusted that arms *N* are then in the horizontal position. At this point in the cycle, there can be no relative motion between block *B* and rack *F*.

Continued movement of block *B* causes spring *J* to compress and the entire assembly to move as a unit farther to the left until the gripper jaws are forced over the tube and it is held under the tension of springs *P*. The driving cam then provides a rest period while the saw *W* cuts the required length from the tubing.

After the saw has completed its operation, bar *A* is reversed and the transfer arms *N* withdraw the work-piece horizontally from guide block *R*. This movement continues until the head of stud *H* again contacts member *D*, at which time rack *F* can no longer move to the right. But as block *B* moves farther in the same direction gear *M* rotates clockwise in mesh with rack *F* until the transfer arms have revolved 180 degrees and deposited the tube in the groove in block *S*. After the transfer arms reach the horizontal position, latch-pin *U* engages the upper side of block *Q*.

The cam then reverses drive-bar *A*, causing the section of tubing to be stripped from between the gripper jaws *O*. This movement of the transfer arms takes place horizontally because they are held in this plant by latch *U*. When the necessary horizontal travel has been completed, block *Q* has moved so that it is no longer under the latch-pin. During this motion, spring *J* is compressed by rack *F*, and in order to prevent sudden expansion of the spring, the upper edge of block *Q* can be shaped to permit a more gradual return. The transfer arms then rotate back to the central position and the cycle is repeated. A plunger (not shown) removes the part from block *S*.

Magazine Feeds Wrappers at Constant Pressure

Wrappers or labels are fed from the top of the magazine shown in Fig. 10. The edge of the wrapper is elevated by vacuum lifter *A* so that it is in position to be picked up by gripper *B* and pulled clear of the magazine.

The most interesting feature of this device is that it maintains a constant pressure between the top of stacked wrappers *C* and the underside of stop-plates *D* and *E*. This is accomplished by means of a rack *F*, pinion *G*, and scroll *H*.

The various parts will assume the positions shown when the magazine is fully loaded. Weight of the wrappers or labels rest-

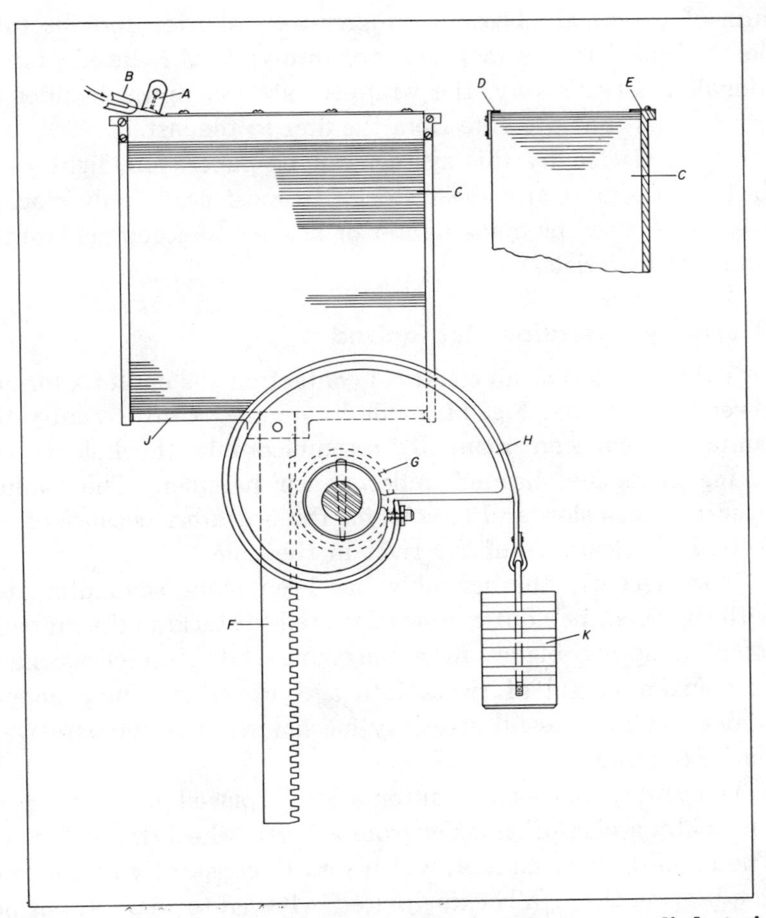

FIG. 10. Rotating scroll H alters the moment of the force applied at pinion G, thereby reducing upward pressure on the stack as the magazine empties.

ing on pressure-plate J is offset by the magnitude of weights K, taking into consideration their distance from the rotational axis of pinion G. These weights are supported by a strap passing around the outer surface of the scroll.

As the wrappers are removed from the stack and the magazine begins to empty, the weight resting on pressure-plate J diminishes. At the same time, the scroll has been rotating in a clockwise direction. Due to the contour of the scroll, the lever arm of the weight becomes progressively shorter and the total force applied to the rack and pressure-plate is reduced proportionally. In this way, the wrappers are fed upward under an almost-constant pressure from the first to the last.

Materials fed by this system can be paper, foil, light cardboard, or almost any sheet stock. In most cases, only stacked weights *K* need be made lighter or heavier to accommodate the different materials.

Assembly Operation Mechanized

A rivet is used as an electrical contact on the end of a formed sheet-brass spring, Fig. 11, made in volume. Until recently the contact rivets were manually assembled into the hole in the spring blade and headed under an air hammer. The manual operation was slow and tedious for the operators because of the 0.001-inch clearance of the rivet in the hole.

More recently the assembly has been made semiautomatic, with the operator's duties now restricted to stacking the properly oriented spring blades in a magazine slide and occasionally dumping a boxful of rivets into a Syntron vibratory hopper feeder. The successful assembly mechanism was comparatively simple to build.

The spring blades are automatically placed over the rivet shanks by a pick-off transfer from a Ferris wheel drum, Fig. 12. The main drive is shaft *A*, which rotates constantly at 100 rpm in a bearing through backing plate *C*. Keyed to shaft *A*, but be-

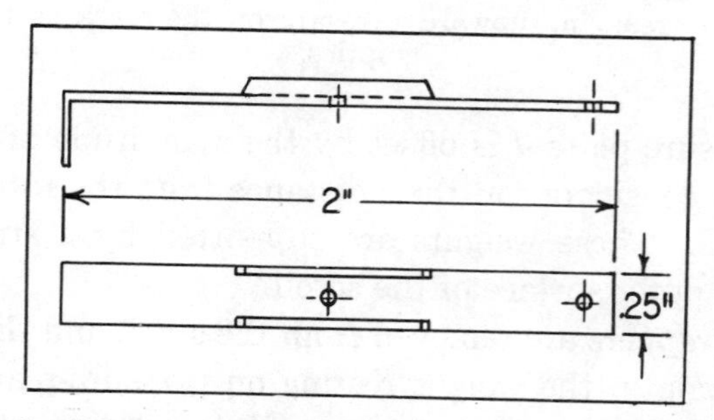

FIG. 11. Work-piece is a formed brass spring. The rivet hole is on the right, and the guide hole in the center.

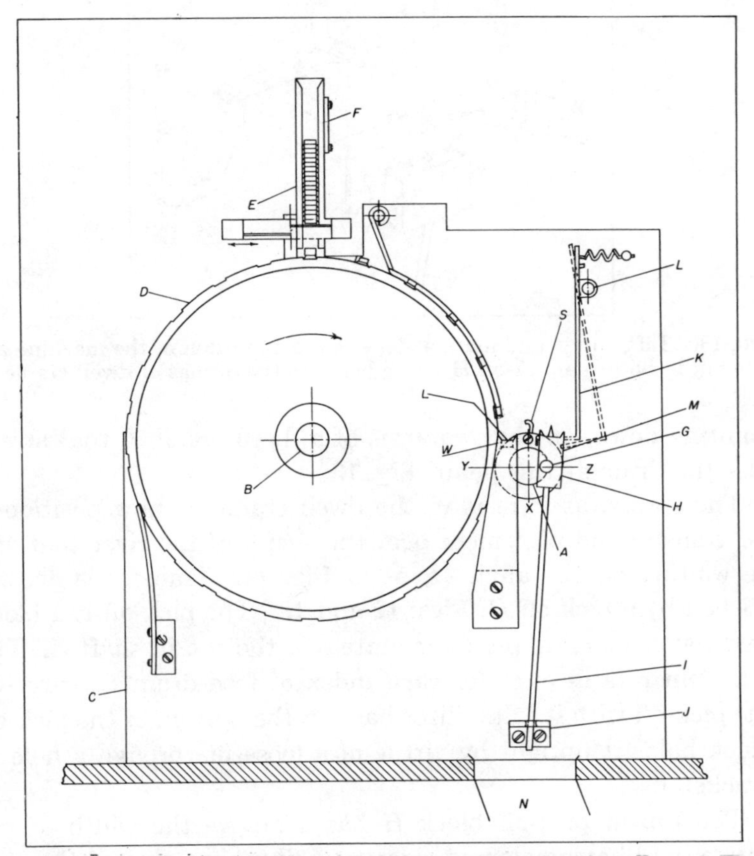

FIG. 12. Positioning and assembly drum indexes by ratchet from shaft *A*. The riveting anvil is contacted with the pick-off *H* in Z position, as shown, but is not included in this diagram.

hind plate *C*, is a cam which, through a lever and pawl linkage, drives a ratchet wheel turning shaft *B*. The rotation of drum *D* is thus intermittent, and each of its twenty-two stop indexes per rotation is aligned with the springs that drop, one at a time, from magazine *F*. The springs lie in the carrier grooves with the bent tab (Fig. 11) pointing upward.

As the drum indexes, the work is held in place by a retaining strap, or guide, that wraps around the outside of the drum. The spring rolls out of the groove, landing horizontal, with the tab

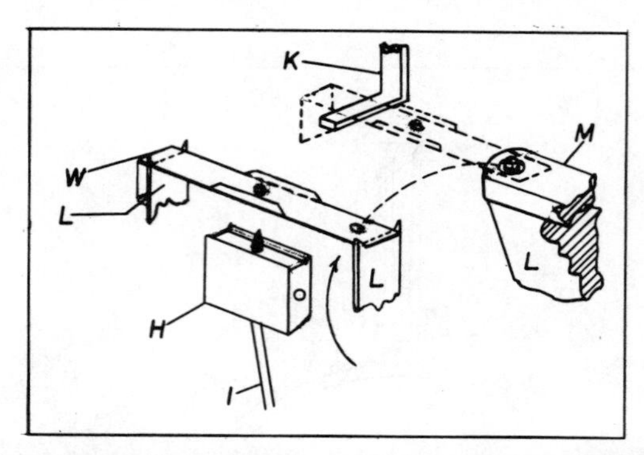

FIG. 13. Lift, carry, and mating steps in the operation of the machine are automatic, with the pick-off *H* rising between the prongs of dwell cradle *L*.

pointing down, on the two arms of a dwell cradle *L* that straddles the drum at this point, Fig. 13.

The spring, as it rests in the dwell cradle, is now positioned for transfer and placement over the shank of the rivet that will be waiting on the anvil *M*, as in Fig. 14. Transfer is accomplished by a pick-off *H*, Figs. 12 and 13. The pick-off is a block that hangs on crankpin *G*, mounted on the end of shaft *A*. The crankpin rotates once for each index of feed-drum *D*, carrying the pick-off with it. Stabilizer bar *I* in the bottom of the pick-off block holds it upright by virtue of a loose fit through a hole in bracket *J*.

The top of pick-off block *H* has a groove the width of the spring, and having sloping sides to facilitate seating of the part

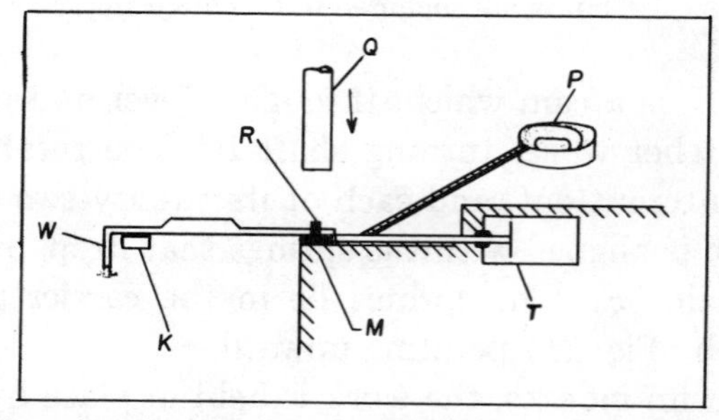

FIG. 14. Mating of spring *W* with hole finding the rivet *R* on anvil *M*. Air cylinder *T* delivers rivet from track of vibratory hopper *P*. Hammer *Q* upsets rivet shank.

by gravity. Near the center of the groove is a pointed stud that "finds" the guide hole in the spring waiting in the dwell cradle.

As the crankpin carrying the pick-off travels upward from position *X*, Fig. 12, the spring is lifted from the dwell cradle at *Y*, with the stud through the guide hole and resting in the groove of the pick-off, Fig. 13. As the crank continues to turn, the spring is carried over to riveting position *Z*, Fig. 12, where the pick-off falls away, leaving the spring with the rivet assembled through the spring's rivet hole and supported by the anvil. The opposite end of the spring sits on rest lever *K*, Fig. 13.

A rivet, meantime, has been positioned by the arrangement in Fig. 13, which also shows the riveting setup. In Fig. 14, rivet *R* from hopper *P* is pushed by air cylinder *T* into position on anvil *M*. Air hammer *Q* upsets a head on the rivet. The hammer is actuated by a micro switch from a cam on shaft *A*, Fig. 12.

As soon as the rivet has been headed, a knockoff finger *S* on shaft *A* swings past, knocking support *K* out from under the work so that it drops into the discharge chute *N*, Fig. 12.

Rotary Work-Transfer Device

An arrangement for transferring work-pieces from a horizontal conveyor to a station for vertical stacking is shown in Fig. 15. The parts, in this case papier-mache egg trays, are each conveyed by one of six pivoting carriers borne on a member rotating at constant speed. An interesting feature of the mechanism is the employment of a stationary cam to control the pivoting motion of the carriers. This device replaced a high-speed oscillating movement.

In construction, the mechanism consists of a rotatable member *A* mounted on a shaft *B* supported in fixed bearings *C*. The six carriers *D* are supported by and pivoted on shafts *E* mounted in bearings *F* secured to the periphery of member *A*. A roller follower *G*, guided by the large stationary plate-cam *H*, controls the movement of each shaft *E* by a connecting arm *J*.

Thus, as member *A* is rotated at constant speed by an appropriate drive, the carriers are each brought by the contour of the cam into the proper position for lifting the work-pieces

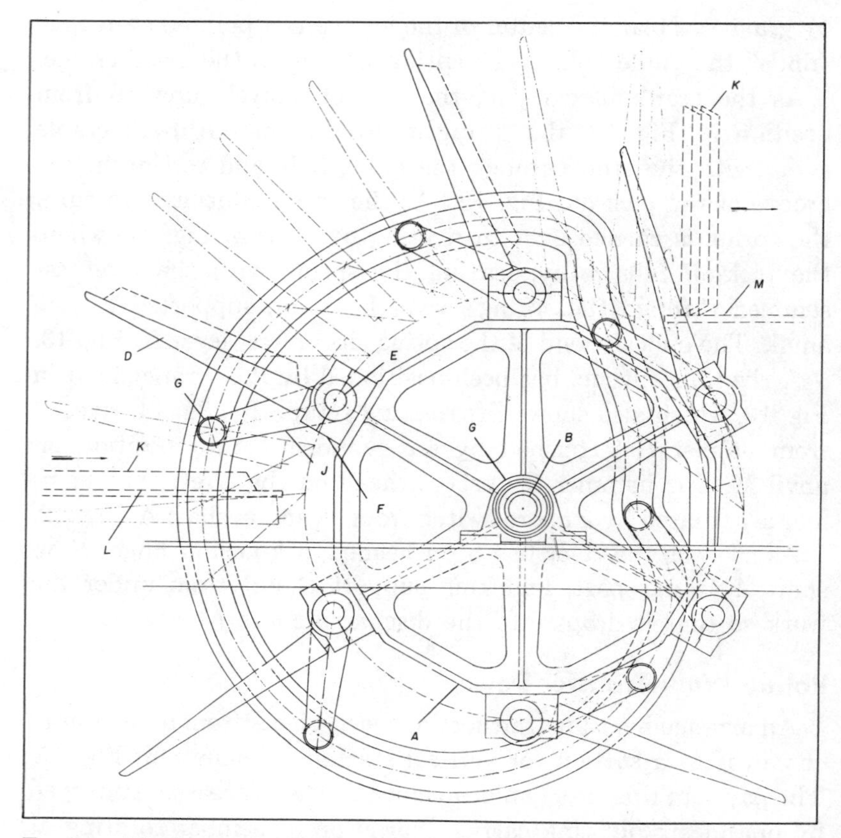

FIG. 15. Work-transfer device featuring stationary cam control of rotating carriers.

K from the conveyor station L and depositing them in a vertical position for stacking at station M. To assure smooth operation, the drive speed is adjusted to synchronize the carriers with the arrival of the work-pieces at the conveyor station L, which should be at a constant interval.

Semi-Automatic Work Feeder Improves Efficiency of Thread Roller

The manual feeding of work-pieces to reciprocating thread rollers has been hazardous both to the operator and the machine. Operators have required considerable skill plus precision

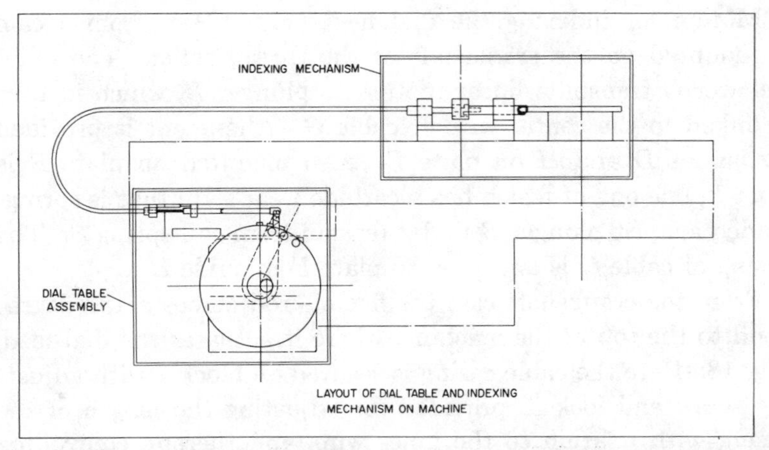

FIG. 16. Thread-rolling machine works on the principle of the automobile choke rod, with the dial-feed mechanism moved by the indexing device.

of hand motion. Careless handling of parts can damage the machine.

Using the attachment (Fig. 16) described and pictured in the illustrations, certain types of straight-shank blanks can now be fed between thread-rolling dies to make the operator's job simply a matter of dropping the work-blanks into holes provided in a dial table. The dial then carries the parts around to loading position above the dies. The construction of the thread roller made necessary the following design for an indexing drive.

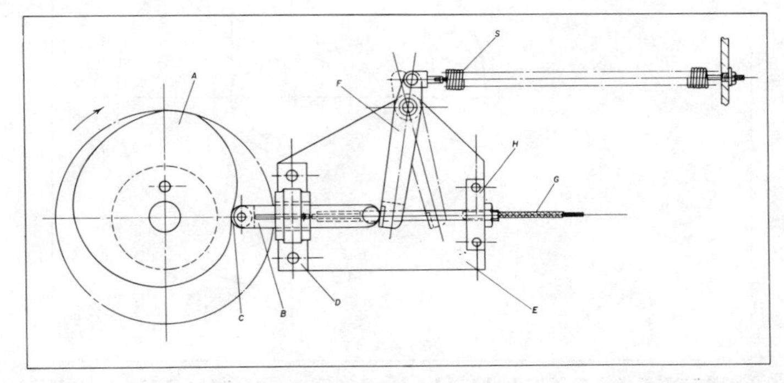

FIG. 17. The indexing mechanism for the thread-roller work feed extends spring *S* which actually drives the dial.

Motion for indexing the dial feed (Fig. 17) is from a cam *A* mounted on the crankshaft of the thread roller. The roller follower *C* transmits linear motion to plunger *B*, which in turn is linked to the center wire of cable *G*. Alignment is provided by guides *D* and *H* on plate *E*. Also mounted on plate *E* is lever *F*, one end of which has a carbide wear strip that is spring-loaded against plunger *B* under tension from coil spring *S*. The casing of cable *G* is anchored to plate *E* in guide *H*.

From the crankshaft cam the flexible rod makes a 180-degree bend to the top of the machine, where it actuates the dial feed, Fig. 18. Here the cable casing is secured on block *I* with adjusting screw and lock *J*, provided for adjusting the length of the casing with relation to the inner wire, and thereby controlling the stroke of arm *X*.

The dial-feed assembly consists of rotating dial *K*, which has thirty equally spaced holes near its outer diameter. Disc *L*,

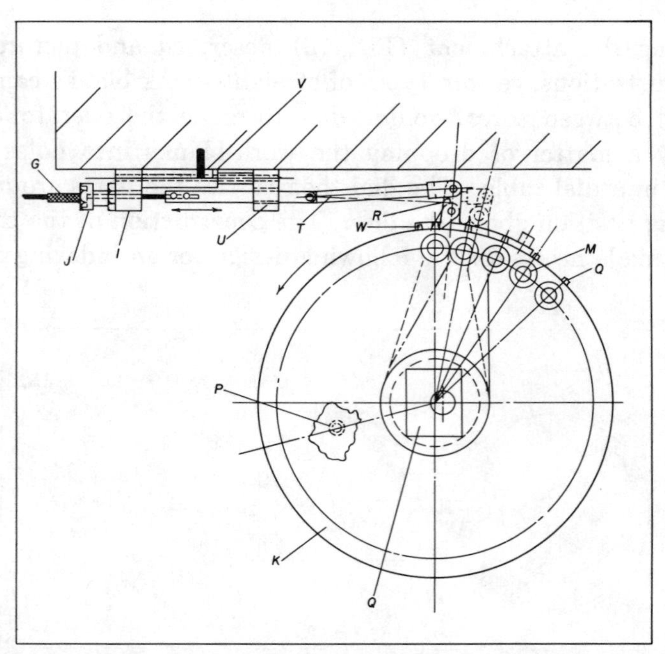

Fig. 18. The dial table feeds blanks to the dies at point *M*, with the other twenty-nine holes open for loading by the operator.

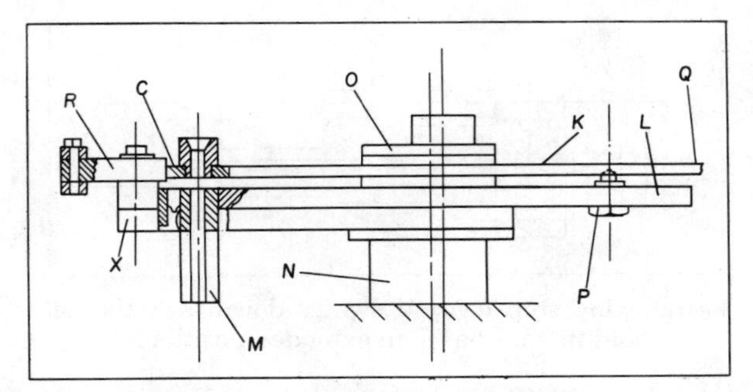

FIG. 19. Elevation view of the dial feed shows feedtube *M* in fixed disc *L* beneath the movable dial *K*.

Fig. 19, is a stationary support for dial *K* and has just one hole through which feed-tube *M* leads to the dies at the loading position.

Both dial *K* and disc *L* are mounted on column *N*, which has within it an eccentric bushing *O* for the dial shift. By turning the eccentric during setup, the position of the feed tube can be varied with relation to the dies to handle a variety of work diameters. Stop location of work-carrying dial *K* is set by an annular row of thirty spherical recesses on its underside. There is a spring-loaded spherical plunger *P* on the top side of fixed plate *L*. The positioning of the recesses is so oriented with relation to the feed tube *M* that when the plunger shoots home in any recess, one of the thirty work-carrier holes in the bottom of moving dial *K* is always aligned with feed tube *M*.

On the periphery of moving dial *K*, Fig. 18, and opposite each work-carrier hole, is a socket-head cap-screw, as at *Q*, which functions as a ratchet tooth for the indexing motion. The arm, swinging on shaft *O*, carries ratchet pawl *R* and at its free end has a linkage for the clevis and link rod *T*, which in turn fastens to the cable wire *G*. Linkage rod *T* is hinged.

A table stop is provided by lever *V*, Fig. 20, which has a slot that will accept inner cable wire *G*. When lever *V* is raised, rod *T* can reciprocate a full stroke, but when it is down on the wire its length holds rod *T* at its full stroke position against the spring *S* (Fig. 17).

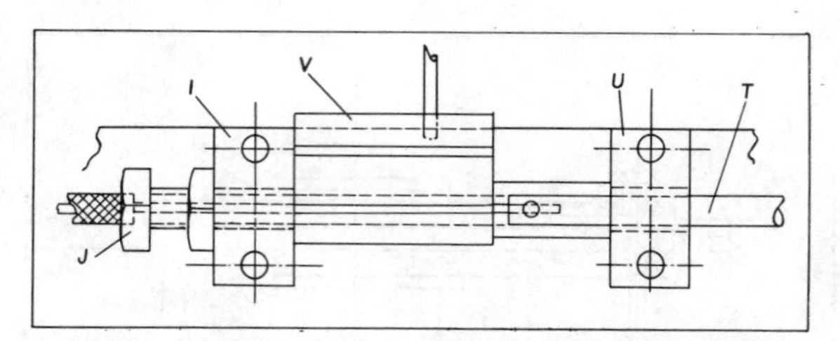

FIG. 20. The indexing stop lever V swings down over the cable wire to hold linkage bar T in extended position.

Indexing table K also has an overshoot control (Fig. 21), which is curved lever W, pivoting on the edge of fixed disc L. The free end rests on indexing arm X. As the indexing arm swings through its arc the contoured bottom of stop lever W causes the far end to rise and fall. The top of its upstroke coincides with the furthest forward motion of the index-arm X and the seating of regular stop positioner P. Thus it catches on its "hook" one of the indexing capscrews Q in a positive mechanical lock.

Straight-Line Motion Through Levers

In processing clutch housings on a transfer machine, movement of the work from station to station is accomplished by the finger type mechanism shown in Fig. 22. Features include simplicity of design, compactness, and a straight-line movement of the work.

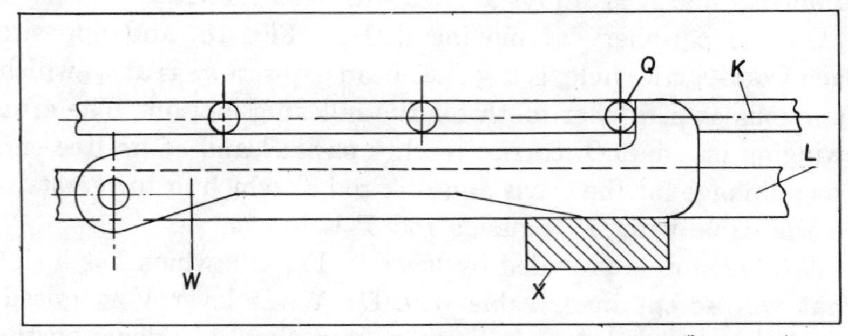

FIG. 21. Table overshoot control lever W has a hooked end that rises to stop bolt head Q when lifted by arm X.

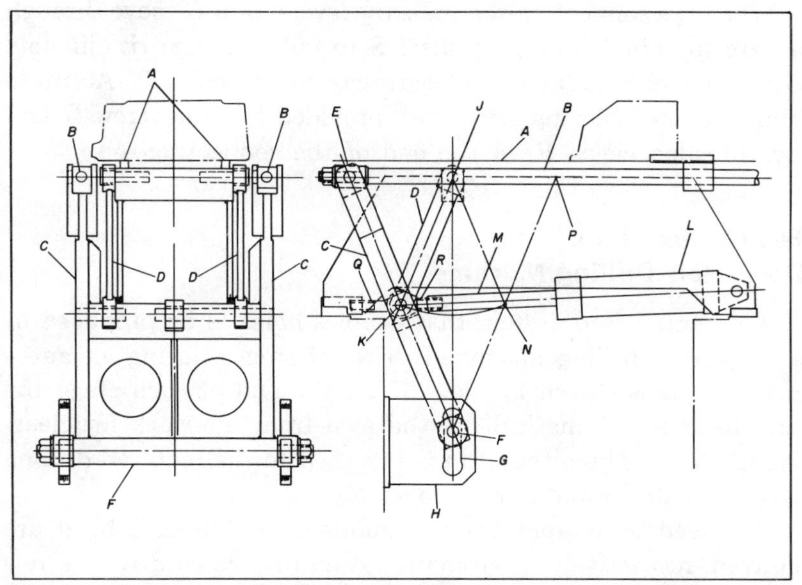

FIG. 22. While the work is indexed in a straight line along rails *A*, shaft *F*, attached to transfer levers *C*, falls and then rises in slots *G*.

Rails *A* support the housings. Along each side are a transfer bar *B*, transfer lever *C*, and link *D*. Trunnion blocks *E* at the top of the levers join them to the bars. The bottoms of the two levers are attached to a common shaft *F* which is contained at each end in a bushing in a slot *G* of a bracket *H* fixed to the frame of the device.

The tops of the two links *D* are fixed to the frame, by pins *J*, which are on the center line of slots *G*. At their bottoms, the links are hinged to the centers of the levers by pins *K*. Links are one-half as long as the levers, with the distance between *K* and *J* equal to the distance between *K* and the center of trunnion block *E*.

To operate the mechanism, air cylinder *L* powers connecting-rod *M*, attached to the levers at *K*. (The rod could be attached elsewhere along the levers, its position depending on the desired ratio of piston stroke to transfer movement.)

When the connecting-rod pulls the levers, pins *K* move through an arc to point *N*, causing shaft *F* to fall and then rise in slots *G* as transfer bars *B* travel in a straight line to point *P*. Accurate limits to the indexing stroke are provided by stop-screw *Q* and by threaded clevis *R* on the end of the connecting-rod.

Feed System for a Deep-Hole Drilling Machine

A patented feed system that enables holes to be produced in a deep-hole drilling machine in several stages during an automatic cycle is shown in Fig. 23. At the end of each stage the drill head is automatically withdrawn from the work for clearing of chips. Then, the head is returned to the required drilling position under rapid power traverse.

The feed and rapid traverse motions of the drill head are derived, respectively, from motors *A* and *B*, which drive a screw *C* through gearing. Upon completion of each drilling stage, the feed motor *A* is reversed, and, simultaneously, the rapid-traverse motor *B* is brought into operation by the action of a torque control system which is not shown. As a result, the drill head is moved away from the work under rapid traverse. At the same time, an electromagnetic type clutch *D* is brought into engage-

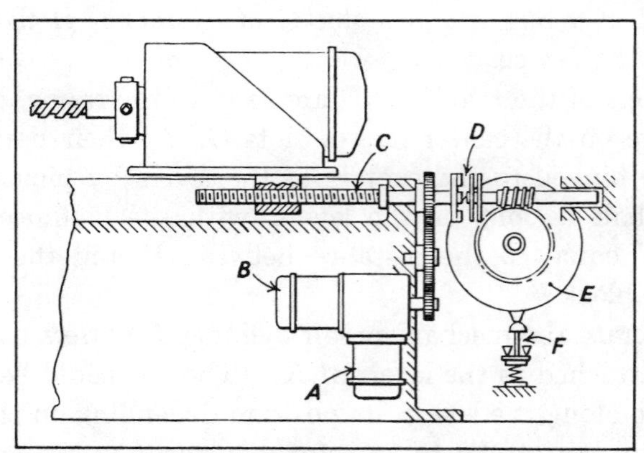

FIG. 23. Feed arrangement for deep-hole drilling that intermittently clears chips.

ment. This causes cam *E* to be driven by means of screw *C* through suitable worm gearing.

At the end of the rapid-traverse motion of the drill head, the feed motor *A* is stopped and the motor *B* is reversed, by a means not shown. In consequence the drill head is rapidly traversed towards the work and cam *E* is driven in the opposite direction. When cam *E* has been returned to its original angular position, it operates a switch *F*, with the result that the rapid-traverse motor *B* is stopped, and clutch *D* disengaged. At this point in the cycle, feed motor *A* is again started in order to perform the next stage of the required drilling operation.

Because feed motor *A* is not running during the rapid traverse of the drill head toward the work, when motor *B* is stopped a small clearance exists between the end of the drill and the bottom of the previously drilled portion of the hole in the workpiece. As a result, the risk of drill breakage due to a slight overtravel of the head during the rapid-traverse motion is reduced.

Automatic Feed Mechanism with Quick-Return Motion

Short metal tubes of various lengths and diameters are polished and buffed on centerless grinding machines. In handling this work on the standard type of centerless grinder, the operator inserts an unpolished tube with his right hand, and as soon as the automatic feed is in action, pushes a handle to actuate the feed-belt pusher. After the tube is polished, the operator inserts a wooden stick in it with his left hand to remove it from the machine. In order to provide for automatically feeding the work to the machine and ejecting it, thus eliminating the need for constant attention by the operator, the regulating wheel was replaced by an endless belt and the automatic feed mechanism shown in the illustration was developed. With this arrangement, the operator merely needs to keep the hopper loaded.

The hopper *A* (see Fig. 24), section *X–X*, is designed for tubes *W*, 1½ and 1¼ inches in diameter, which are loaded parallel to each other. A hinged bottom *B* is swung on its pivot by an eccentric *D*, below *B*, which receives its rotary motion from a

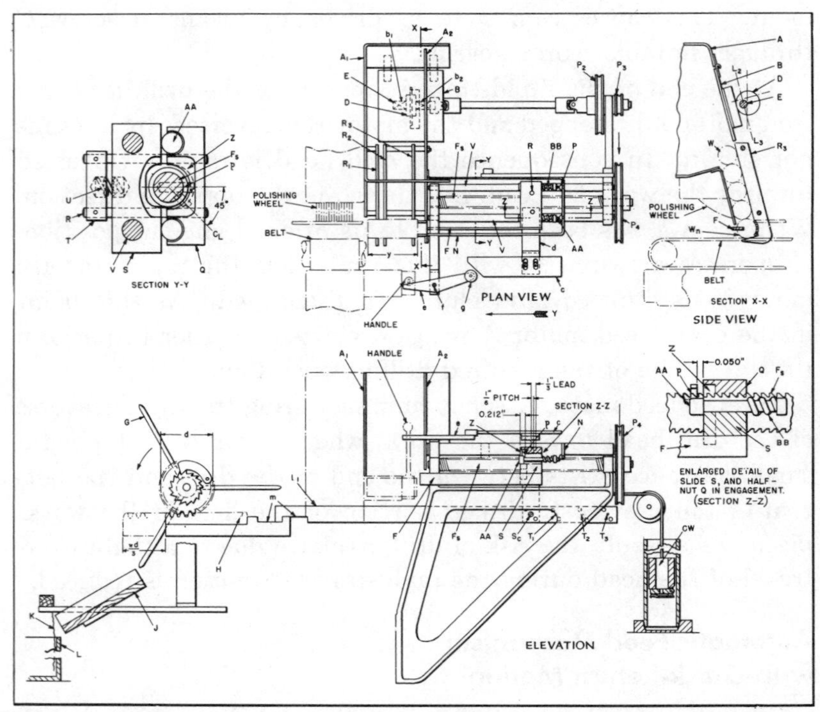

FIG. 24. **Feed mechanism used with centerless grinder for automatically loading and unloading tubes during a polishing operation.**

constantly rotating shaft E. This shaft is supported in bearings b_1 and b_2 (see plan view), which are adjustable in the direction indicated by line L_2–L_3 (section X–X), so that the amount of movement of hopper bottom B can be regulated. Shaft E is driven by pulley P_2. Pulley P_3, driven by P_2 drives P_4 which drives feed screw F_s.

Referring to the plan view and to the elevation, A_1 is that part of the hopper which is fastened to the machine and has a rail R_1; A_2 is an adjustable section of the hopper, designed to accommodate any length y of tube W, and has rails R_2 and R_3 on which the work rolls into position W_n.

The adjustable wall A_2 has an extension N provided with two stop-buttons P which limit the return stroke of slide S (section Y–Y) as will be described. The slide mechanism, which has a pusher F, also includes a half-nut Q into which two pins R are pressed. These pins fit loosely in holes through slide S and are fastened to a spring support T. A spring U keeps half-nut Q and slide S in contact, or closed, in which position Q is engaged with feed-screw F_s. The right-hand buttress thread of the feed-screw feeds slide S in the direction of arrow Y, on fixed rods V, when the half-nut and feed-screw are engaged. When they are disengaged, as shown in section Y–Y, where slide S and half-nut Q are separated, slide S returns to the starting position by means of the pull provided by counterweight CW, which is essentially a dashpot, since its outer casing is partly filled with oil.

The disengagement of the half-nut Q and the feed-screw for the return stroke is accomplished by means of pin p in feed-screw F_s, which enters a space Z milled in the half-nut. When members Q and F_s are engaged, pin p is at line C_L, shown in dot-and-dash lines in section Y–Y, and a clearance of 0.050 inch exists between the pin and the front of the half-nut, as shown in the enlarged detail, section Z–Z. The lead of the feed-screw is ⅓ inch per revolution; hence, as the feed-screw turns through 270 degrees, or three-quarters of a revolution, the nut will advance $\frac{3}{4}\ (\frac{1}{3} - 0.05) = 0.212$ inch.

At this point, the pin will have entered the half-nut a distance of 0.212 inch, as shown at section Z–Z in the front elevation, and will have raised it against spring U to separate it from slide S and the feed-screw. The distance the half-nut is raised is sufficient to clear the heads of bolts AA from slide S so that two springs BB can pull them along the top of the slide, as will be understood by reference to sections Y–Y and Z–Z. The two parts remain disengaged while the bolt heads rest on top of slide S, instead of being engaged with it, as shown in the enlarged detail Z–Z. The counterweight CW then pulls the slide (and the half-nut) in the return-stroke direction.

Since the amount of weight CW is adjusted to overcome friction plus the compression of springs BB for the length of the heads of bolts AA, when the ends of the bolts strike stops P, springs BB are compressed and the bolts back up so that their heads drop down into engagement again with slide S. Spring U

then exerts pressure against slide S, so that the half-nut and feed-screw engage again, thereby repeating the reciprocating motion. The working area between these elements of the feed mechanism is small, but it is sufficient for the light work it is required to do.

This reciprocating movement of any predetermined length y is repeated automatically, as long as the feed-screw rotates. Incidentally, this mechanism also acts as an automatic timer, the time duration of the cycles being proportional to the set length y.

As shown in the plan view, the hopper is set up for a certain length y. If a different length of tube is to be polished with the same equipment, the adjustable section A_2 is reset to the new length y and a screw S_c (front elevation) is reset to any of the tapped holes T_1, T_2, etc. The two holes on each side of S_c or T_1, etc., are for dowel-pins which are not removed.

A ratchet wheel, mounted on a pinion, carries three metal rods G that pick up polished tubes and carry them to a chute J, from which they slide into a receptacle L after passing a cloth apron K. A connecting arm L_1 on slide S engages a rack H during the forward stroke, after an idle period provided by open sections m. This idle movement is used to obtain a fast return stroke while the counterweight CW falls freely before reaching the oil cushion.

On the forward stroke, arm L_1 moves the rack a distance $\frac{\pi d}{3}$ (since the pitch diameter of the pinion is d) to position it for turning the ratchet wheel 120 degrees on the return stroke. During this forward movement, the ratchet wheel and rods or arms G are stationary, due to the action of a pawl that pivots on the pinion and slides over the teeth of the ratchet wheel. This permits the automatic pick-up of a polished tube on one of the stationary arms.

The first part of the return stroke is fast, as previously mentioned, and amounts to the idle motion of the rack plus 55 degrees of the rotation of arms G, when the pawl engages the teeth of the ratchet; the remaining 65 degrees is slowed up by the cushioning effect of the dashpot.

A cam *c* is fastened to a flat spring *d*, see plan view; the latter, in turn, is fastened to slide *S*. A rocker arm *e* swivels on a fixed pivot *f*. It has a roller *g* at one end and pushes the machine handle at the other end. When the slide moves in the direction of arrow *Y*, cam *c* exerts pressure on the handle to apply pressure to the belt. The amount of pressure is adjustable by means of two elongated slots in cam *c*.

CHAPTER 16

Feeding and Ejecting Mechanisms for Power Presses

The use of a properly designed feeding and ejecting mechanism is an important factor in power press operation for maintaining a low percentage of spoiled work and a relatively high production rate. The mechanisms described in this chapter were designed for a wide variety of press functions.

Other similar feeding and ejecting mechanisms will be found in Chapter 16, Volumes I and III and Chapter 15, Volume II of "Ingenious Mechanisms for Designers and Inventors."

Mechanism for Rotary Positioning of Work-Transfer Arm

In order to reduce costs in two stamping operations on a sheet-metal part, existing dies were mounted in tandem on a mechanical press. It was intended that the operator load the work manually in the first die and that the part be transferred automatically to the second die. This transferral is accomplished by the mechanism shown in Fig. 1. The parts are ejected from the second die by air.

The feature of the mechanism is a swinging arm *A* which incorporates a vacuum system. The arm swings through 120 degrees in transferring a part from the first die to the second. In addition, the arm must lower a slight amount on the die to pick up the work, rise to a height that will insure clearance of the work over the two dies, lower a second time to place the part in the second die and again rise to the swinging position. Figure 1 shows the arm in the neutral position it occupies when the ram is at the bottom of its stroke.

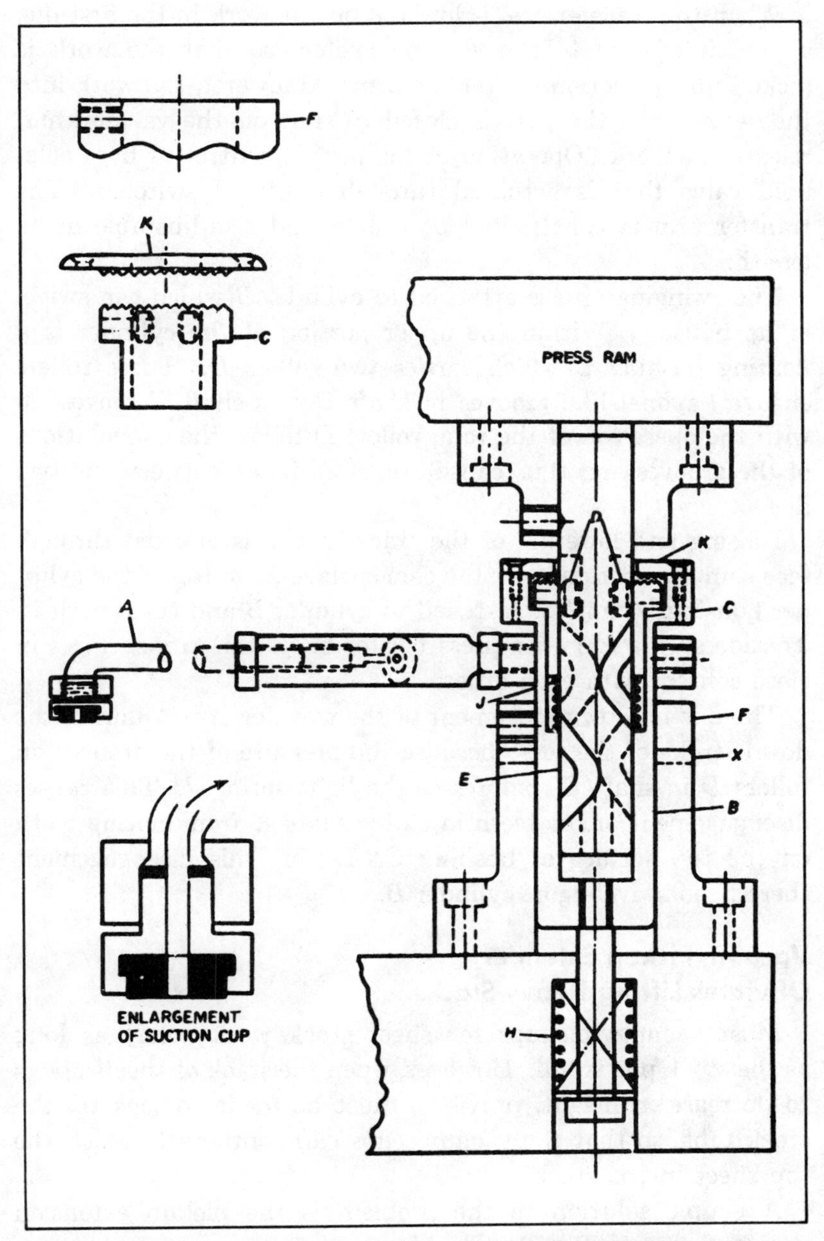

FIG. 1. Swiveling vacuum arm designed for automatically transferring parts from one die to a second die.

When the transfer arm is lowered on the work in the first die, a port is opened to the vacuum system so that the work is picked up by suction. After the arm has lowered the work into the second die, the port is closed to shut off the vacuum and release the work. Operation of the port is controlled by a solenoid valve that is actuated through electrical switches. The transfer arm is constructed of tubing and is adjustable as to length.

The swinging arm is attached to cylinder *B* which can swivel in its housing. Within the upper portion of the cylinder is a floating bushing *C* which carries two rollers *D*. These rollers engage bayonet-like grooves in shaft *E*. As shaft *E* moves up with the operation of the ram, rollers *D* follow the convolutions of the grooves and thus cause arm *A* to index between the two dies.

Raising and lowering of the transfer arm is effected through face cam *F* as it rides over the cam surface *X* on top of the cylinder housing. Cam *F* is fastened to cylinder *B* and turns with it. Tension spring *H* in the press bolster keeps the cam surfaces in close contact with each other.

There is no down movement of the transfer arm *A* during the down stroke of the press because the pressure of the grooves on rollers *D* in shaft *E* compresses the light spring *J*. This causes disengagement of the teeth in clutch plate *K* from mating teeth on the top surface of bushing *C*. During this disengagement there is no swiveling of cylinder *B*.

Vacuum Pickup Extender Obviates Lifting Heavy Stacks

Most vacuum pickups for sheet stock work nicely as long as the stock pile is full. However, when the stack of sheets starts to decrease in height, provision must be made to jack up the stock table so that the vacuum cups can continue to reach the top sheet in the stack.

A simple solution to the problem is the pickup extension device shown in Fig. 2. The shank of the vacuum cup moves downward automatically as the pile is used by the press. The

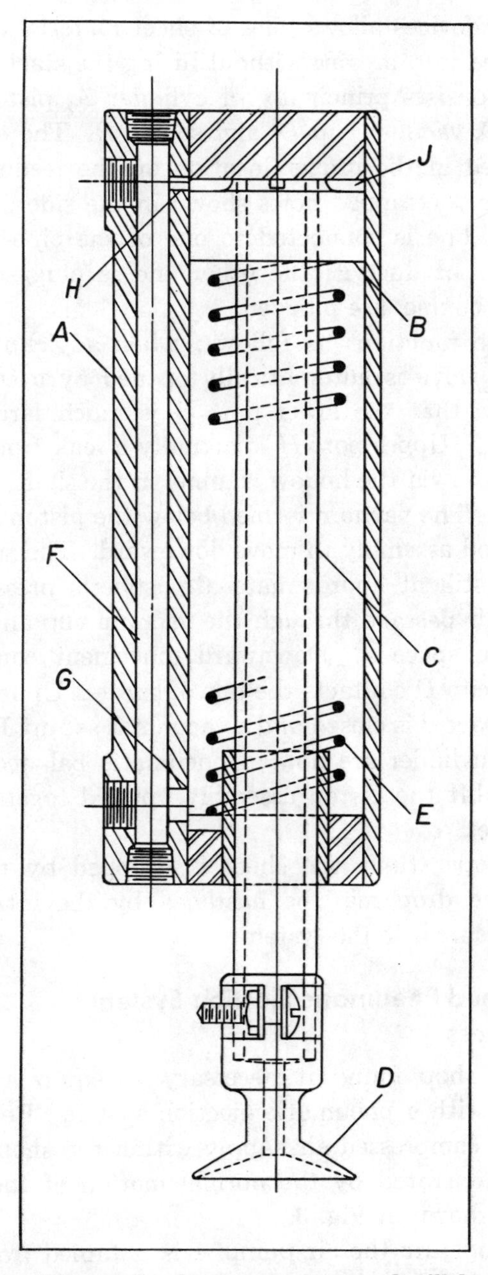

FIG. 2. Vacuum lifter grabs sheet on contact, and will handle the whole of a big stack.

use of these devices allows piles of sheet material over 4 inches high to be fed into presses without lifting the stack.

The lift consists principally of cylinder *A*, piston *B*, hollow piston-rod *C*, vacuum cup *D*, and spring *E*. The entire assembly is secured in the proper location on the feeding device by cap-screws in the tapped holes shown in the side of cylinder *A*. The vacuum line is connected to one of the pipe-tapped holes at either end of duct *F*; the other end is plugged. All other openings, of course, are plugged.

The device functions as follows: when a pickup is desired, the vacuum valve is automatically opened by a cam, or other means. Note that the lower port *G* is much larger than the upper port *H*. Upper port *H* is actually a leak from space *J* to the atmosphere via the hollow channel in the shaft, at this stage of operation. The vacuum formed below the piston *B* causes the piston and rod assembly to move downward. The space *J* above piston *B* is still at approximate atmospheric pressure because there is an air passage through the vacuum cup and the hollow piston-rod to space *J*. Downward movement continues until the vacuum cup *D* contacts the pile of sheets. Upon contact, air passage to space *J* is closed and a vacuum is set up. Because both ends of the cylinder are now in pneumatic balance, the spring *E* is free to lift the piston assembly upward together with the adhering sheet.

At the proper time, the sheet is dropped by releasing the vacuum. The drop may be hastened by the introduction of some air pressure into the system.

Self-Contained Pneumatic Ejection System for Punch Press

A certain shop found it necessary to equip a crank type punch press with a pneumatic ejection system. Because of the absence of a compressed air supply within the shop, a self-contained unit, operated by the normal motion of the press, was installed, as shown in Fig. 3.

Power to operate the air pump *A* is obtained from the drive mechanism of the press. Connecting-rod *B* is extended at its

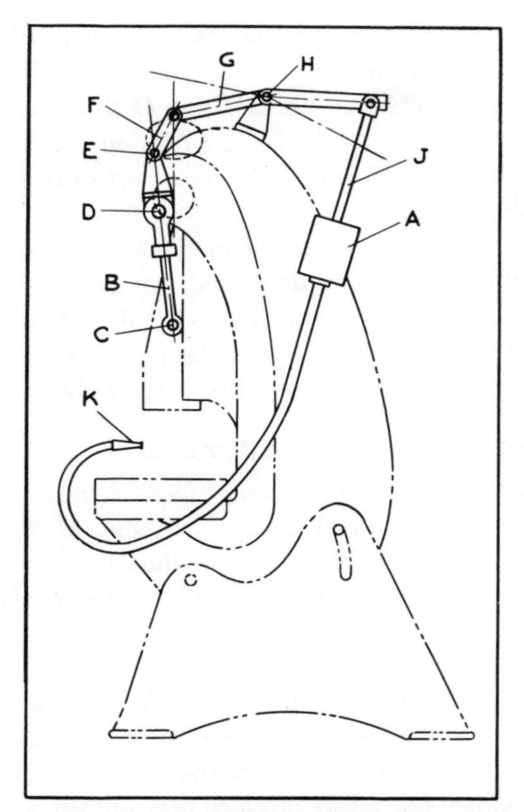

FIG. 3. Ejection system for punch press utilizes air pressure from a self-contained unit.

upper end so that cross-head pin C, crankpin D, and pin E of the extension arm lie in a straight line. The path of pin E, which connects the extension arm to free link F, is a curve which, for practical purposes, can be considered as an ellipse. The minor axis of this curve is equal in length to the stroke of the crank, and the major axis is equal in length to the stroke of the crank multiplied by $\frac{CE}{CD}$.

To achieve maximum efficiency of the ejection system, the stream of air should be discharged during the latter portion of the up stroke only. This was accomplished by making the length of free link *F*, which actually consists of two parallel links, equal to the approximate radius of curvature of the lowest part of the oval path traversed by pin *E*. Link *G* thus remains motionless during the initial part of the press up stroke, but pivots rapidly around shaft *H* during the final part of the same stroke, as can be seen from Fig. 3.

As link *G* pivots, piston-rod *J* of the air pump, which is secured to the press frame, is depressed. In this way, an air blast is discharged between the punch and the die. A nozzle *K* is attached to the end of a flexible air hose and directs the stream of air to any desired point in the work area of the press.

An Intermittent Feed for Strip Material

An arrangement for intermittently feeding the required length of strip material to a press is shown in Fig. 4. This patented mechanism is actuated by a pin *A*, which is attached to a slotted plate fixed to the bottom of the vertically reciprocating member of a press, or similar machine.

Pin *A* moves in the vertical plane and pivots a bellcrank *B*. This member, in turn, has an open-ended slot which engages with a pin projecting from one side of a sliding block *C*. Block *C* is guided and can only move in the horizontal plane. Thus, vertical movements of pin *A* produce horizontal reciprocation of

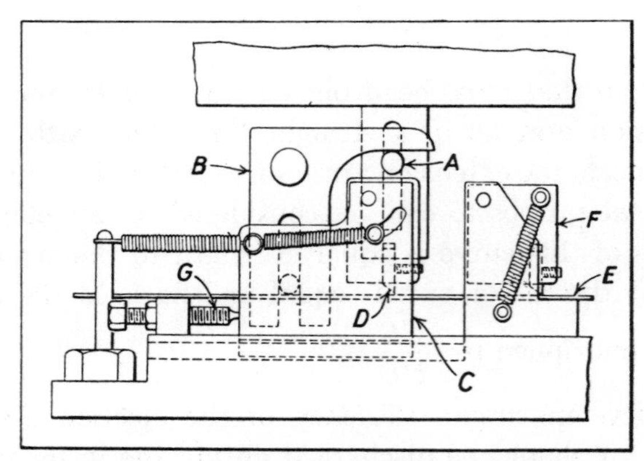

FIG. 4. A device for intermittently feeding lengths of strip stock to a die in a press.

the block. A pivoted holder and blade *D* are housed within block *C*, the blade being held in contact with the strip material *E* by means of a tension spring secured at one end to block *C*.

In operation, as pin *A* travels upward, block *C* moves to the right and, through the gripping action of the blade *D*, the strip material is fed in the same direction. A second spring-loaded and pivoted blade-holder *F* prevents the strip from moving back when block *C* is returned by an associated tension spring seen at the left in the illustration. An adjustable screw *G* limits the return movement of block *C* and the extent of the forward stroke is determined by the position of pin *A* in its slot, permitting the arrangement to be used for various sized parts.

Instant-Release Latch Mechanism

On an automatic machine, a bowl-shaped part must be tipped 90 degrees from a horizontal resting position over transfer bars to a vertical position in a cradle on the same transfer bars. The part is first moved off the bars into position over a gate which picks up the part and rotates it 90 degrees into the bars.

The problem in designing this equipment was to provide some means of latching the part to the tipping gate during rotation and releasing it into the transfer cradle at the precise moment that the part reaches its vertical position. The tipping gate then returns to its starting position. Such a latch mechanism is shown in Fig. 5.

The work-piece *A* is moved from Position 1 over transfer bars to Position 2 above tipping gate *B*. The tipping gate permits work-piece *A* to drop a few degrees below the horizontal plane, so that the work-piece may pass over button *C*, which prevents the part from sliding down the tipping gate during indexing. The tipping gate is rotated by shaft *D*, which passes through a latch operating cam *E*, that is fixed to a frame member. Shaft *D* also passes through elongated slots in the clevis end of the latch-operating link assembly *F*, which straddles cam *E*. The cam follower *G*, located between the clevis legs of link assembly *F*, is allowed to pivot freely on pin *H*, but cannot pivot farther than stop *J* in a clockwise direction.

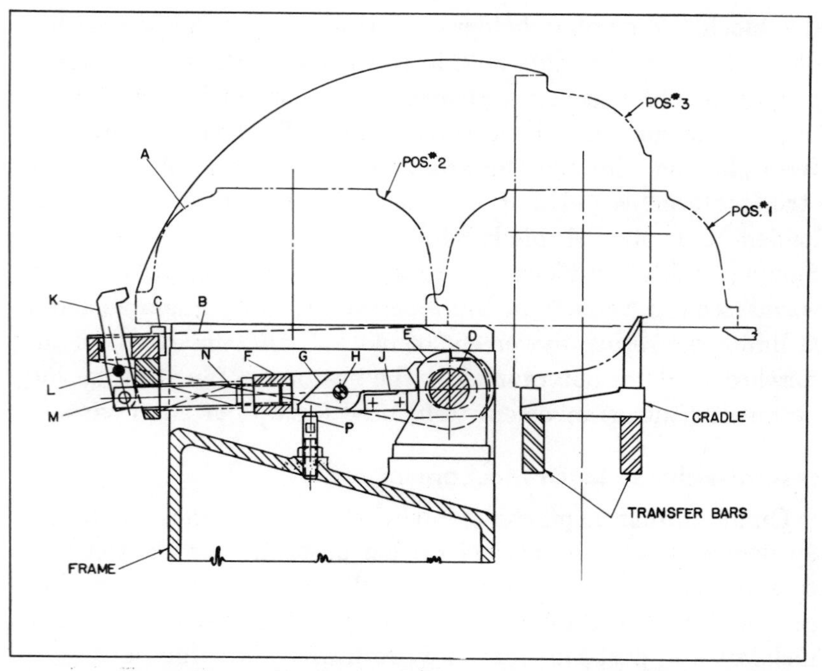

FIG. 5. An instant-release mechanism which facilitates a work-indexing movement.

Latch *K* pivots on pin *L* which is fixed to the tipping gate. The end of the link assembly passes through a clearance hole in the tipping gate, is forked to straddle latch *K* which pivots on pin *M*. The rotating motion of the tipping gate is imparted to the link assembly through pin *M*. As the tipping gate rotates, cam follower *G* backed up by block *J* rides up on the high lobe of cam *E*. The resulting motion compresses spring *N* *and* closes latch *K*.

Part *A* is now latched securely to the tipping gate *B* during the dwell portion of cam *E*. At the precise moment the part reaches its vertical position, the cam follower reaches a step in cam *E*, and the compressed spring *N* forces open latch *K*, instantly freeing the part from the tipping gate. The part is left resting in its transfer cradle (Position 3), as the tipping gate returns to its starting position. However, the cam follower *G*, which had hooked itself over the stop in cam E, pivots on pin H, as the tipping gate rotates. This allows it to pass over the high lobe of the cam. As the tipping gate finally comes to rest at its starting position, the cam follower G is reset by button P. The latch mechanism is now ready to start its next cycle.

High-Speed Punch Press Feed Mechanism

Unlike the conventional feeding mechanisms provided on punch presses that utilize one-half the crankshaft rotation for their operation, the device illustrated in Fig. 6 functions through a span of 240 degrees. This feed mechanism enables a strip-fed, short-stroke punch press to be converted to a high-speed automatic machine.

With power feeding mechanisms, the intermittent motion as well as the necessity of overcoming the inertia of the coiled stock

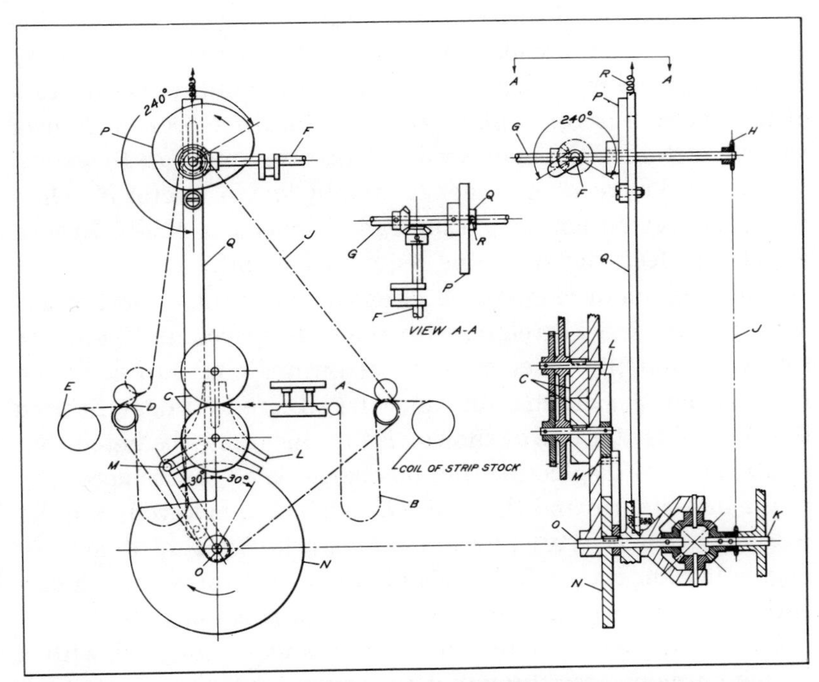

FIG. 6. Cam-controlled differential mechanism combined with Geneva movement to provide a high-speed feed for punch presses.

each time a feed movement occurs, limits the maximum speed at which the material may be fed. These speed-retarding factors have been eliminated in the device here described.

The strip stock is threaded between sprocket-driven rollers *A*, leaving loop *B*, and then over a small idler pulley and through the die (see Fig. 6). From the die, the material passes between feed rollers *C*, leaving a second loop, then through another pair of sprocket-driven rollers *D*, to be finally wound on friction-driven reel *E*. The length of each of the two loops is slightly greater than the required feed per stroke. The circumference of rollers *A* and *D* are equal to the length of feed. The circumference of feed rollers *C* is three times as large.

A pair of miter gears transmits the rotary motion of crankshaft *F* to drive-shaft *G*. This drive-shaft is positively linked, by means of sprocket *H* and chain *J*, to the sprockets on rollers *A* and *D*, and also to shaft *K* of a differential mechanism. All four sprockets are of the same diameter.

The mechanism was designed so that feeding occurred during 240 degrees and punching or forming occurred during 120 degrees of the cycle. Intermittent rotation of the three-branch Geneva wheel *L* actuates the main feed rollers. Movement of the wheel is controlled by arm and roller *M*, and by locking sector *N*. One-third of a revolution, or a 120-degree movement of the Geneva wheel, will feed the stock the required amount.

The roller on driving arm *M* enters a slot of the Geneva wheel at a right angle. Therefore, with the slots spaced at 120-degree intervals, and the roller entering at an angle of 90 degrees, it follows that the driving arm must be situated 30 degrees from the vertical center line as shown. From this it may be noted that a total angular displacement of 60 degrees is all that is necessary of the driving arm and the shaft *O*, to which it is keyed, in order to complete one feeding cycle of the stock. To spread this 60-degree motion over 240 degrees of crankshaft rotation, a differential gear mechanism, coupled to a cam, is employed.

Cam *P* is keyed to the drive-shaft and is designed with a simple harmonic rise through 240 degrees, and a return through the remaining 120 degrees. As the drive-shaft rotates in the direction of the arrow in the illustration, shaft *K* is driven at the same speed. In coordination with this movement, the differential housing is rotated in the same direction as shaft *K* by the action of cam *P* on rod *Q*. The rod is attached directly to the differential housing. By designing the cam to move the housing 90 degrees during a 240-degree rotation on shaft *K*, a total of 180 degrees of rotation will be lost to shaft *O*. In this way, only the 60 degrees necessary to index the Geneva wheel will be transmitted through the differential mechanism during the 240 degrees of crankshaft rotation used for feeding.

The opposite effect is produced by the differential during the last 120 degrees of cam rotation during which the punching or forming operation is being performed. As rod *Q* rises — due to tension spring *R* forcing the roller to follow the cam contour — the housing is moved in the opposite direction and shaft *O* is accelerated. This brings driving arm and roller *M* around to the starting position for the next cycle.

Sorting and Feeding Shells Closed End First

Hollow-drawn cylindrical shells often present handling difficulties when they must be fed rapidly and continuously to an automatic machine with the closed end foremost. This is especially true when the shells are fed to a sorting mechanism from a hopper.

The hole in the press-drawn shell *W* extends approximately two-thirds the length of the piece. Rapid delivery of this shell with its closed end foremost was required to insure economical production. The sorting mechanism for this job had to be of simple design, mechanically actuated, and relatively free from clogging tendencies.

The shells fall unassorted into a large hopper from a drawing press. By simply agitating the hopper, the shells pass down the vertical portion of a chute which feeds them into the short horizontal section *A* leading to the sorting and feeding mechanisms. The hopper and the vertical portion of the chute which holds about twelve shells are not illustrated. The sorting or feeding mechanism is interposed in the horizontal portion of the chute.

Chute *A* (see Fig. 7), is a cylindrical steel tube with an inside diameter of sufficient size to permit an easy flow of parts *W* down its vertical portion and through to the horizontal section. The end of chute *A* is fitted into a hole drilled through a boss in the left-hand wall of the cast-iron sorting body *B*.

Body *B* has flange extensions *D* with hold-down bolt holes *E* for fastening the sorting mechanism to a bracket extension on the press frame. A deep slot *F* is the sorting chamber of the

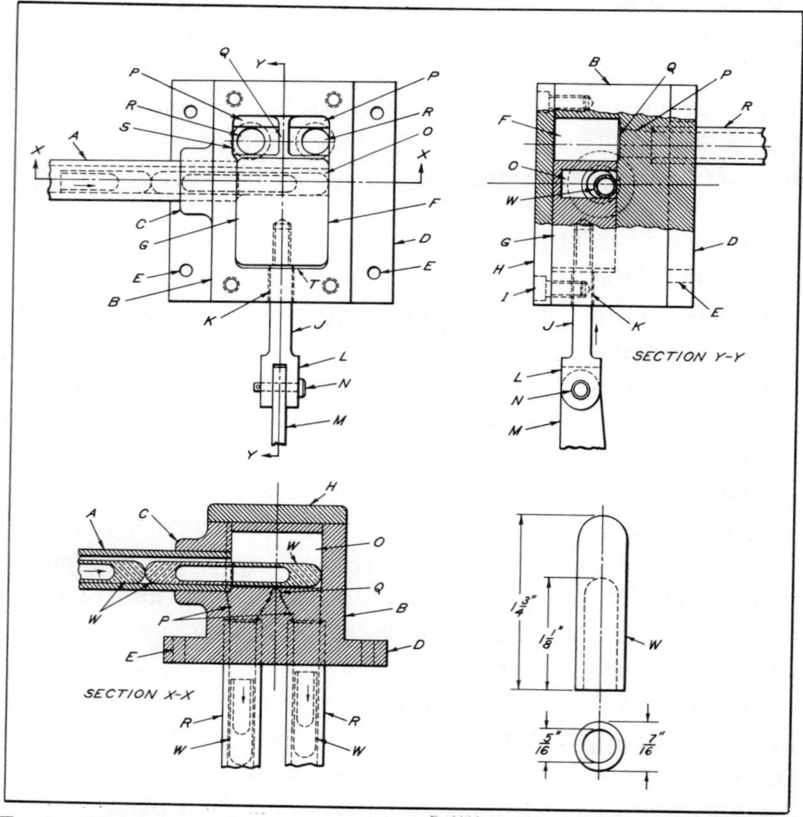

FIG. 7. Mechanism by means of which shells that are fed either open end or closed end first into a feed chute are all sorted and fed closed end first into chute leading to press for succeeding operation.

device into which the shells must pass. The floor of the slot is approximately 1/32 inch below the bottom edge of the bore of chute *A*, as shown in the lower view. Slot *F* is about 0.050 inch wider than the over-all length of the shell *W*.

The walls at each end of slot *F* serve as positive stops to control the sliding movements of the short hardened and ground steel slide *G* which moves easily in the slot. This slide is considerably shorter than the slot length to allow a movement of about 7/8 inch. The depth of slot *F* and the thickness of slide *G* are equally important, and are made not less than three-quarters of the over-all length of the shell.

The slide thickness should be about 0.005 inch less than the depth of the slot to allow a free working fit. The slide is actuated by a simple rotating barrel cam (not shown); linkage between the cam and slide being provided by the short connecting-rod *J*. The shank of rod *J* is screwed permanently into the end of the slide and passes through clearance hole *K* in body *B*.

Each revolution of the barrel cam causes slide *G* to make forward and return movements of 7/8 inch. The cam is designed to provide a short dwell period for the slide at each point of reversal. These reversal points coincide with the positions in which the shell is loaded into slot *O* and ejected from it. The width of slot *O* is slightly greater than the outside diameter of the shell, and the depth is made not less than seven-tenths of the over-all length of the shell, although the diameter may influence the depth dimension.

When slide *G* is in the retracted position, a shell will pass readily from chute *A* into slot *O* so that one end is in contact with the right-hand side of the slot *F*. As slide *G* moves forward within slot *F*, the shell will be carried along, and the plain end of the slide will automatically close off the mouth of chute *A* to prevent another shell from passing into slot *F*.

In the floor of slot *F* are two outlet wells *P*. The sides of these wells are inclined, as shown in heavy broken lines below bridge *Q* in section *X-X*. The minimum width dimensions of each well should be made appreciably greater than the outside diameter of the shell. An allowance of 1/8 inch was found satisfactory.

Bridge *Q* lies between the wells and central with the width of the slot *F*.

The two tubes *R* converge, forming a single tube of slightly larger diameter which feeds the shells directly to the dial mechanism of the press. The junction point of tubes *R* is omitted from the illustrations.

Slide *G* is shown in the loading position and near the extreme end of its stroke. In this position, slot *O* will be in alignment with the outlet of chute *A*, and since a considerable number of shell components will be carried in the chute, the weight of those in the vertical portion will exert enough pressure to push the first shell into slot *O*.

The illustration shows the work or shell located with its domed end foremost and lightly pressed against the right-hand wall of slot *F*. As soon as the shell is positioned in slot *O*, slide *G* is moved forward by the link-cam drive, for a distance of approximately ⅞ inch. This movement will also be imparted to the shell contained in the slide. The same movement of slide *G* serves to close up the end of the chute *A*, thus preventing the next shell in the bore of the chute from entering slot *F*. The blank side of slide *G* passing across the mouth of the chute *A* will hold the line of shells in position throughout the remainder of the slide movements.

When *G* is in its most extreme position the shell will be supported only at its mid-point by the very narrow bridge *Q*, and since the domed end is much heavier, it cannot remain balanced on the bridge *Q*, and will immediately tilt over in a clockwise direction and pass into the right-hand well *P*. It then passes down the vertical outlet tube *R* with its domed end foremost as shown by the broken lines in the section *X-X*.

A shell advancing to slot *F* with its open end foremost will enter slot *O* and be traversed to the opposite end of the sorting chamber in exactly the same manner as a shell entering dome end first. But this time the heavy end of the shell will fall in the left tube *R*.

CHAPTER 17

Hoppers and Hopper Selector Mechanisms for Automatic Machines

Tool engineers and machine designers are often faced with the problem of designing mechanisms to pick up parts from hoppers for delivery to the assembly machines. By "hopper feeding" is meant the indiscriminate dumping of a load of parts into a hopper of suitable size and shape, from which the parts are picked up, in the proper position, and deposited in a track for feeding to a machine by gravity. Ordinarily, the pick-up member is so shaped that the parts cannot enter the track if they are not in the right position, and therefore are dropped back into the hopper. Occasionally, the shape of the part and the speed requirements of the machine make it necessary to pick up parts that are not all in the same position. In that case, prior to going into the assembly machine, the parts are required to pass through an auxiliary mechanism, or separator, which arranges them all in the required position.

Many types of hoppers have been designed and built with varying degrees of success. One type of hopper may work successfully for a part of a certain shape, but may prove entirely unsuitable for pieces of a different contour. A great deal of thought must be given to the selection of a hopper for any particular job. Every new problem is unique in some respect, and will necessitate variations in the type of hopper selected.

Other hoppers and hopper selector mechanisms are described in Chapter 17, Volume III of "Ingenious Mechanisms for Designers and Inventors."

Designing Hopper Feeds for Square and Hexagonal Nuts

Nuts — both square and hexagonal — are employed in such large quantities as fastenings that the problem of automatically feeding parts of this type to various machines often confronts tool engineers and machine designers. Occasionally, the nut blank must be transferred to machines that perform secondary operations. In other cases, the finished nut must be automatically delivered to a particular location for assembly to other components.

A number of hopper designs have proved successful for handling work of this kind. Among these are the centerboard, the paddle wheel, and the rotary hopper types. Choice of the correct form of hopper to use will depend upon production requirements and, to a certain extent, upon the size of the nuts.

The centerboard type of hopper is widely employed for small and medium size nuts, either square or hexagonal in shape. Because of its low cost, large capacity, and excellent operating efficiency, this type of hopper should receive first consideration when a part delivery problem is met. A centerboard type of hopper and separator mechanism for feeding square or hexagonal nuts, properly positioned, to an automatic machine is illustrated in Fig. 1.

The hopper consists of a body, usually made in two parts as shown in section *X–X*; and a blade, fastened to an arm, which oscillates through the mass of parts that have been previously placed in the hopper. Parts that happen to be in the correct position drop into a groove machined in the top edge of the blade, and are raised by the blade in its upward travel. At its uppermost position, this groove is in line with a track, down which the nuts slide toward the machine. An actuating rod transmits motion to the centerboard blade from a cam or crank, not shown in the drawing. The centerboard arm is fastened to and pivots about a shaft.

A lever projecting from the opposite side of this shaft operates a knock-out slide. The knock-out slide advances, upon

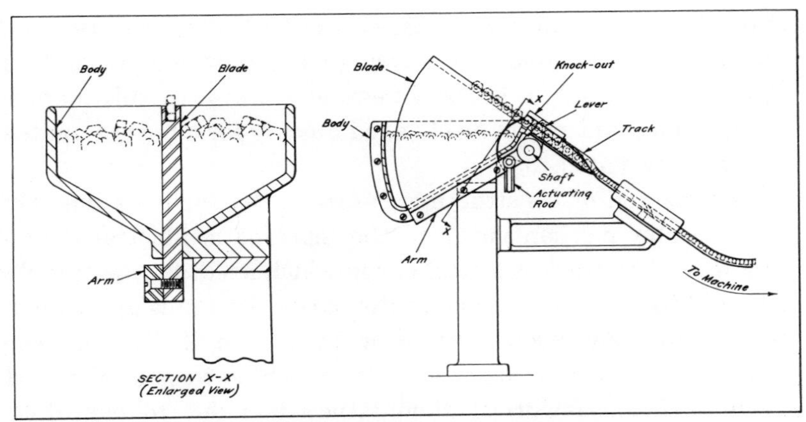

FIG. 1. Centerboard type of hopper and separator mechanism for feeding properly positioned nuts to an automatic machine.

descent of the center-board blade, and clears the mouth of the track of any parts obstructing it. An enlarged sectional view through the knock-out mechanism is seen in Fig. 2.

In cases where it is planned to use the hopper for more than one size of nut, the centerboard blade should be constructed sectionally, as shown in Fig. 3. The two positions in which it

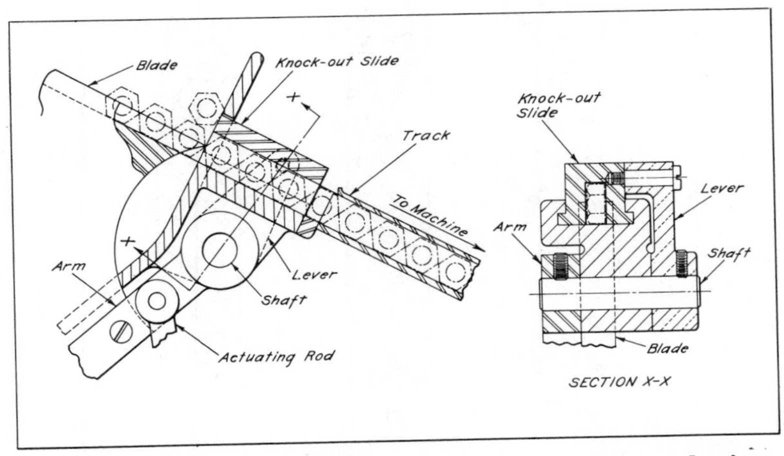

FIG. 2. Enlarged sectional view through the knock-out mechanism of hopper seen in Fig. 1. Oscillation of the knock-out slide clears the track mouth of improperly positioned parts.

is possible for the nuts to be raised by the blade are shown at *A* and *B*. A replaceable cap, keyed to the centerboard blade, is fastened by means of socket-screws as shown. In this manner, the feed mechanism can be changed quickly for a part of another size, as shown at *C*.

One method of constructing the track that delivers the parts to a selector mechanism or to the machine is illustrated at *A* in Fig. 4. The track is a steel bar in which a slot has been milled longitudinally to accommodate the parts. Two rails are fastened to the track by screws, as shown in section *X–X*. This construction allows the operator of the machine to push the parts along with a screwdriver should they jam due to one of the parts being slightly over size.

The end of the track is welded to a fastening plate, which is secured to the hopper with screws. At the opposite end of the track, a similar plate is fastened to the selector mechanism in the same manner. This allows quick removal of the track when changing the setup for other sized nuts.

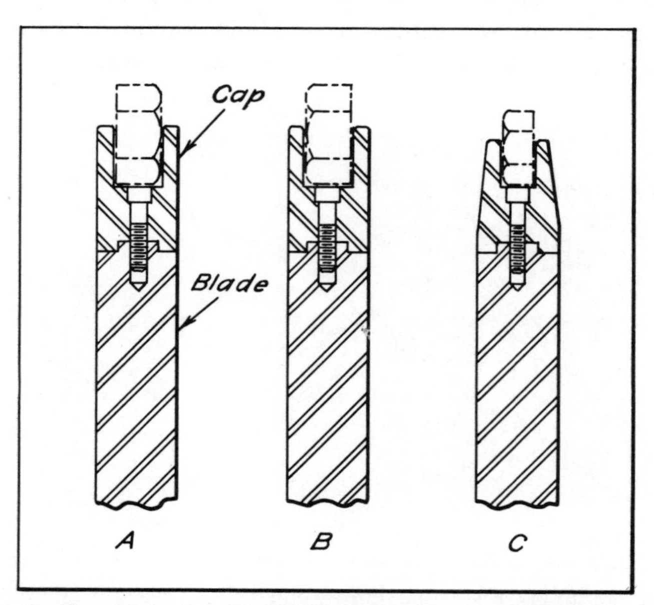

FIG. 3. Sectional construction of the centerboard hopper blade permits rapid change-overs for different sizes of nuts.

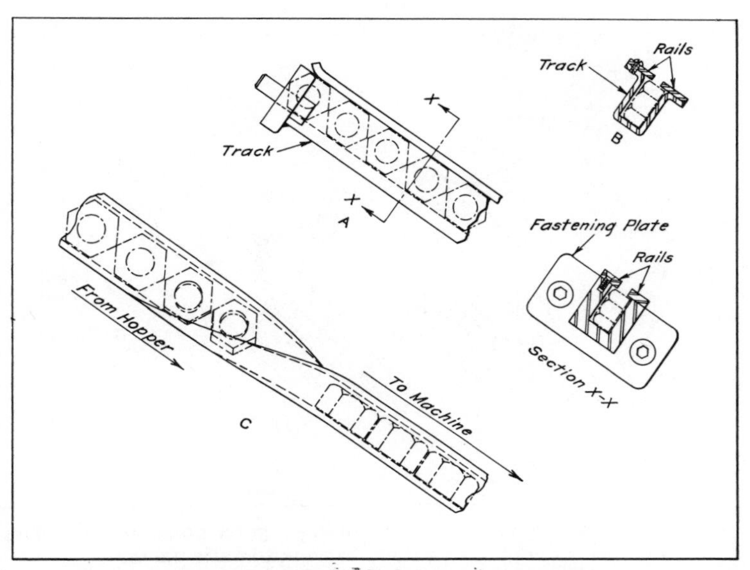

FIG. 4. Details of track construction shown in their position relative to the hopper. The track shown at *A* is machined from bar stock while that at *B* is formed from sheet metal.

A less expensive type of track construction is illustrated at *B*. In this case, the track is made of heavy-gage sheet metal, formed as shown. Rails, fastened to the track with small screws, permit dislodgment of jammed parts through the central gap between them.

A quarter twist may be applied to the track at a position between the hopper and the selector mechanism, or between the hopper and the machine, when the parts must be delivered in a horizontal position. Such a twist in the track is illustrated at *C*. Generally, a considerable amount of hand filing or grinding must be done on the walls of the track at this point to enlarge the sides for proper clearance of the parts.

The construction of one type of selector mechanism employed to turn improperly positioned hexagonal nuts upside down is illustrated in Fig. 5. The sectional view at the right shows the relative position in which the mechanism is mounted on the machine.

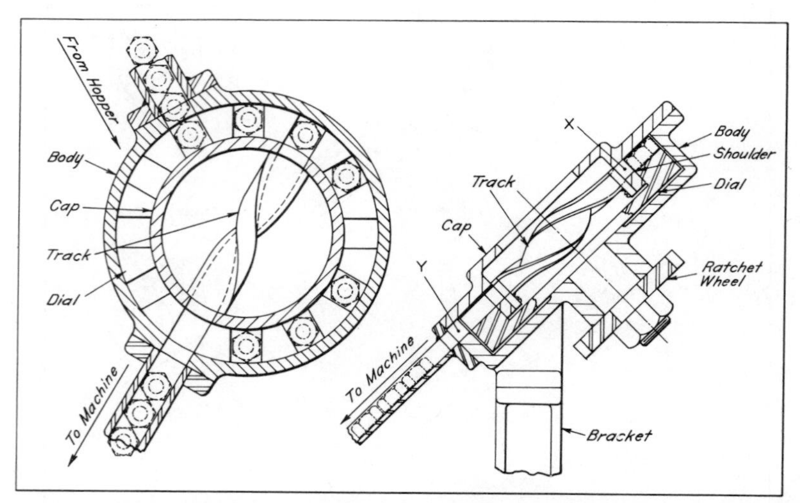

FIG. 5. Selector mechanism for inverting improperly positioned hexagonal nuts prior to delivering them to an automatic machine.

The selector mechanism consists of a cast body, which is fastened in a stationary position to a bracket on the machine; a cap, screwed and doweled to the body, which holds the twisted track of the mechanism; and a dial, free to rotate intermittently within the body and driven by a ratchet wheel.

In operation, the nuts enter the dial from the hopper track, as seen at the left. As the dial indexes, the nuts are carried to a position opposite the mouth of the twisted track. At this point, an opening X has been machined in the stationary cap in such a manner that a shoulder is formed between the cap and the body.

If the nuts are properly positioned (that is, if the flat side of the nut is down), this shoulder prevents them from entering the track, and they are carried around by subsequent indexes of the dial to point Y. Here they pass through an opening in the body, enter another track, and slide down to the machine.

However, when nuts enter the selector mechanism incorrectly positioned (with their curved side down) and reach the point opposite the track mouth, they slide over the projecting shoulder and enter the twisted track. In sliding through this track, they are turned over and enter the track leading to the machine correctly positioned for assembly.

Owing to the extremely slight projection of the shoulder, all parts of the selector mechanism must be made sufficiently strong are fitted precisely to eliminate vibration. Improper fitting or vibration might allow properly positioned parts to jump over the shoulder and slide toward the machine resulting, of course, in incorrect assembly. The ratchet that indexes the dial should be driven by a crank. If a cam is employed, it should be of the harmonic motion type to give gradual acceleration and deceleration.

The same basic type of selector mechanism can be used for square nuts, as seen in Fig. 6. However, in this case, the selector shoulder is formed differently, due to the altered contour of the part. It will be noted that the shoulder projects upward at the corners, thereby preventing properly positioned nuts from enter-

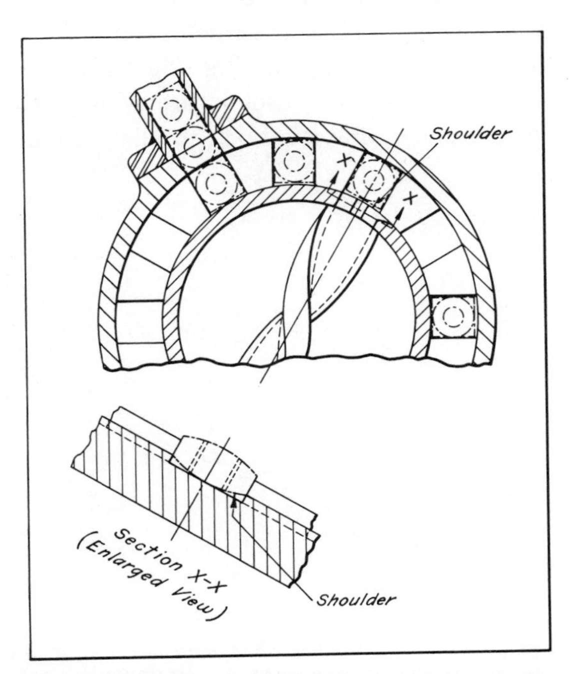

FIG. 6. Selector mechanism for correctly positioning square nuts before delivering them to the machine.

ing the twisted track. Incorrectly delivered nuts, however, slide over this shoulder and are inverted by the twisted track prior to entering the machine.

Another type of hopper frequently employed to feed nut blanks to automatic machines is the paddle wheel type shown in Fig. 7. This consists of a body which is made in two halves that are fastened together with screws and dowels. A spacer, slightly wider than the nuts, is provided between the body halves to allow the nuts to slide freely between the confining portions of the body. When parts are dumped indiscriminately into the hopper, a few will fall into the track formed in the bottom. The rotating paddle wheel pushes the nuts upward, and when they reach the top of the hopper, they enter a track and slide down to the machine.

A retaining dog, which pivots loosely on a pin, prevents the nuts from sliding back every time one of the blades of the paddle wheel clears the track and before the next blade has advanced more blanks to push the row upward. While it is not absolutely necessary, such a retaining dog adds to the smooth operation of the hopper. An offset is provided at the pivoting end of the dog, as shown in view Y–Y, to allow the nuts to slide

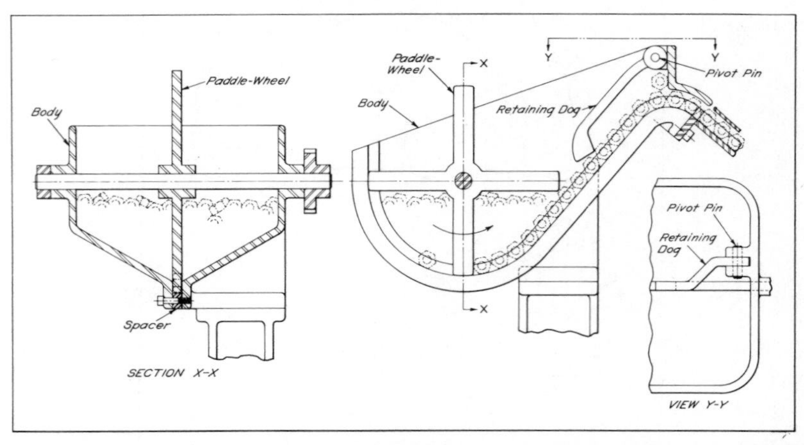

FIG. 7. The paddle wheel type of hopper is an effective means of feeding square or hexagonal nuts to automatic machines.

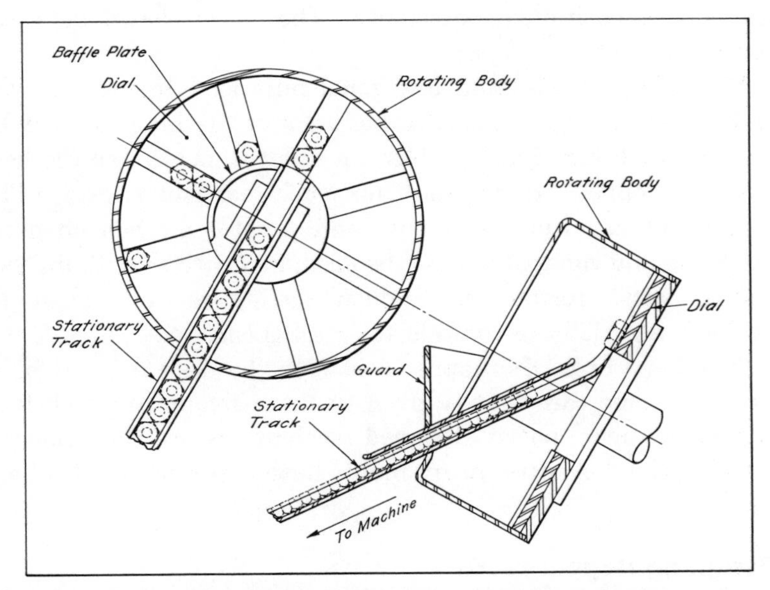

FIG. 8. Rotary type hoppers consist essentially of a rotating body and dial, a stationary track, a baffle plate, and a guard.

up the face of the hopper when the track is full. These parts then fall back into the hopper.

Rotary hoppers, such as the one seen in Fig. 8, can also be arranged to feed nuts to automatic machines. The hopper shown consists of a rotating body attached to a dial. The dial has radial slots milled in it, as shown. A stationary baffle plate prevents the nuts that are being carried upward in the slots from falling back into the hopper. As these parts reach a position opposite the mouth of the stationary track, they enter the track and slide toward the machine. When the track is full, the nuts go past the track mouth and drop back into the hopper.

As in the selector mechanisms described, a shoulder can be arranged at the track mouth to allow only correctly positioned parts to enter, if this feature is desired. In such cases, of course, the track must be given a half twist at some point between the selector and the machine, so that the nuts will be delivered in the correct position for assembly. The guard shown increases the capacity of the hopper.

While both square and hexagonal nuts lend themselves well to hopper feeding, a considerable amount of thought must be given to each individual problem in order to determine the best possible hopper arrangement for each particular case. The component parts of the hopper and selector mechanism must be strong and rigid, and must be machined with smooth finishes and to close tolerances. Sudden jarring motions must be avoided, especially in the selector mechanism.

In other words, the hopper feed must be designed and built with the same care that is given to the machine to which it is applied. Flimsy, poorly designed mechanisms are the principal cause of the difficulties many plants have experienced with hopper feeds.

Designing Hopper Feeds for Bottle Caps – I

The "double-disc rectifying hopper" shown in Fig. 9 is designed to feed bottle caps to automatic machines. It operates on an entirely different principle from the pin type hopper. In the double-disc hopper, the two discs *E* and *F* are fastened to the shaft *G* and kept in continual rotation in their housings *A* and *B*, respectively. The two housings are separated by four spacers *C*. A thrust bearing *H* is provided to give easy and smooth rotation. Each disc is provided with four drivers, each driver consisting of a plunger *O*, a spring *K*, and a split washer forced on the plunger. This unit of the hopper is supported on the column *D*.

As the caps fall on the disc *E*, they are thrown by centrifugal force to the inner side of the housing where they finally take a vertical position. The drivers then push the caps to the slot *I*, through which they are guided to the rectifying chute *L*. This chute is provided with two ribs which direct the caps facing one way down the chute *M*. The caps facing the opposite way fall on disc *F*. Therefore, all the caps falling on disc *F* are in one position; and when the operation of placing these caps in a

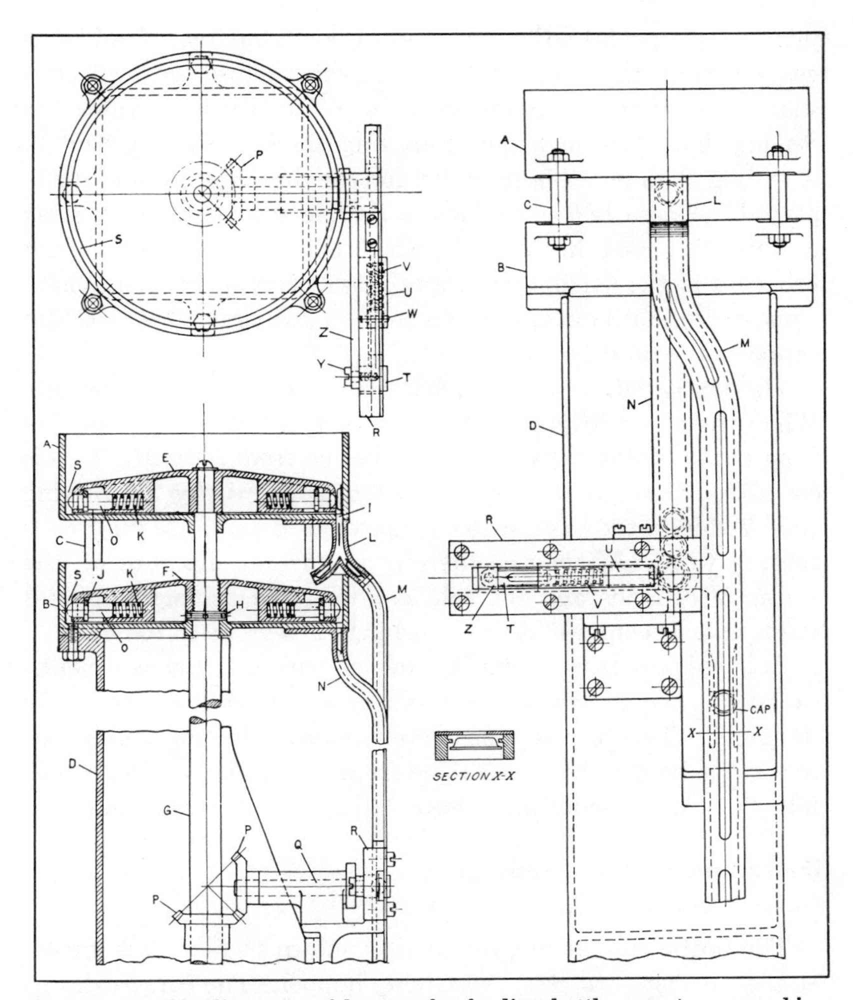

FIG. 9. Double-disc type of hopper for feeding bottle caps to cap making or bottling machines. Caps falling from disc *E* into chute *L* which are facing the desired way, fall into chute *M*. Caps facing the opposite way fall on disc *F*, and their positions are reversed before they enter chute *N*.

vertical plane is repeated (the caps being pushed into chute *N*), they are in the same position as those caps which passed down chute *M*.

Thus there are two lines of caps which must be combined to form one line for feeding the machine below. This is accomplished by a pusher mechanism operated through a pair of bevel gears *P* from the main shaft *G*. The mechanism consists of a shaft *Q* provided with an eccentric plate which operates the driving slide *T* through the connecting bar *Z* pivoted on stud *Y*. A sliding chamber *R* is provided for the driving slide and push-plate *U*. These two parts have a flexible connection consisting of a pin *W* forced into plate *U*, which is free to operate in a slot provided in the driving slide, and a light compression spring *V*. This entire mechanism is supported on the bracket which also supports the rotating shaft *Q*.

The operation of the mechanism needs little explanation. When the chute *M* is full of caps to a point above the opening from chute *N*, the caps should not be fed from *N* to *M*. This is controlled through the compression of spring *V* by the driving slide *T* due to the resistance encountered in trying to push caps from *N* to *M*. As soon as there is an opening in chute *M*, the spring *V* expands and the full length of the driving slide and push-plate becomes effective, pushing caps from *N* to *M*.

This hopper arrangement is provided with a safety slot identical to the one provided in the pin type hopper feed previously described. This slot is even more necessary in the double-disc rectifying hopper here described because of the greater possibility of caps passing into chute *M* in the wrong position.

Designing Hopper Feeds for Bottle Caps – II

The bottle-cap feeding mechanism shown in Fig. 10 is known as the quarter-turn chute rectifying hopper. The convex disc *B* is kept in rotation at a moderate speed in the housing *A*. The disc is supported by a thrust bearing *F*. A bronze bushing *G* serves as a bearing for the main shaft *E*, to which the disc *B* is fastened.

Four buttons *D*, pressed into the disc *B*, serve to agitate the caps until centrifugal force carries them to the inner side of the housing, where they take a vertical position in the annular slot *W*. The caps pass from slot *W* down through the opening *Z* into the rectifying chute *C*, where they encounter the two ribs

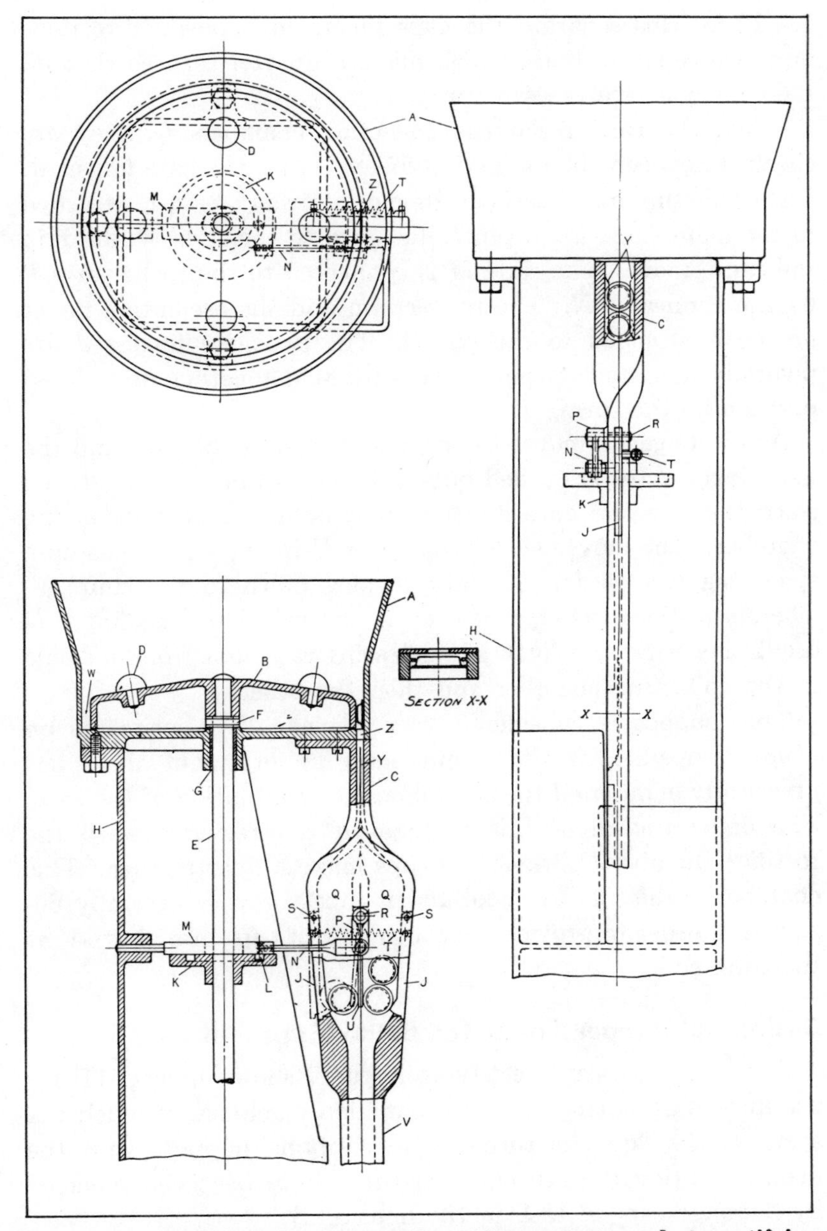

FIG. 10. Bottle-cap feeding mechanism with quarter-turn chute rectifying hopper which is so designed that the delivered caps all face the same way.

Y. These ribs separate the caps facing in opposite directions into two separate lines, which make a quarter-turn so that all the caps will face the same way.

When the caps reach the collecting chambers *Q*, they are checked between the oscillating finger *U* and the gate fingers *J*. The oscillating finger receives its motion from a cam *K* attached to the main shaft *E*. A pin *L*, fastened to the slide *M*, fits into the cam groove. The slide *M* is connected to the rocker arm *P* by the connector *N*. The rocker arm and the oscillating finger are both attached to the pin *R*. The two gate fingers *J* are pivotally mounted on pins *S* and are held together in a closed position by the spring *T*.

As the finger *U* moves to one side, it causes the caps and the gate finger *J* to be pushed outward. The spring *T* is stretched since the opposite gate finger cannot enter into the collecting chamber. The movement of the finger *U* to one side opens up a space big enough for the caps to pass down to the chute *V*. Therefore, the discharge of caps is controlled by the finger *U* oscillating back and forth and releasing caps first from one side of the collecting chamber and then the other.

This hopper is considered very efficient. The principle on which it operates is simple and most important of all, it has practically eliminated the scratching and mutilation of the caps. The one criticism of this mechanism is directed toward the rectifier chamber *C*, where the caps make a quarter turn. This chamber is difficult to make and it must be very carefully designed in order to provide a smooth flow of caps into chute *V* at the lower end.

Designing Hopper Feeds for Bottle Caps – III

The "double-escape rectifying hopper" shown in Fig. 11, for use in feeding bottle caps to production machines, is much the same as the "quarter-turn chute rectifying hopper" with the main exception that the quarter-turn chute has been replaced by double-escape slots *S* in the housing *A*.

Each slot *S* is machined so that it will permit bottle caps to pass through into the chutes *C* and *D* only when they face one

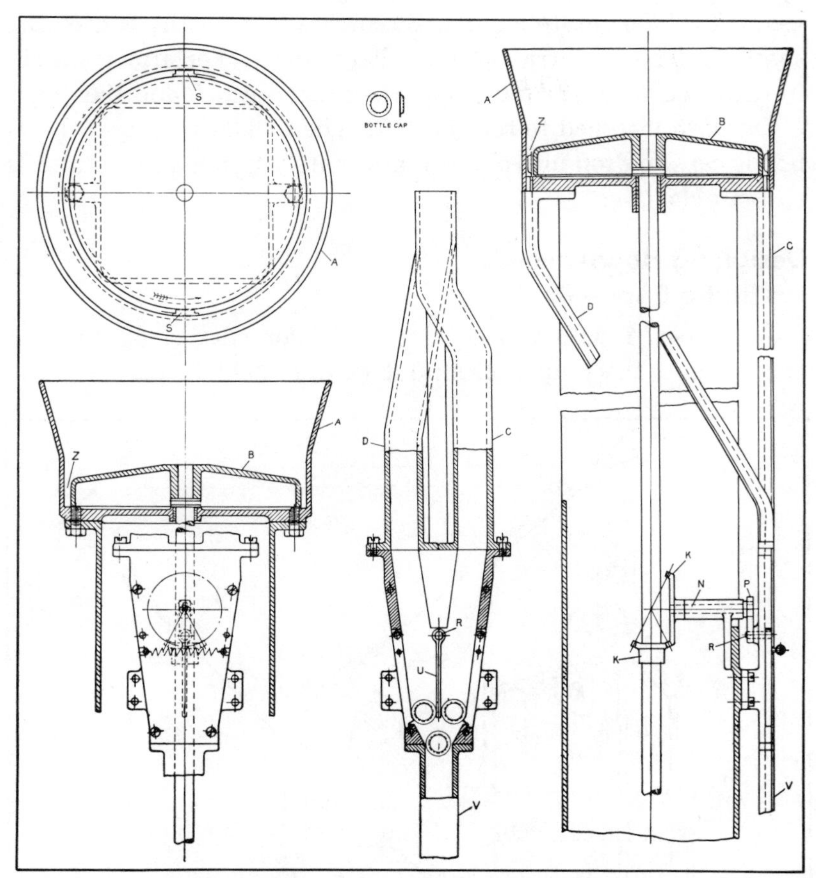

FIG. 11. **Hopper feed with "double-escape" slots for automatic feeding of bottle caps.**

way. The result is that the caps are divided between chutes *C* and *D* according to the positions in which they drop into the slot *Z* from the rotating disc *B* in the hopper. By placing these escape slots *S* 180 degrees apart, the necessity of making a twist in one of the chutes in order to bring all the caps to the collecting chamber in one position has been eliminated.

The operation of the collecting chamber is the same as that of the quarter-turn rectifying hopper. A slight change is shown in the operating mechanism. This consists of a pair of bevel gears K which rotate a small eccentric plate on the end of the shaft N. The eccentric plate oscillates the rocker arm P which, in turn, oscillates the finger U. The rocker arm P and the finger U are both fastened to the shaft R. The oscillating finger allows bottle caps to drop into chute V alternately from chutes C and D without clogging.

Designing Hopper Feeds for Bottle Caps – IV

A hopper A with the usual plate B for cutting off the cap supply, and the hinged clean-out door C held in place by the

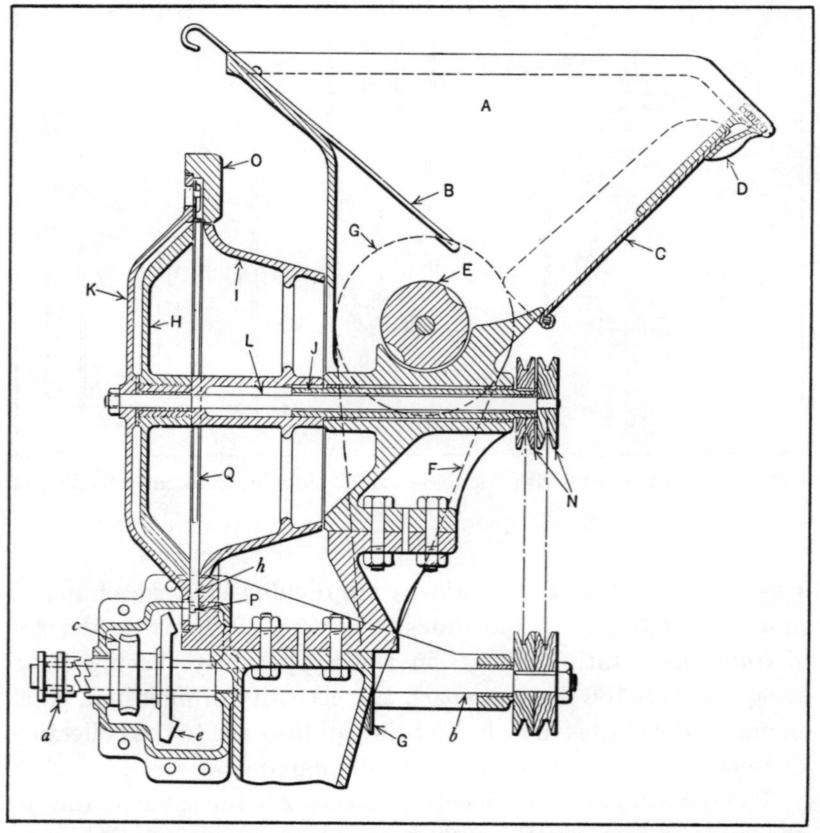

FIG. 12. Cross-section through hopper and driving members of bottle-cap feeding mechanism shown in Fig. 13.

latch *D* is shown in Fig. 12. A fluted roller *E* constitutes the first means for obtaining cap control in the hopper. The fluted roller is driven by the belt *F* which connects pulleys *G*. The rotation of the roller at a steady speed controls the feeding of caps into the revolving chamber so that the caps are discharged in small numbers.

The revolving chamber consists of two dish-shaped members *H* and *I* fastened together and keyed to the hollow shaft *J*. The pin-wheel *K* is fastened to an independent shaft *L* passing through the hollow shaft *J*. Each shaft is provided with a pulley *N*, and is so mounted that the two shafts can be rotated independently. This form of construction permits regulation of the speed of the revolving chamber relative to the pin-wheel *K*, and also permits changing the direction of rotation of the pin-wheel relative to the revolving chamber.

In one case, it was found that best results were obtained when the pin-wheel *K* was rotated in the opposite direction to the revolving chamber and at the same speed as the chamber. It was observed that there was then a minimum amount of churning and agitation of the caps, and that they were removed rapidly from the chamber by the pin-wheel.

The pin-wheel *K* is provided with ordinary straight pins *P*. A housing ring *O* and the guide ring *Q* are provided to form a chamber for the caps. The ring *Q* is fastened to the housing ring *O* and serves to guide the caps until they pass out to the open cap-rectifying chute *R*, Fig. 13. This chute is a very clever though simple device. It is provided with ribs which, aided by the fact that the location of the center of gravity of the cap causes it to fall upward, serve to place all caps in one position on the disc *S*. A cover *T* prevents the caps from flying out of the chute before they reach the point to which they are being directed.

After reaching the horizontal disc *S*, which revolves continuously in its housing, the caps have about completed their passage. Centrifugal force throws the caps to the inner edge of the housing. From here they are directed to the chute *U*, seen in Fig. 14, which conducts them to the machine below.

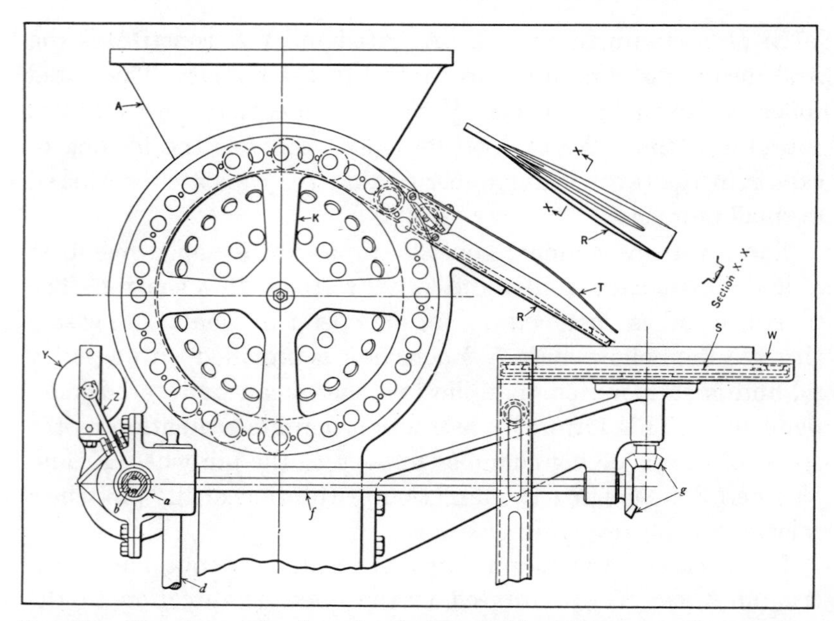

FIG. 13. Hopper mechanism for feeding bottle caps to production machines.

It is apparent that some means must be provided to control the flow of caps from the pinwheel *K* to the disc *S*. This is taken care of by an electrical arrangement consisting of contactors *V*, which are fastened to the disc cover *W*. One of these contactors is provided with the finger *X* which extends through a slot in the disc cover so that it can come in contact with the caps passing under the cover *W*. A solenoid *Y* is shown in Fig. 13.

The plunger of the solenoid *Y* is attached to one end of the rocker arm *Z*. The other end of the rocker arm is shaped into a yoke and provided with two pins which fit into a groove machined in the clutch member *a*, Figs. 12 and 13. This clutch member is attached to the shaft *b* by means of two keys, thus permitting a sliding motion for the clutch *a* under the action of the rocker arm *Z*. The second member of the clutch is part of the worm-wheel *c*.

The entire hopper receives its power from the shaft *d* through a worm not shown in the drawing, and the worm-wheel *c*. The disc *S* is rotated from a pair of bevel gears, one of which (not shown) is in contact with the bevel gear *e*, through the shaft *f* and the pair of bevel gears *g*. The bevel gear *e* is fastened to the worm-wheel *c*, forming a single unit free to rotate on the shaft *b*.

Caps from the hopper *A*, Fig. 12, pass through the fluted roller and, entering the revolving chamber, immediately drop to chamber *h*, where they fall between the pins *P* on the pin-wheel *K*. The pin-wheel carries the caps up between the guide-bar *Q* and the ring *O*, and discharges them onto the open cap-rectifying chute *R*, Fig. 13. Here they are all caused to face upward and

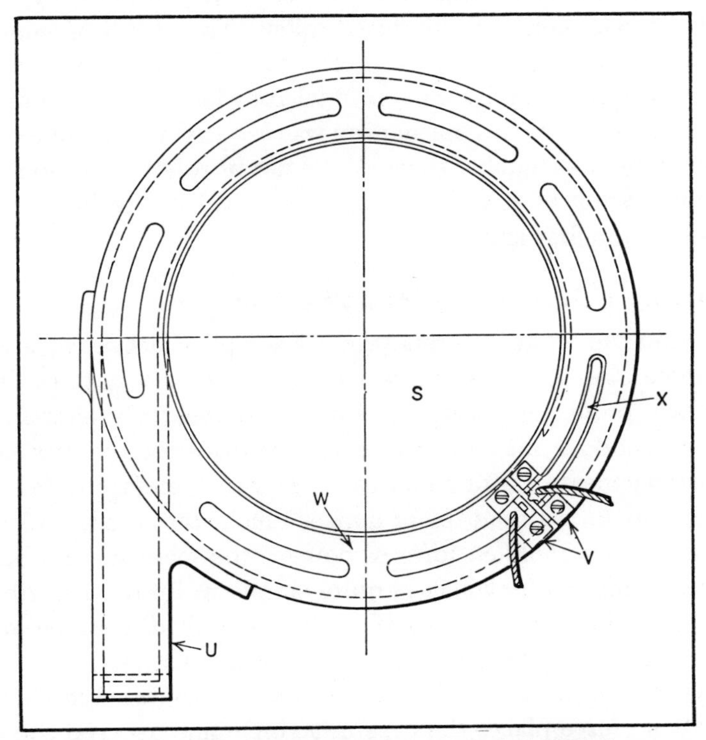

FIG. 14. Plan view of disc *S*, Fig. 13, showing finger *X* that completes electrical circuit, stopping feed when disc *S* becomes overloaded due to clogging of machine.

slide down to the disc *S*, from which they are discharged to the chute *U*.

Now, if the disc *S* is crowded to capacity, the feeler of the electrical contactor *V*, Fig. 14, is raised so as to complete the circuit and energize the solenoid. If the contactor is allowed to make contact for a short period, nothing happens at the solenoid because of the lag in the time it takes to completely energize the solenoid. However, if the period of contact is continued, as would be the case if there were a full line of caps passing under the feeler *X*, the solenoid becomes effective.

Upon being energized, the solenoid plunger is drawn in, causing the clutch members to be separated through the rocker arm *Z*. This immediately stops the rotation of the pin-wheel *K*, the revolving chamber, and the fluted roller. The disc *S* remains in rotation because it is actuated by the gear *e*, which is fastened to the worm-wheel *c* and is in continual rotation from the shaft *d*. Churning action in this mechanism was reduced to a great extent in the revolving chamber through the use of reverse rotation and the elimination of the pin arrangement used in the hopper described in the first article.

Designing Hopper Feeds for Bottle Caps – V

Certain cap manufacturers produce a type of bottle cap about twice the size of the ordinary cap used for beer bottles. The machines producing these large caps are relatively slow-speed machines and do not require as elaborate designs of hopper feeding arrangements as for small caps. It was therefore found advantageous to use the type of hopper illustrated in Figs. 15 and 16. This is a simplified form of hopper, the main feature being the rectifying chute, which is quite simple in itself. The rib *A*, Fig. 15, divides the caps so that they are all discharged with their open face up on the horizontal rotating disc *B*.

The charge of caps is thrown on the horizontal disc *C*, and centrifugal force places the caps in a single line between the bar *D*, Fig. 16, and the inner edge of the housing *E*. From this disc the caps pass down the rectifying chute *F* to the second horizontal disc *B*, where centrifugal force again is used to form

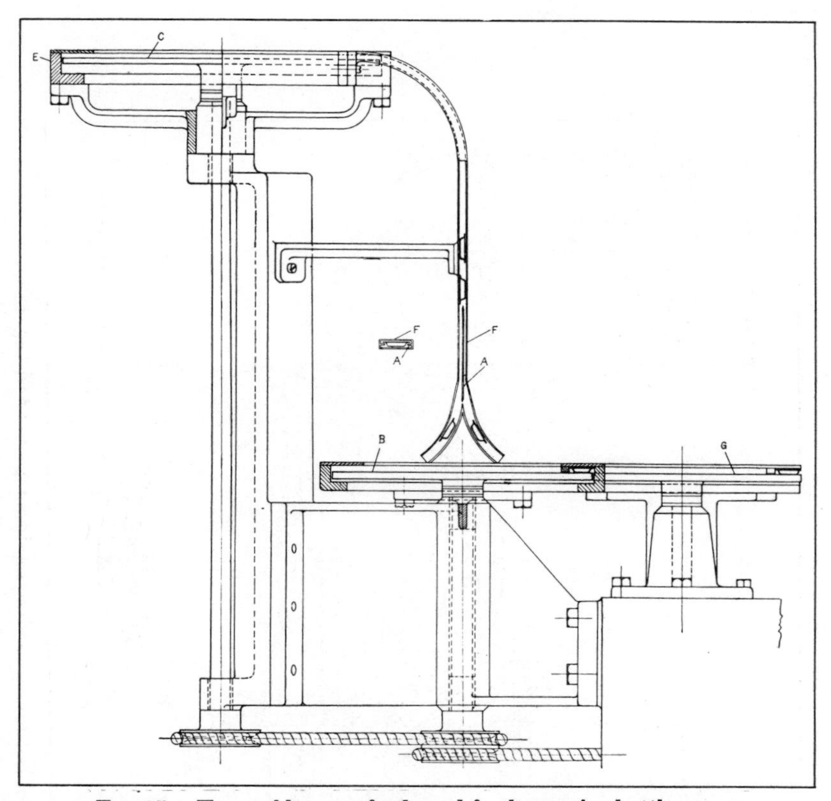

FIG. 15. Type of hopper feed used for large-size bottle caps.

a single line of caps. From disc *B*, the caps pass to disc *G*, from which they are fed into the feeding mechanism.

It is obvious that this hopper is very much simplified both in design and construction. Nevertheless, it has maintained excellent performance. However, it is restricted to a slow-speed feeding and has no means for controlling the flow of caps other than that obtained by limiting the number of caps thrown on the first disc from the bin overhead.

Designing Hopper Feeds for Bottle Caps – VI

Bottle cap manufacturing machines and bottling equipment are among the most essential elements of the brewing industry.

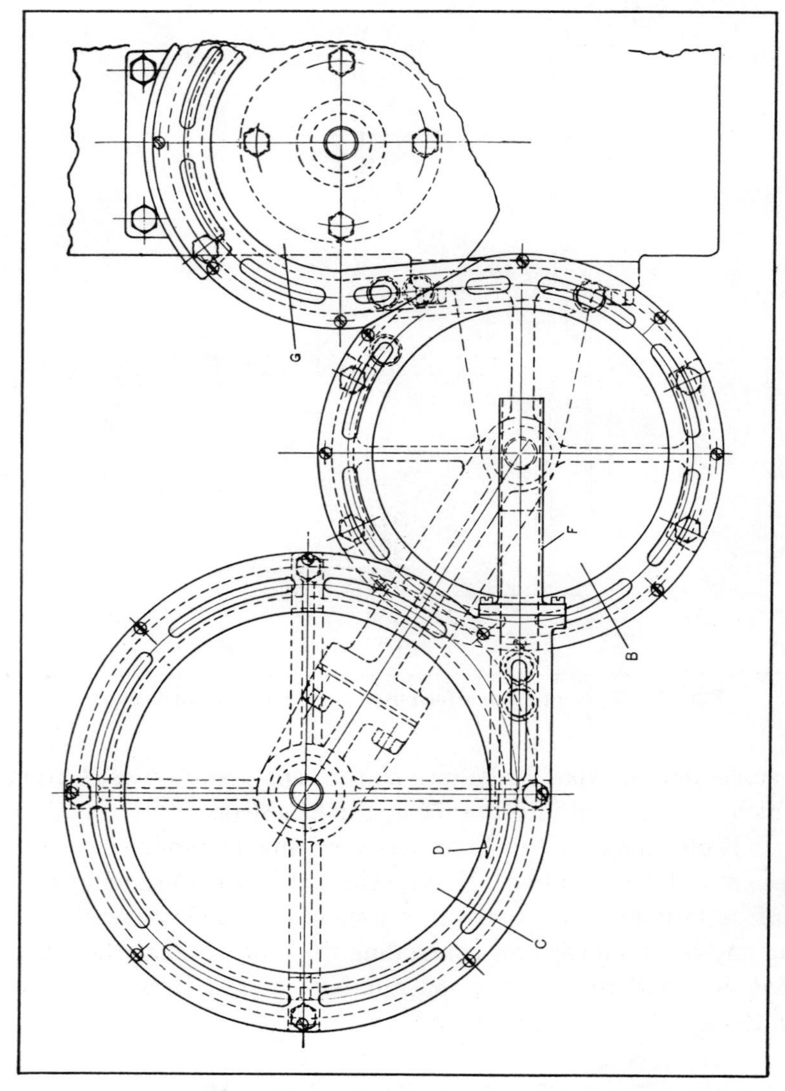

FIG. 16. Plan view of bottle cap feeding equipment shown in Fig. 15.

Remarkable progress has been made in the development of these elements along lines intended to increase production and achieve automatic operation. The hopper feed requirements of the bottle cap manufacturing machines and the bottling machines are very much alike and present designing problems requiring much the same treatment.

The word "hopper" suggests simply a deep receptacle or container for holding a supply of caps for the machine. However, there are many details connected with these hoppers that make them more than a simple receptacle. The most important of these details is the mechanism for arranging the caps so that they will all be face up or all face down, depending upon the kind of machine under consideration. In the bottle cap manufacturing machine, the caps are generally delivered face up, while in the bottling machine, they are usually required to be face down.

The ordinary bottle cap is usually made of sheet tin of a relatively thin gage and can be distorted under light pressure. It is provided with a cork disc, retained in place by an adhesive such as gutta-percha or albumen. An appropriate design is usually lithographed on the top of the cap. Therefore, any device that handles the caps must be so designed that it will not disturb or mutilate the lithographed design.

The general construction of a so-called "pin rectifying hopper" for delivering the caps to a machine in an orderly manner, all with the same side up, is shown in the accompanying illustration. The chute *B*, see Fig. 17, shown by the dot-and-dash lines, supplies caps to the main hopper *A*. The plate *M* serves as a means of checking the flow of caps, in case too large a quantity is placed in the chute. A clean-out door *N* is provided, which is held in place by clip *P* and latch *R*. The rotating member consists of the dish-shaped plate *C* and the ring *D*, which are held together as a single unit by the pins *E*. The ends of these pins are riveted over, as shown in the enlarged section *Y–Y*. The rotating member is fastened to the shaft *F*, supported in bearings *H*, and is kept in continual rotation by a belt drive to the pulley *L*. It is desirable to have a minimum amount of

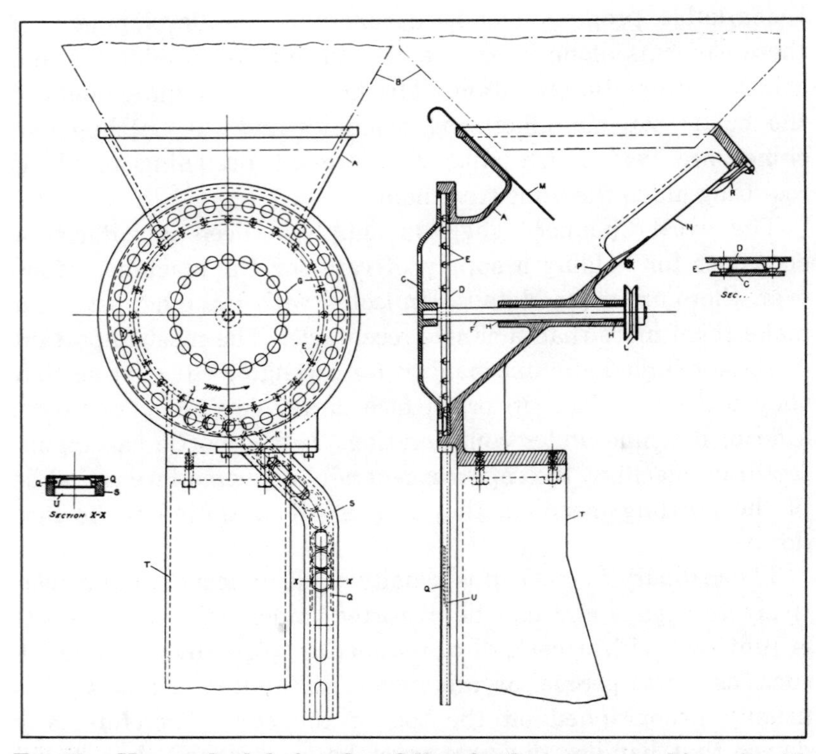

FIG. 17. Hopper feed designed to deliver bottle caps to machine, all with the same side up.

clearance between the ring D and the hopper A. A large number of holes G are provided in the plate C and slots are cut in the cover of chute S for the purpose of observing the movement of the caps. The entire hopper is supported by column T.

When the caps fall into hopper A, they are kept in a state of continual motion by the pins E. From section Y–Y it will be seen that the caps can pass between the pins E only when they are in one position. It is, therefore, the purpose of the rotating member to keep the caps in motion until they are so positioned that they will pass between the pins to the chamber J. From this chamber the caps fall into the slot K and thence down the chute S, which feeds them into the machine in an orderly manner.

As a precaution, chute S is provided with a safety slot which prevents any cap that may fall in the wrong position from passing to the machine below. The safety slot is shown in section X–X. It consists of two ribs Q, which hold the cap in the chute if it is in the proper position, or permit it to drop out of the chute through the opening U if the position is reversed. This safety device becomes effective when a cap is crushed, so that it passes between the pins E to chute S.

The main advantage of this arrangement is its simplicity. Its few operating parts and low construction cost make it very desirable from the manufacturing point of view. On the other hand, the rotating member causes the caps to turn around in the hopper and often results in scratching the lithographing. This condition is aggravated by excessive pressure when there is no automatic action available for regulating the quantity of caps admitted to the hopper. Therefore, the main disadvantage is in the churning action, coupled with excessive pressure, which results in serious abrasion and scratching.

"Jamproof" Feeding Mechanism

Figures 18 and 19 present an interesting materials handling system for advancing small parts onto a feed track. The system possesses merit because of the manner in which the feed rate is controlled to prevent any parts from jamming on the track. In this instance, small round-head screw blanks are being fed into a threading machine.

A shallow pan A, Fig. 18, holds a batch of screws. The pan rotates at about 40 rpm around an inclined axis. As the pan rotates, the screws are directed onto a fork B, where some of the screws will align themselves and hang by their heads. At its other end, the fork is hinged to the feed track C, the slot of which forms a continuation with that of the fork.

The fork is intermittently raised to the position shown by the broken lines, where the screws on the fork can slide onto the feed track. This movement of the fork originates from a cam (not shown) rotating on the bottom of a vertical stem D joined to the fork by a link E. As the cam slowly rotates, it intermit-

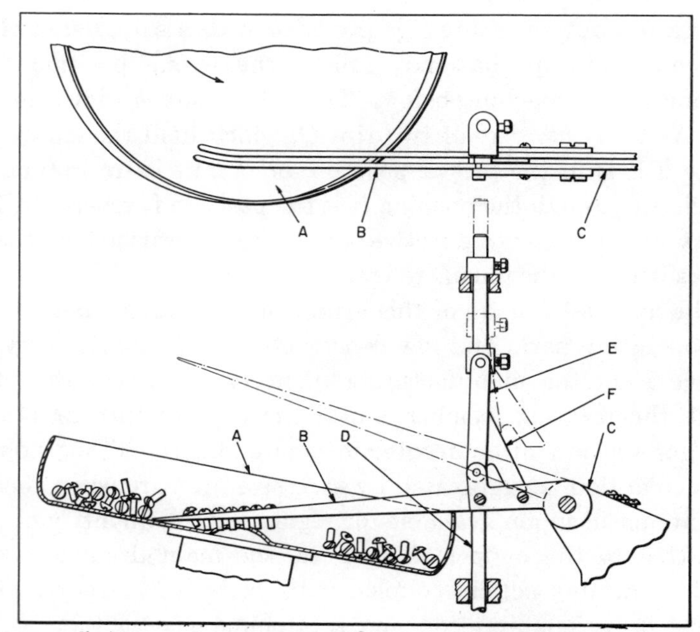

Fig. 18. When the track is not congested, the fork can drop into the pan to pick up screws.

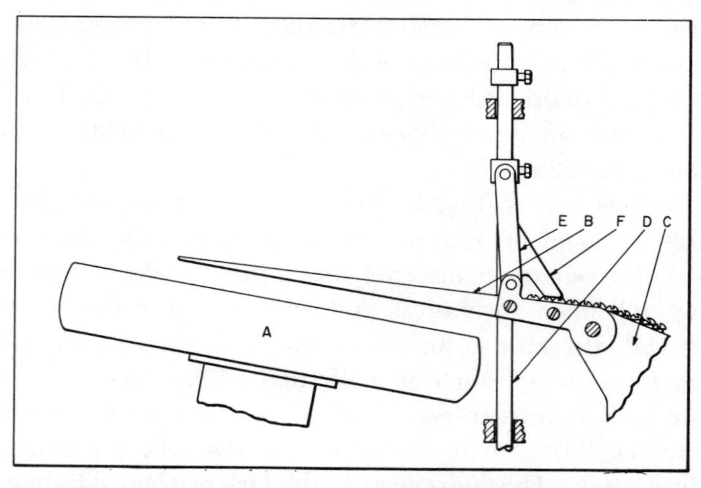

Fig. 19. When the track is congested, the feeling finger prevents the fork from dropping into the pan.

tently raises the stem, then lets the stem fall down. In this manner, the fork is also intermittently raised or lowered.

When the pan is full, each time the fork is raised it may deliver more screws than can be accommodated on the feed track. Consequently, there may be a serious jam on the track, if it were not for a feeling finger *F* which is incorporated in the mechanism. This finger is fixed to the stem, and moves up and down with it.

If the line of screws in the track extends back onto the fork, as in Fig. 19, then the down stroke of the stem is arrested when the finger strikes the heads of the screws. Since the stem and fork are linked together, the fork does not swing completely down into the pan until the congestion on the track has ended. Then, as in Fig. 18, the finger can enter the track during the reciprocation of the stem, and the fork can drop into the pan in order to pick up more screws.

Disc-Stacking Device

The power-operated device shown in Fig. 20 is used for stacking discs like the one shown at *A*. The same arrangement can also be employed for other forms of discs or shallow shells. It is entirely automatic in operation, except for the removal of the complete stack of discs. The particular device shown has been in use for several years and has never given trouble or required repairs of any kind.

Discs *A* are stacked in piles of varying numbers after being punched out of sheet metal on an inclined press. The stacking device consists essentially of two feeding or work-traversing screws *E* and *F* with a special form of square thread. This thread is cut very thin to provide a large space between adjacent threads. The screws are identical in size and shape, with the exception that one has a left-hand thread and the other a right-hand thread.

The screw *E* fastened to the shaft *J*, is driven by the main shaft *L* through a pair of spiral gears *K*. Two spur gears *H* transmit power to the screw *F*, which rotates on stud *G*. The

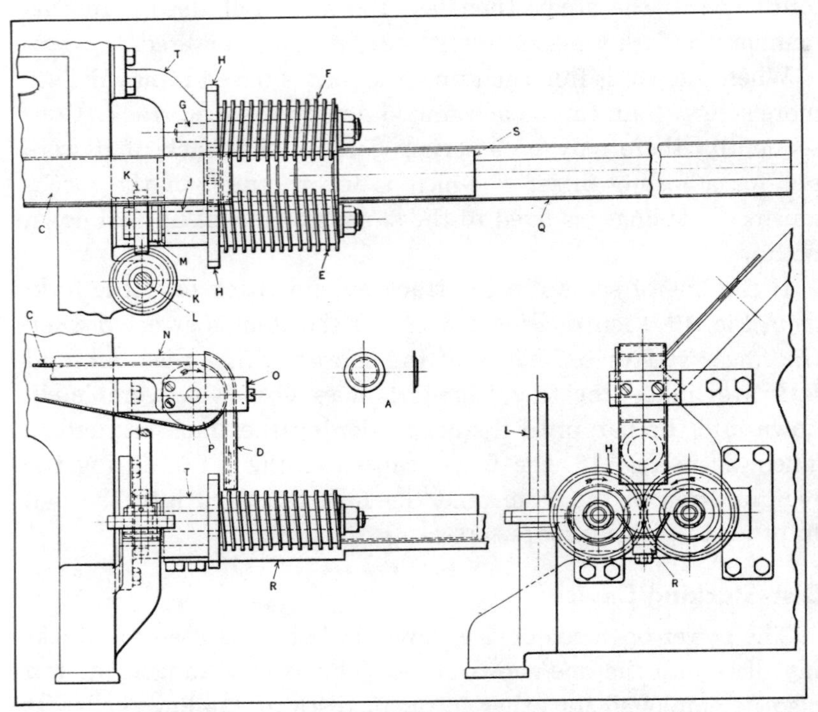

FIG. 20. Device for stacking discs such as shown at *A*.

entire mechanism is carried by a bracket *T* fastened to the press. Thrust bearings *M* are provided for the spiral gears.

A conveying arrangement consisting of a belt *C* and a pulley *P* delivers the disc to the chute *D*. The bracket *O* supports this conveying equipment. A guide strip *N* is located at the side of the belt. The vertical chute *D*, which directs the discs from the belt to the screws *E* and *F*, is fastened to bracket *O*.

The operation of the device is very simple. The press, being inclined, discharges the discs on the conveyor belt *C*, which, in turn, carries them to the vertical chute *D*, through which they drop to a position between the threads of the revolving screws *E* and *F*. The screws advance the discs until they are discharged into the trough *Q*, which is supported by a bar *R* fastened to the bracket *T*. As the discs are deposited in the trough in an orderly fashion, the slug *S* is gradually pushed back. The trough is graduated so that the attendant can see when the desired number of discs has accumulated. They can then be removed and the slug *S* pushed back to its original position.

As the stacking device receives its supply from a conveying belt, it can be located at any required point. The best results are obtained when the screws *E* and *F* revolve one revolution for each disc produced. At this rate, a smooth and orderly stack is obtained.

Crank Principle Controls Shuttle Between Supply and Discharge Chutes

Figure 21 shows a device for directing small objects — ball screws, rivets and the like — from one chute to two chutes. Initially, nest *E* in slide *D* is located beneath chute *A* allowing an object to fall into it. The nest is then caused to move to one of the chutes *B* or *C* by crank or cam. After the object has fallen into the chute, the nest moves back to chute *A*, picks up another object, then moves to the opposite chute and drops the second object.

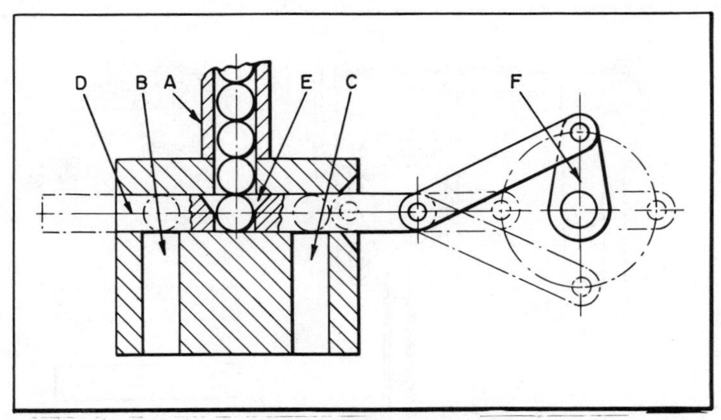

FIG. 21. Transfer of parts from one vertical plane to another is made possible by utilizing the principle of the crank.

Mechanism for Orienting Pins in Assembly Machine

Prior to pressing pins into side plates of a cotter-pin roller chain it is necessary to position them so that all pins going through a tube *A* leading to the assembly machine will have the cotter-pin hole at the bottom end. These pins are of the type shown at *X* in Fig. 22. They are ½ inch in diameter by 2¾ inches long.

In the mechanism shown a tube extending from a rotating hopper carries pins fed by gravity. However, the pins are not oriented as to position of the cotter-pin hole. As drum *B* rotates, gravity forces each pin into a slot on the drum. Each pin is subsequently dropped on rollers *C* and *D*, which turn as indicated by the arrows.

The distance between adjacent high points of the rollers is 0.015 to 0.020 inch less than the diameter of the pins. As the rollers revolve the pin also rotates. When the center line of the cotter hole in the pin is horizontal, the hole end of the pin falls through and the opposite end slides off, as indicated by the dotted lines at *Y*. Thus all pins leave the mechanism with the cotter hole at the bottom end.

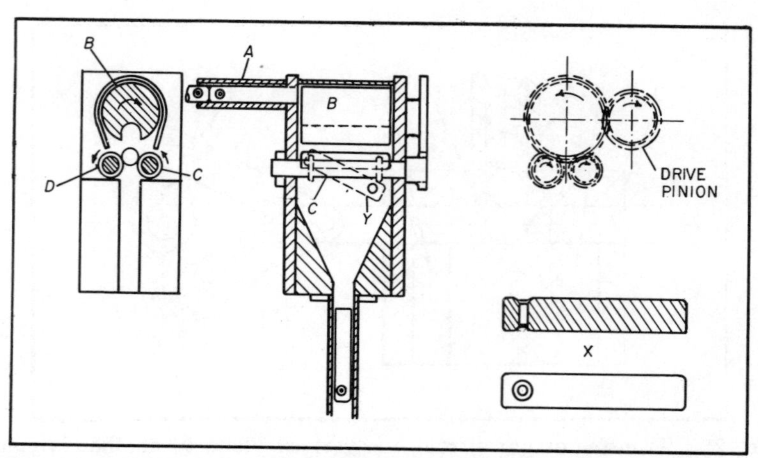

FIG. 22. Mechanism for feeding roller pins to assembly machine with drilled end downward.

CHAPTER 18

Varying Continuously Rotating Output

Some machines require mechanisms that convert uniform rotary motion into variable rotary motion. The mechanisms in this chapter cover this and provide for varying continuously rotating outputs.

Two-Gear Drive Produces Variable Output Motions

Machines often need a drive in which the input shaft turns with uniform angular velocity and the output shaft rotates at a different velocity. Examples are an alternately increasing and slowing of rotation, with stopping; the same type of motion but with the shaft coming to a prolonged dwell; and a varying rotation, interrupted by a reverse oscillation. The two-gear drive shown can produce any of these motions, depending on the motion of a pin that can be readily changed to give the desired output.

Essentially, the mechanism consists of a four-bar linkage A_oABB_o, as shown in Fig. 1. Points A_o and B_o are pivot points on a stationary frame. The input crank rotates around point A_o and carries pivot point A, which is the center of an internal gear. Attached to the internal gear is a pin at point B. A rocker arm — and a pin in the frame at point B_o — pivot on this pin.

The internal gear meshes with an external gear attached to the output shaft.

Center A of the internal gear rotates around A_o, which is the same axis on which the external gear rotates. Consequently, the two gears are always in mesh.

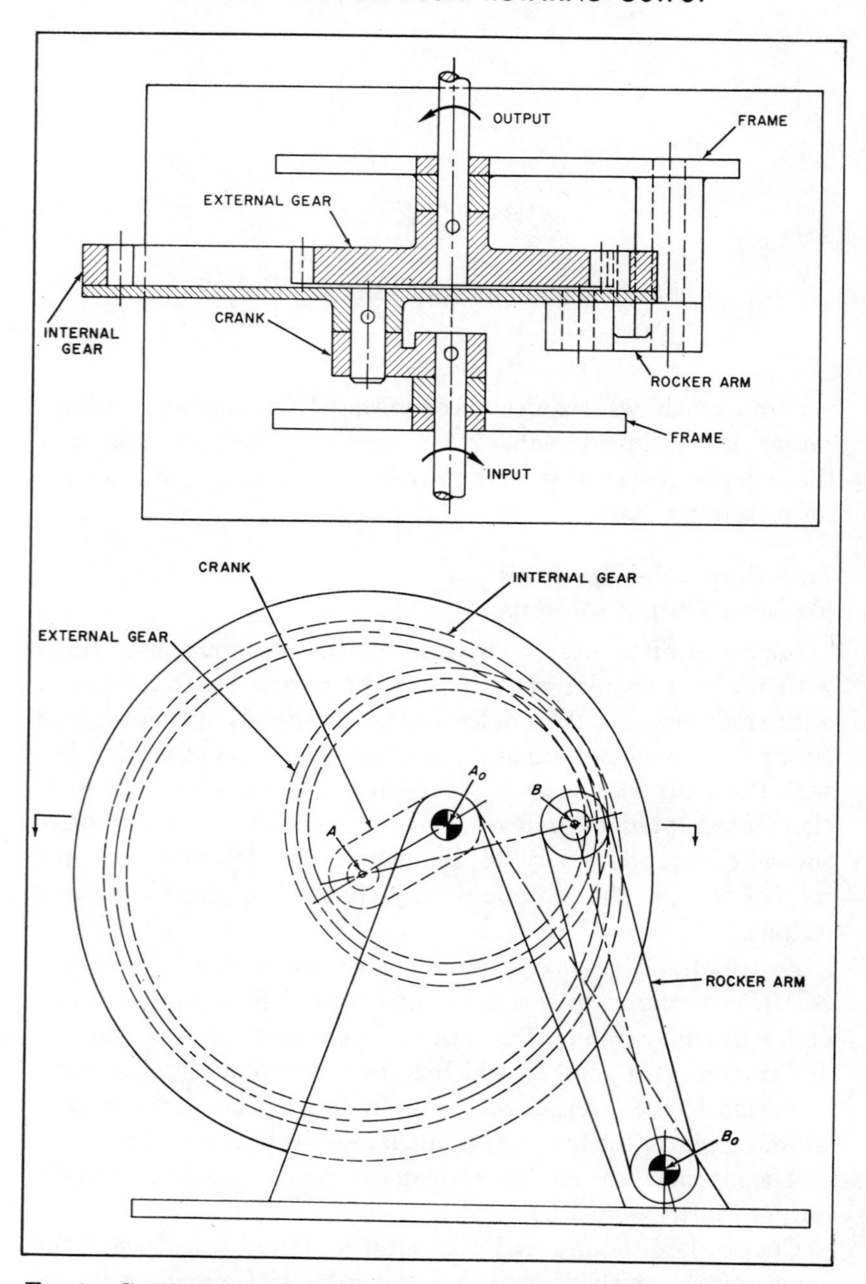

FIG. 1. Crank translates internal gear, causing external gear, with which it meshes, to rotate.

As the crank rotates, the center of the internal gear rotates around point A_o. The pin attached to the internal gear moves in a circular arc around point B_o. The combined motion of points A and B causes the internal gear to rotate and oscillate around the external gear.

Location of the pivot on the internal gear relative to its pitch circle is important. If the pin is outside the pitch circle, the output motion of the external gear is a rotation interrupted by an oscillation. If the pin is on the pitch circle, an instantaneous dwell results. If it is inside the pitch circle, the output motion will still be a rotation but it will alternately speed up and slow down.

Theoretically, dwell is instant. Practically, dwell can be considered prolonged. Two units in series will give an output motion with a larger dwell than one unit. Two units can also give an output with two dwells.

Cam and Link System Produces Varying Rotation Rate

Figure 2 shows two views of a mechanism which converts uniform rotary motion into variable rotary motion, maintaining

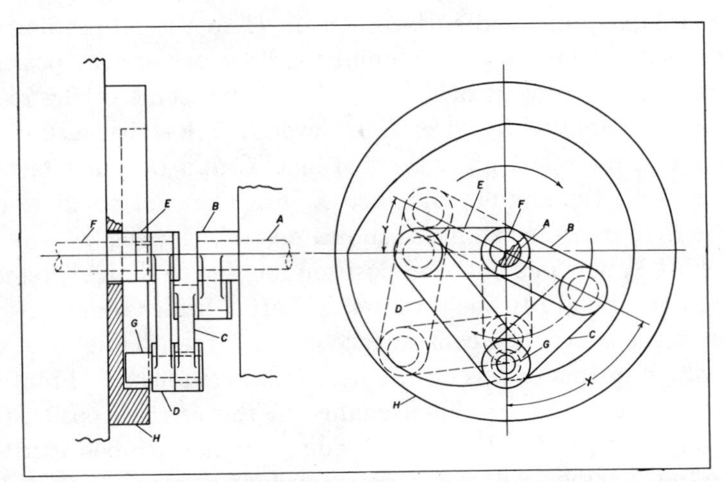

FIG. 2. The result of this link and cam system causes driven shaft F to slow down and then pick up speed with relation to drive-shaft A.

required relationships between the positions of the drive-shaft and the driven shaft. This mechanism was designed for use on a machine on which it is necessary: (1) for the mechanism carried on the driven shaft to have its angular position changed relative to the mechanism carried on the driving shaft; (2) to maintain the changed position for a required portion of the cycle; and then (3) to gradually return the driven shaft to its original relative position.

The driving shaft *A*, rotating uniformly in the direction indicated by the arrow (clockwise), carries the lever *B* keyed to it. Lever *B* carries the link *C*, which is connected to link *D*. Link *D* is connected to lever *E*, which is keyed to the driven shaft *F*. Link *D* carries, at its lower end, the follower roller *G*, which is in contact with the internal cam *H*, attached to a stationary part of the machine with its center coinciding with the centers of shafts *A* and *F*.

Referring to the diagram at the right, the linkage drawn with solid lines indicates the positions of the elements when the rotation and relative angular positions of shafts *A* and *F* coincide. At this point, roller *G* has arrived at the point terminating the concentric surface of the smaller diameter on the inside of cam *H*. Continued rotation of shaft *A* causes roller *G* to ride down the circularly inclined surface of cam *H* to the larger-diameter concentric surface, at which point the links occupy the positions shown by the broken outlines. In so doing, some of the rotary movement applied by lever *B* to lever *E* is lost because of the change in the relative positions of links *C* and *D*. Thus, lever *B*, in rotating the angular distance *X*, has transmitted motion to lever *E* to a lesser extent, as indicated by the angle *Y*.

Shaft *F*, now in a changed position relative to shaft *A*, remains in this position until the rotation of shaft *A* brings roller *G* to the end of the concentric cam surface of the larger diameter of cam *H*, which in this case is on the horizontal center line. From this point on, there is a gradual change in the relative positions of shafts *A* and *F* due to roller *G* riding up the inclined cam surface to the larger-radius concentric surface. This change in shaft positions is in the opposite direction to that previously produced, so that when roller *G* reaches the vertical center line, shafts *A* and *F* are again in their original relative positions, remaining so for a half-cycle when the levers and links will again be in the position shown by the solid outlines.

Adjustable Drive Produces Continually Variable Rotation

A chain-operated drive which converts uniform rotation into variable rotation and can be adjusted while in operation is shown in Fig. 3. The purpose of the drive is to increase the rotative speed of the driven shaft for a portion of each revolution so that the dwell of a cam may be changed.

Driven shaft *A* is keyed to a gear *B* and passes through the bore of a disc *C*. The hub of member *C* is free to rotate in a pillow block *D*. A collar *E* retains disc *C* in the pillow block, and a second collar *F* holds shaft *A* in disc *C*. Mounted on the same disc is a sprocket *G* that is rotated uniformly by a drive chain. In addition, disc *C* has a groove that carries a plate *H*, which is free to slide in this groove.

Gear *B* rotates in a rectangular section removed from plate *H*. One side of this cutout in plate *H* has rack teeth which mesh with gear *B*, while the opposite side of the cutout is machined to clear the gear. Plate *H* also carries a roller follower *J* at one end. A block *K* is mounted on the bed of the machine and is grooved to accommodate a bar *L*. This bar is free to slide in block *K* but is retained in the groove by two blocks *M*.

A stud *N* is secured to bar *L*, and is drilled and tapped to receive a screw *O* which, in turn, is supported by blocks *M*. Bar *L* also carries a ring *P* and a disc *Q* mounted concentric with the ring. These members are so spaced that the roller follower *J* can move freely around the circular path between them. In the position shown, bar *L* with ring *P* and disc *Q* has been moved horizontally to the right by means of screw *O*, placing the center of disc *Q* a distance *X* from the center of shaft *A*.

If disc *Q* and ring *P* are moved to the left so that their common center coincides with the center of shaft *A*, the rotation of sprocket *G* is transmitted through disc *C* and plate *H*, causing

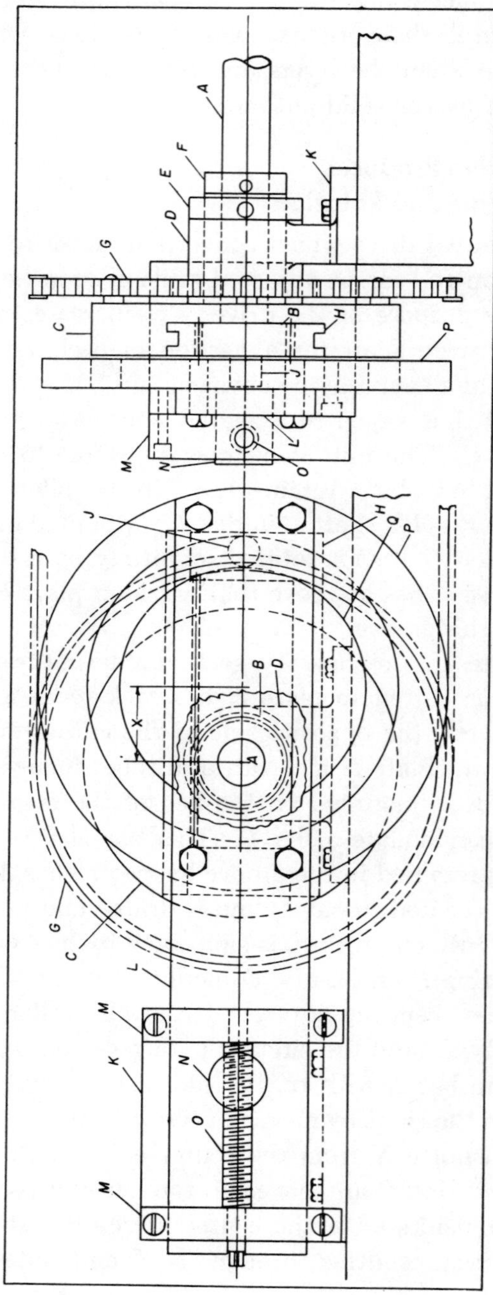

FIG. 3. This drive which produces continually variable rotation can be adjusted while in operation. When center of disc Q coincides with that of shaft A, however, no variation in the drive is obtained.

roller follower *J* to travel in a circular path concentric with the center of shaft *A*. The rack teeth on the inside of plate *H* that mesh with gear *B* simply act as a key to transfer the rotation to gear *B*. Thus, the rotation of gear *B* and shaft *A* are then synchronized with that of sprocket *G*.

When plate *L* is moved horizontally to the right toward the position shown in the illustration, roller *J* no longer travels in a path concentric with the center of shaft *A*. Instead, the roller alternately approaches and recedes from the center of shaft *A*, thus imparting a reciprocating motion to plate *H* and its rack. The magnitude of this reciprocation is controlled by the adjustment of screw *O*. Reciprocation of the rack produces a gradual and alternating increase and decrease in the rotational speed of driven gear *B* relative to that of the driving member. In this manner a continually varying rotational motion is imparted to shaft *A* by the uniformly rotating sprocket.

This mechanism is not limited to producing a uniformly variable rotation of the driven shaft. By changing the shape of disc *Q* and ring *P*, a variety of movements may be obtained in the same manner as from any cam.

Coupling Connects Displaced Shafts

Rotary motion can be transmitted between shafts with a considerable degree of displacement, using the coupling shown in Fig. 4.

This coupling can be used for high-speed and high-torque applications. Inherently, it is dynamically balanced. Angular velocities of the driving shaft are identically reproduced in the driven shaft and no phase shift is produced when shaft displacement is changed.

The drive-shaft is keyed to a disc. Three eccentric, equally spaced pins extend from the face of the disc. The driven shaft has an identical disc and pin arrangement, but is parrallelly displaced from the drive-shaft. A connecting or intermediate disc has three cylindrical pins that extend from each side face and have the same spacing as pins in the other two discs.

Two sets of three links act as connecting members between the discs. These links are of the same length and are less the diam-

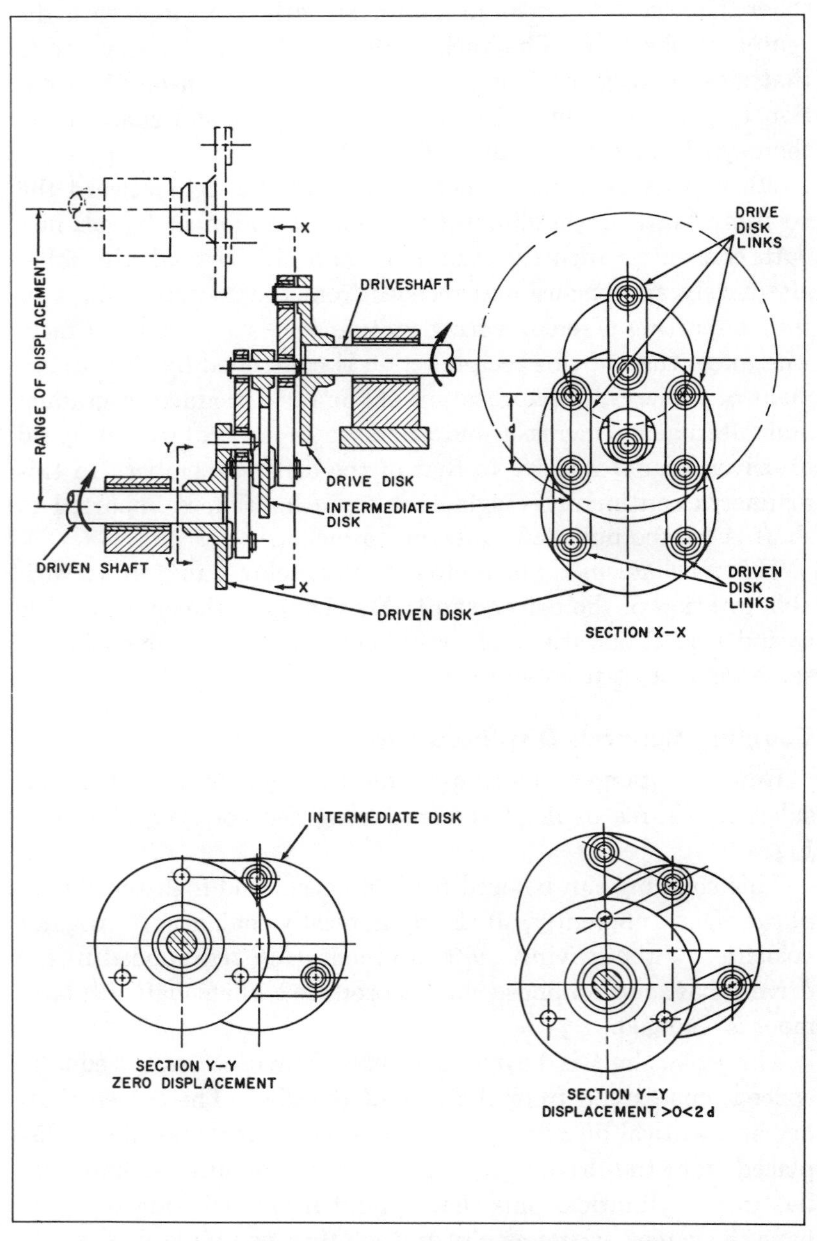

FIG. 4. Constant rotation is transmitted through coupling connecting displaced shafts.

eter of bolt circle formed by the pins. Each has two equally spaced bearings that engage the corresponding pins in adjacent disc faces.

Maximum radial displacement of the shafts is determined by the distances between bearings in the links, multiplied by the number of link sets. Since two sets of three links each are used, the maximum possible radial shaft displacement is $2d$. If the shafts are in the fixed position shown in section Y–Y and the shaft displacement is greater than zero and less than $2d$, the intermediate disc is suspended from the links between the other discs.

The pivot points of the two sets of links form three triangles which are geometrically determined by the length of the links and the fixed parallel displacement of the shafts.

Since the intermediate disc is suspended by three pins located at geometrically determined points, the center of the intermediate disc is also geometrically determined and cannot be moved unless the parallel shaft displacement becomes zero and the centerlines of the pins in the outer disks fall together. In this case the intermediate disc is free to swing about the centers of these pins. For practical applications, the zero-displacement position should be avoided.

When varying the displacement of the shafts, the intermediate disc automatically adjusts its position in compensation. The shafts can be displaced under loads.

Cam-Controlled Differential Mechanism

In a punched-card machine it was necessary to include a mechanism that would convert a uniform rotary input motion into a varying rotary motion. The mechanism shown in Fig. 5 proved very successful in comparison with other mechanisms designed to perform the same job.

In the preferred mechanism, the two shafts are geared together in the ratio of 1 to 1 by an arrangement that is not shown. On shaft A is fastened a cam C and its complementary cam D. These two cams drive arm T through the rollers R and S. Arm T carries gear E, which is driven by gear I. The arm also carries

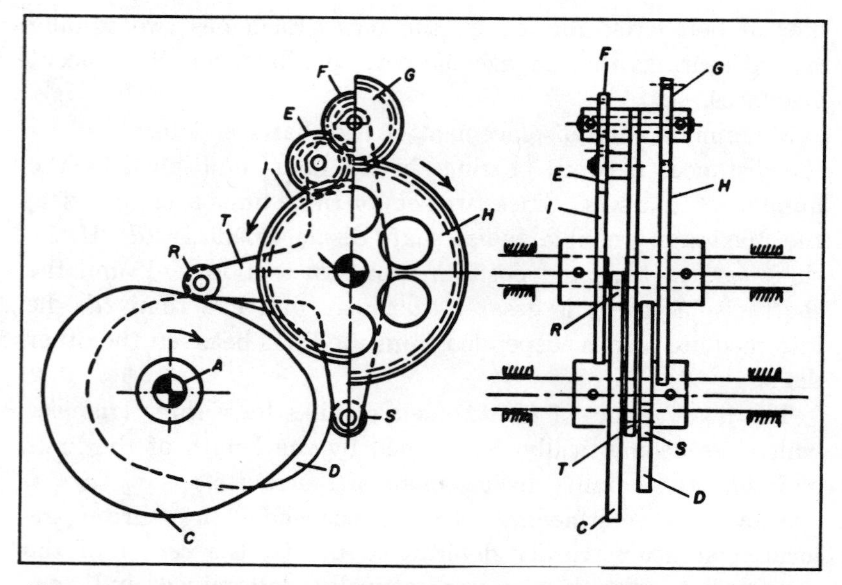

FIG. 5. Derivation of varying rotary motion from uniform rotary motion.

gear *F*, which is in mesh with gear *E*, and gear *G*, which is keyed to the same shaft as gear *F*.

As gear *I* rotates, it transmits motion through the gear train *E*, *F*, *G*, and *H* with a constant angular velocity. However, because of the oscillating motion of arm *T*, gear *E* will roll on gear *I*. The result is that gear *H* will move with a nonuniform velocity according to the shape of the curves on cams *C* and *D*.

Cams *C* and *D* impart a constrained movement to arm *T*. If only one cam were employed, springs would be necessary to hold the follower on arm *T* in contact with the cam.

Timing of Cam Changed While Machine Is in Motion

It was required that the timing of cam *A* (see Fig. 6), be changed while in motion. Drive-shaft *B* rotated the keyed gear *F*. Gear *F* rotated pinion *G*, supported by the normally stationary gear *C*. Pinion *G*, in turn, rotated gear *L*, through pinion *J*, in the opposite direction to shaft *B*. The timing of cam *A* is adjusted by rotating pinion *D*, resulting in gear *C* rotating, say *X* degrees. The attached pinions will rotate X degrees and the cam *A* which is attached to gear *L*, will rotate 2X degrees as gear *L* has the same pitch diameter as *F* and the pinions are of equal pitch diameter.

Unidirectional Rotation Regardless of Changes in Drive Direction

Occasionally it is necessary to drive a shaft in only one direction even though the driving member may alter its own direction of rotation. In designing the drive mechanism for a recorder drum, such a situation arose. The drum was required to maintain a single direction of rotation regardless of the fact that its driving member fluctuated between a clockwise and a counterclockwise rotation.

A clutch and bevel gear arrangement, Fig. 7, consists of three bevel gears; gear *A* being the driven member, and gears *B* and *C* being the driving members. Both driving gears are mounted on roller clutches *D* which are, in turn, mounted on a common drive-shaft *E*. The clutches provide positive drive when rotated

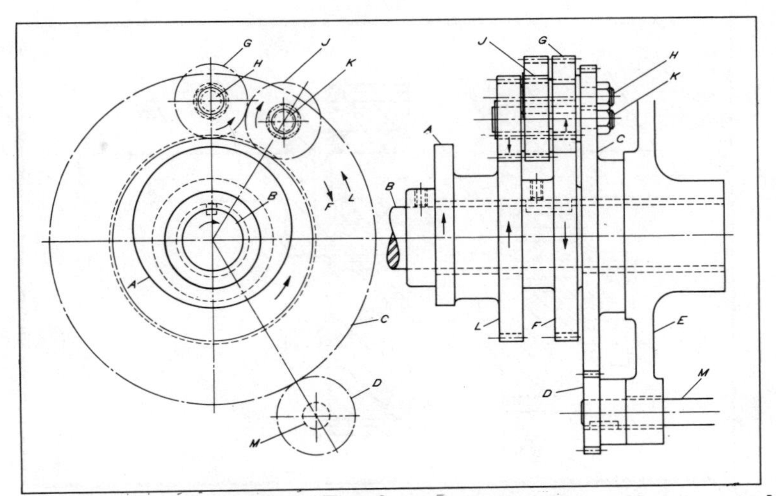

FIG. 6. The gear train from *F* to *L* permits the timing of cam *A* to be changed while it is rotating.

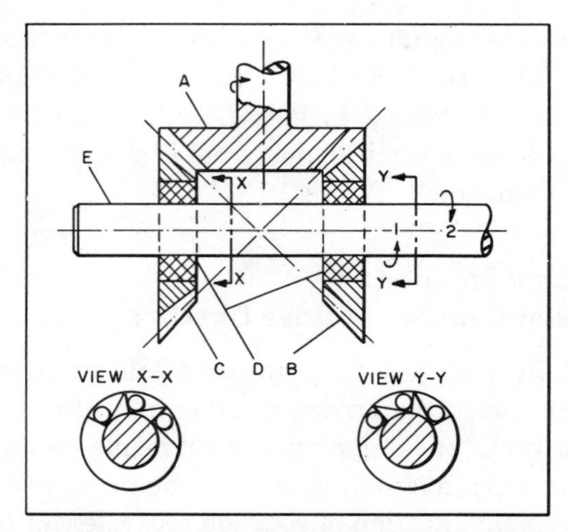

FIG. 7. Two bevel gears mounted on roller clutches *D* provide unidirectional rotation for driven gear *A* regardless of rotational direction of driveshaft *E*.

in one direction, and overrun, or "free-wheeling," with reverse rotation. Each of the two clutches in this set-up is mounted in opposite directions from one another; hence bevel gears *B* and *C* cannot drive in the same direction.

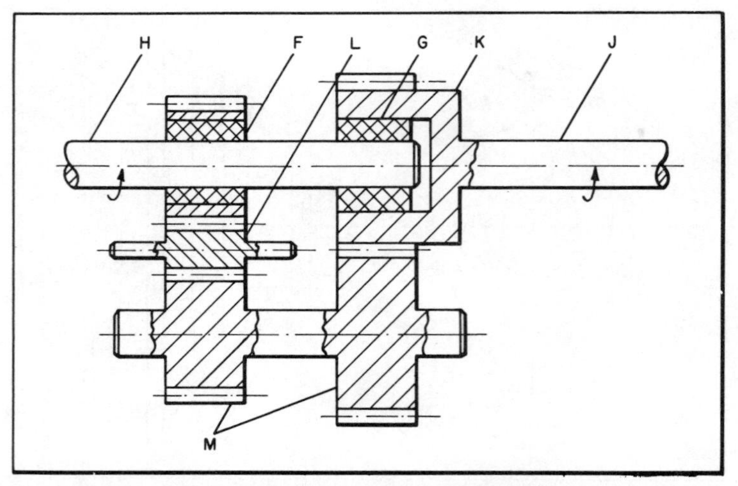

FIG. 8. Spur gear set-up designed to accomplish the same purpose as bevel gearing shown in Fig. 7.

When the drive-shaft rotates in the direction indicated by Arrow 1 (at the right-hand end of drive-shaft), the clutch in gear *B* engages while the clutch in gear *C* slips, thus driving gear *A* as indicated. Upon reversal of the direction of drive-shaft rotation to that indicated by Arrow 2, the operation of the clutches is also reversed. In this way, the direction of gear *A* is unchanged.

A set-up employing spur gears to obtain the same end result may be seen in Fig. 8. Here again, two roller clutches *F* and *G* are mounted on common drive-shaft *H*. In line with this is driven shaft *J*, on the end of which is an integral external spur gear *K*. Gear *K* is counterbored to fit over the outside diameter of clutch *G*.

As the drive-shaft rotates in the direction indicated by the arrow, clutch *F* slips while clutch *G* engages, thus forcing the driven shaft to rotate as shown. When drive-shaft rotation is reversed, clutch *F*, which is mounted within the bore of a spur gear, engages as clutch *G* slips. Motion is transferred through idler gear *L* to intermediate gears *M*, and finally to driven gear *K*, thereby causing the direction of rotation of drive-shaft *J* to remain the same.

Timing of Driven Shaft Changed While Running

A power train through sun gears which rotate within a stationary, but adjustable ring gear permits the timing of a driven shaft to be changed while it is running. The arrangement is used in positioning revenue stamps correctly in a tobacco-packaging machine.

Ordinarily, driven shaft *A* (see Fig. 9), and driving shaft *B* revolve in unison. The transmission is as follows: Pinion *C*, integral with the continuously revolving driving shaft *B*, meshes with two sun gears *D*. These sun gears are the same size as the pinion. They are free on studs *E* fixed in diametrically opposite points in disc *F*. Near the center, the disc forms a hub which is pinned to driven shaft *A*. Both sun gears mesh with stationary ring *G*. Thus, as shaft *B* revolves, the sun gears re-

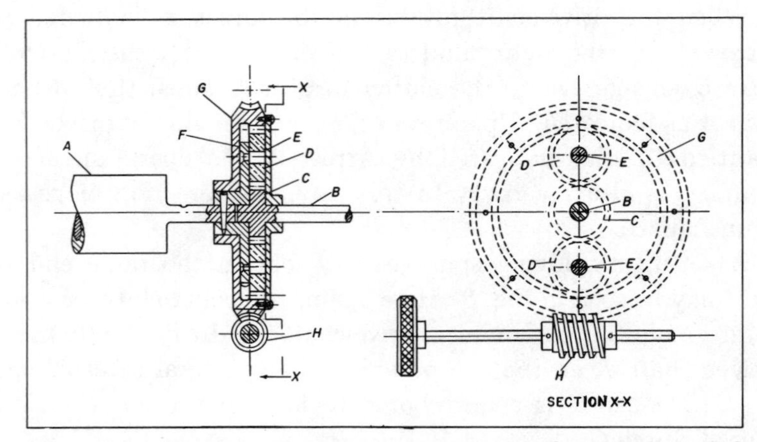

FIG. 9. When the ring gear G is rotated by worm H, the timing of driven shaft A changes with respect to the driving shaft B.

volve on their studs, and in addition they rotate within the ring gear, causing the disc to turn shaft A.

For changing the timing of shaft A with respect to shaft B, the outside of the ring gear is machined to a worm-wheel. The changes of angular displacement of the disc with respect to the pinion, and advances (or retards) the driven shaft the required amount. No stoppage of the mechanism is made, and since the worm is self-locking, it will preserve the timing, once set.

Mechanism for Varying Rotation of Winding Mandrels

The mechanism illustrated in Fig. 10 is designed to vary the rotation of two mandrels on which paper tubes are wound. Paper from the web is wound first on mandrel A and then on mandrel B. The tube thus produced on one mandrel is stripped off while a tube is being wound on the other mandrel. A completed tube consists of a given number of full turns of paper plus a fraction of a turn as indicated at a in diagram X. The means for applying adhesive to the proper portion of the web for securing the lap section of the rolled tube is not shown.

In order to remain synchronous with the other parts of the machine, the mandrels can only make a specified number of full turns. Therefore, a driving mechanism for the mandrels had to be designed which would give the required number of full turns plus a fraction of a turn a on the winding cycle of each roll without increasing the total number of complete revolutions. To accomplish this, it was necessary for the mandrel being stripped to have its rotation slowed to compensate for the fractional rotation provided for the lap a before starting to wind a new tube.

The mandrels A and B, and their respective worm-wheels C and D, are driven by worms E and F. The worm-wheel C and worm E are right-hand, while the worm-wheel D and worm F are left-hand. The two worms E and F are made integral, and are slidably mounted on their driving shaft M.

The extra rotational movements of the mandrels are obtained by axially reciprocating the worms. A shaft G, which makes one revolution for each complete winding cycle of the two mandrels,

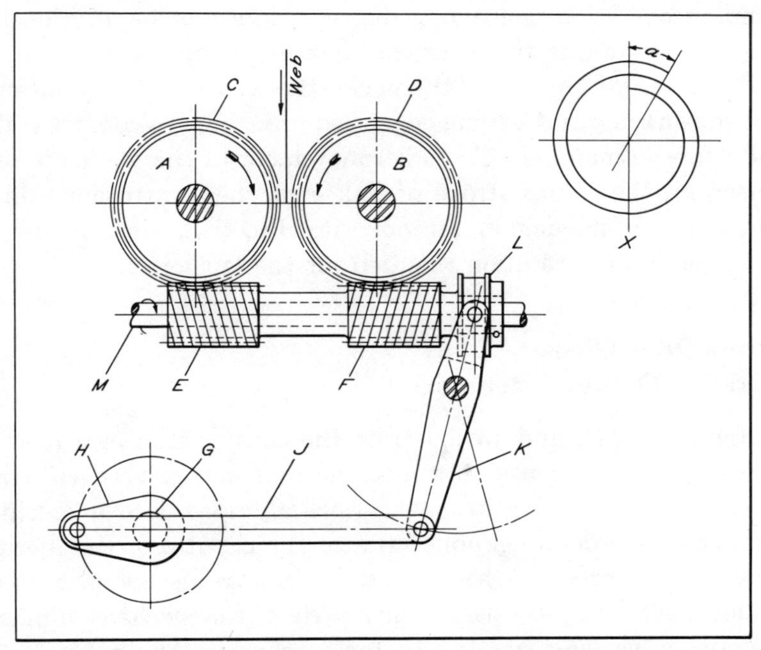

FIG. 10. Mechanism designed to vary rotation of mandrels used to wind paper tubes.

provides the drive for the worm-reciprocating movement. On shaft *G* is a crank *H* of appropriate length, which is connected to lever *K* by link *J*. The forked end of lever *K* carries shoes that ride in a groove provided for them in a collar *L*. Collar *L* is pinned to a hub at the outer end of worm *F*.

When the machine is in operation, the web is introduced between the mandrels, as illustrated, and is brought into contact, by means not shown, with, say, mandrel *A*, to which it is held by vacuum. Mandrel *A* rotates in the direction indicated, winding the paper around it the required number of laps. At the same time, the worms move to the left a sufficient distance to give the required extra part of a revolution. It should be noted that although mandrel *A* is given the additional rotational movement, this same amount is subtracted from the rotational movement of mandrel *B*. Thus it will be apparent that the overlap *a* is the sum of the under speed of the beginning of the winding cycle and the over speed at the end. This does not affect the winding operation, since *B* is being stripped of its previously wound tube during this decelerating part of the cycle.

At the proper point in the cycle, the web is cut (by means not shown) and the leading edge then brought into contact with the opposite mandrel. The axial movement of the worms is reversed on the return stroke of link *J* at the proper time; and the worm *F*, moving in the opposite direction, gives mandrel *B* the additional rotation required for the lap joint.

Worm Drive Gives Variable Output Automatically

Figures 11, 12 and 13 illustrate the construction of a mechanism designed for use on a machine that makes a woven wire product in which the strands of wire are spaced to a specific pattern. In order to produce the required pattern, the mechanism which processes some of the wire strands operates at a uniform rate of speed, while other parts of the mechanism must operate at varying speeds. In the mechanism shown, a shaft rotating at a uniform speed transmits motion to another shaft,

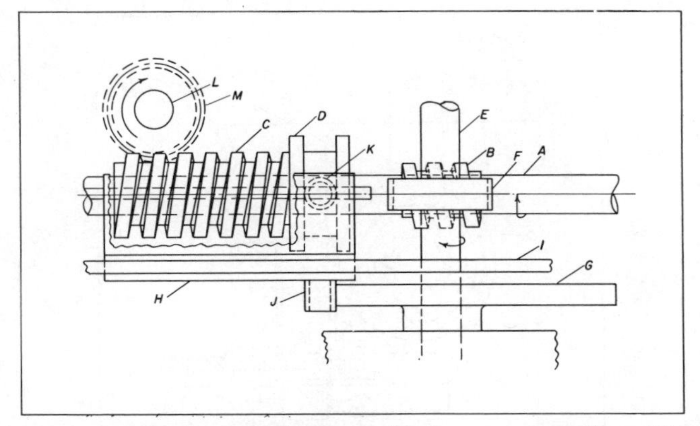

FIG. 11. Front view of speed changer with the cam G at the mid-point of its dwell.

in synchronism during part of the cycle, followed by an increase of speed, and then by a period of rest, which returns the two shafts to the original relative rotation. Figure 11 is a front view, Fig. 12 a plan view, and Fig. 13 an end view of the mechanism.

Drive-shaft A rotates in the direction indicated by the arrow and carries worm B, attached to it. Grooved collar D and worm C are fastened to each other as shown and are arranged to slide axially on shaft A, their rotation being provided by an internal

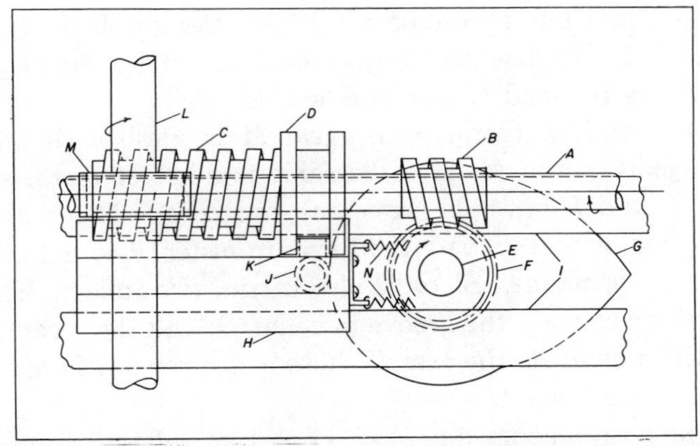

FIG. 12. Variable-speed worm-gear drive, plan view, showing slide-block H, which moves pusher worm C to left in a rack action.

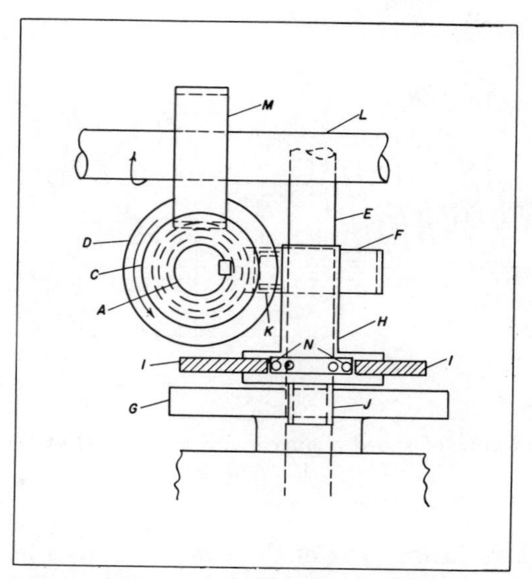

FIG. 13. The relationship of the slide-block H and its springs N is shown in this end view of the device.

key sliding in a groove in shaft A. Worms B and C have the same lead, except that B is right-hand and C is left-hand.

Shaft E, which is free to rotate, is mounted on a stationary part of the machine and carries worm-gear F, in mesh with worm B, and cam G. The contour of cam G is shown in Fig. 12. It is laid out to produce a 240-degree dwell on the base circle, 60-degree rise and 60-degree fall, with a rise equal to three times the lead of worms B and C.

Driven shaft L carries worm-gear M in mesh with worm C. Worm-gears F and M have the same number of teeth, so that shafts E and L have the same rotative speed. Block H has a sliding mount on two bars I, which are fastened to a stationary part of the machine. Block H also carries two rollers. The first is K, which engages the groove in collar D; and the second roller is J, which contacts the cam G. Two coil springs N hold roller J in contact with cam G.

In the position shown in Figs. 11 and 12, roller J is in contact with cam G at the mid-point of the dwell period. Rotation of shaft *A* transmits rotation to the shaft *E* in the direction shown. At this stage in the cycle, no motion is transmitted to slide-block *H*. Shaft *A* also transmits rotative motion to shaft *L* through worm *C* and gear *M*, in the ratio of the number of *M*'s teeth.

With the continued rotation of shaft *E*, the rise section of cam *G* contacts roller *J* and motion is thus imparted to the slide-block *H*. The movement of slide-block *H* is transmitted to collar *D* and worm *C* through the roller *K*. In addition to transmitting rotary motion to shaft *L*, the axial movement of worm *C* causes it to act as a rack to increase the speed of rotation of shaft *L*. Because the rise of cam *G* is equal to three times the lead of worm *C*, gear *M* turns through three teeth in addition to the normal worm rotation in the same period of time.

After cam *G* has reached the peak of its rise, springs *N* draw slide-block *H* back to the starting point at a rate of speed controlled by the contour of cam *G*. This movement is at a rate which permits slide-block *H* to move a distance equal to one lead of worm *C* with each rotation of shaft *A*. Therefore, no rotation is transmitted to shaft *L* on the return movement of block *H*.

With this mechanism, the relative rotation of shafts *A* and *L* remains unchanged over a 240-degree rotation of cam *G*. The rotation of shaft *L* is doubled over the 60-degree rise of cam *G*, and remains stationary over the 60-degree fall of cam *G*, so that when the cycle is completed drive-shaft *A* and driven shaft *L* again assume their original relative rotation.

CHAPTER 19

Clutch and Disconnecting Devices

Clutches and disconnecting devices which play an important role in automatic and semiautomatic machinery are described in this chapter.

Slip Clutch Suitable for Tape Drives

Many applications present themselves for a slip clutch that will not change its rating, can be infinitely adjusted, is not sensitive to speed changes, and will not wear out. One use for such a clutch is in a tape drive. Drives for both punched-paper and magnetic tapes can use this clutch. Clutches of various sizes and ratings can be built in the design shown in Fig. 1.

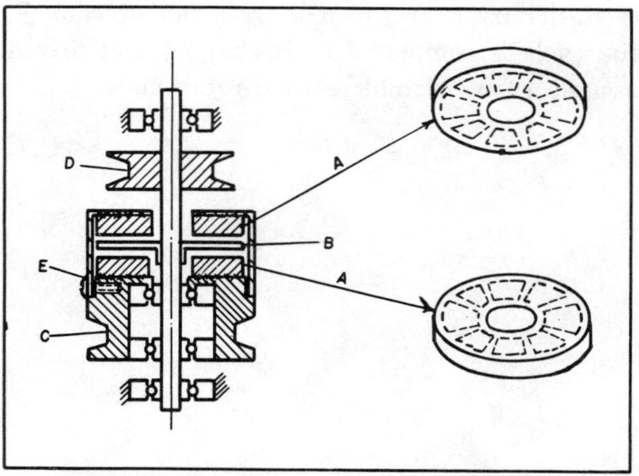

FIG. 1. Nonwearing slip clutch especially suitable for tape drives.

The clutch consists of two ceramic magnets *A* with eight separate poles in each of the opposing faces. North and south poles are alternated on the magnet faces. These magnets are mounted on the driver pulley as indicated. A disc *B* of high-hysteresis cobalt steel runs between the two ceramic magnets on suitable bearings. The output pulley *D* is driven by the disc. Power is, of course, applied through input pulley *C*.

Adjustment is achieved by rotating one of the magnets in relation to the other. Screw *E* passes through a slot in the shell of the input pulley. The highest torque is obtained when unlike poles of the magnets are opposite each other. The drive results from the magnetic path, established in disc *B*, resisting change.

Driving Mechanism Prevents Reverse Movement of Driven Shaft

Feed arrangements for a special wire-forming machine required intermittent, unidirectional rotation of the feed-drive shaft. Any shaft reversal during dwell periods in the feed cycle had to be avoided. To meet these demands, the drive mechanism, shown in Fig. 2, was developed.

Wire feed spool *A* is keyed to driven shaft *B*, as shown at X (section Z-Z). The shaft runs freely within the stationary, circular bearing bracket *C*. A clockwise rotation (indicated by the arrow) is imparted to the spool by pulley *D*, keyed to the shouldered end of the driven shaft. A nut and washer retain the pulley in position.

A counterbored recess is machined in the upper surface of bracket *C*. Cast-iron circular plate *E* is mounted at the base of this recess, a counterbore in the bottom face of the plate forming a running fit over an internal hub on the bracket. The through hole in plate *E* is bored to be a running fit with driven shaft *B*.

A deep slot is cut diametrically across the upper surface of plate *E* to receive sliding bar *F*. Two diagonally opposed sides of this slot are relieved for clearance. A short elongated slot, located approximately in the center of the sliding bar, a slip fit over the driven shaft. Headed steel pin *G* is driven through a

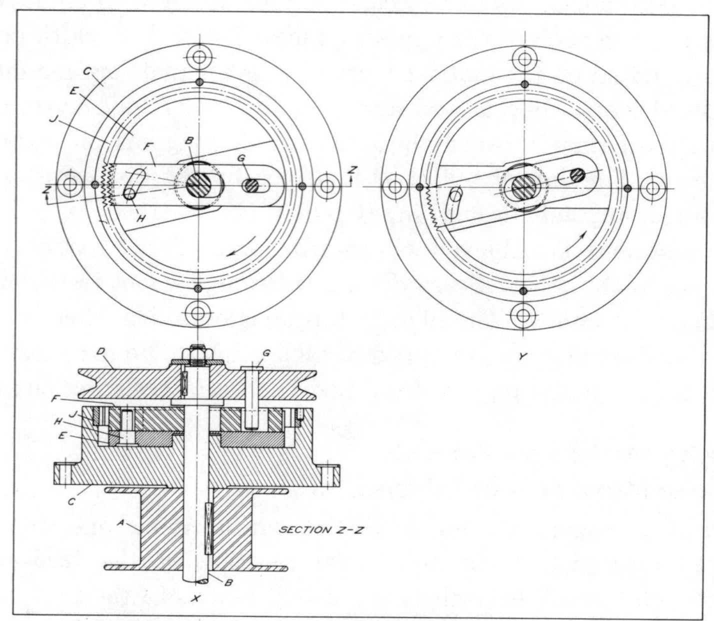

FIG. 2. Drive mechanism arrests any reverse movement of driven shaft *B*.

hole in the web of the pulley and projects into an elongated slot in the right-hand end of sliding bar *F*. At the opposite end of the bar is a short arcuate slot. Engaging this slot is the projecting portion of shoulder-pin *H*. This pin is pressed into a hole that has been drilled in circular plate *E*.

Cut across the left-hand end of the sliding bar is a series of hardened, fine-pitch vee serrations. A case-hardened steel ring *J*, having internal vee serrations matching those of the sliding bar, is pressed and pinned into the largest diameter step of the recess in bearing bracket *C*.

The diagrams at X in the illustration show the relative positions assumed by the various components while shaft *B* is being driven in its normal (clockwise) direction of rotation. Pulley *D*, sliding bar *F*, and circular plate *E* rotate in unison, being connected by pins *G* and *H*. The curved slot bearing against pin *H* serves to hold the sliding bar to the right, thus allowing free movement of all the components.

When pulley D is at rest, any movement of spool A in the reverse direction will be instantly arrested. Such a movement will be transmitted to the pulley and thus to pin G. The pin, in turn, will cause bar F to rotate counterclockwise for a short distance. As this occurs, and the left-hand slot is caused to slide in contact with pin H, the bar will be forced to the left until it engages the vee serrations on ring J. In this way, the drive mechanism is positively locked, as shown at Y, and will remain so until normal rotation is resumed.

Two-Revolution Clutch

In designing a certain mechanism, provision had to be made to rotate a drive-shaft two complete revolutions and then to stop the shaft. The accompanying drawing shows the means by which this action was obtained. All rotating parts are of low inertia. The full lines in Fig. 3 show all parts in their locations when the mechanism is stopped. However, the driving member A rotates constantly in the direction indicated by the arrow.

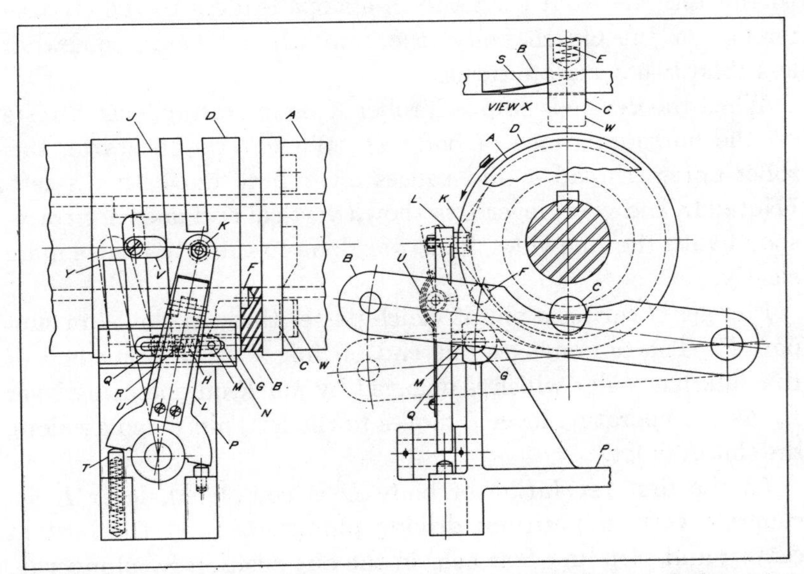

FIG. 3. Clutch which makes two complete revolutions and then stops accurately.

When the mechanism is in motion, a link, which is not shown and which is attached to the left-hand end of stop link *B*, moves the latter from the full-line position to the dotted location. This movement permits driving plunger *C*, carried in clutch body *D*, to move forward under the pressure of spring *E* (view X) until it bears against the face of driver *A*. This position is maintained until one of four cam-like recesses registers with plunger *C*, whereupon spring *E* forces the plunger to enter the recess as indicated at *W*. Clutch body *D* and driving member *A* then revolve together.

An outstanding feature of the device is the means employed to hold link *B* out of the path of plunger *C* until the second revolution has started. To hold link *B* in the lowered position, a hole *F* is provided in the side of the link that faces the clutch body. When link *B* has been lowered to release plunger *C*, the hole registers with plunger *G* and the latter is snapped forward into the hole by means of spring *H*.

Clutch body *D* now revolves with member *A*. A right-hand square thread is cut in the outer surface of body *D* at *J*. This thread terminates at each end in a slope leading to the circumference of the clutch body, the total thread being somewhat less than two complete turns.

When the device is stopped, roller *K* on operating lever *L* rests on the outside surface of body *D*. When rotation starts, this roller enters thread *J* and causes lever *L* to move to the left. (Note: In end view, lever *L* is shown vertical and does not correspond with the side view. This was done to show the parts more clearly.)

Plunger *G* carries a pin *M* which passes through slot *N* in support *P*. The pin engages one end of link *Q*. The other end of this link has a slot which is engaged by pin *R* on operating lever *L*. As the operating lever *L* moves to the left, pin *R* moves along the slot in link *Q*.

As the first revolution of body *D* is completed, lever *L* assumes a vertical position; driving plunger *C* is at the bottom center; and stop link *B* is held in the out position by plunger *G*. This permits the start of the second revolution.

Sometime after the second revolution starts, pin *R* will reach the end of the slot in link *Q*. Further movement of the pin and link to the left will move plunger *G* until it disengages from hole *F*. Link *B* is thereupon returned to its original position by the action of a spring. Then, when plunger *C* reaches link *B*, a wide slot on the plunger engages the thin portion of the cam surface *S* (view X) on the link, forcing the plunger out of engagement with the recess in body *A*. Inertia will carry the rotating parts until plunger *C* engages the stop portion of link *B*.

Upon the clutch body *D* having completed two turns, operating lever *L* will have moved to the extreme left as shown by dotted lines at *Y*, and roller *K* will have risen to the outside surface of body *D*. Operating lever *L* is now free to be moved by means of the spring plunger *T* to its original position.

In order to permit roller *K* to rise from the engaged position with the thread to the outside surface of body *D*, the upper portion of lever *L* is hinged as shown. This upper portion is forced into its operating position by flat spring *U*. Swinging block *V* permits roller *K* to pass over the central portion of the thread groove in returning to the starting position. This block is held in position across the thread groove by a torsion spring that surrounds its pivot screw. When roller *K* passes this point, riding in the groove, block *V* is rotated about its pivot screw into a pocket, as shown by dotted lines, permitting the roller to pass. All parts are now in position for another two-revolution cycle. Stop link *B* and bracket *P* are carried on a frame not shown.

Intermittent Rotation for Instrument Pointers

A mechanism that converts the continuous rotation of a shaft to the intermittent rotation of a stop-wheel is shown in Fig. 4. In the application illustrated, the angular position pointer *A* follows the shaft *B* with a quick motion each time the shaft completes a 60-degree rotation.

Basically, the mechanism consists of a clutch that engages and disengages automatically by the axial movement of one of

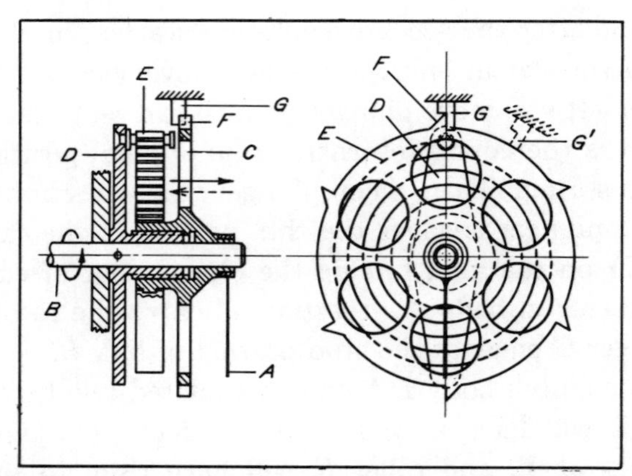

FIG. 4. When shaft *B* rotates, stop-wheel *C* moves axially until lobe *F* clears stop *G*. The stop-wheel then rotates 60 degrees.

the parts. This axial movement is derived from the continuous rotation of the shaft by a screw and nut action.

Stop-wheel *C*, carrying the pointer, is mounted freely over shaft *B*. Also over the shaft, but pinned to it, is an arm *D*. An integral stud on the arm is threaded externally, and engages an internal thread on the stop-wheel. Fastened to, and wound around the hub of the stop-wheel is a flat coil spring *E*. The outer end of this spring is connected to the arm. As the shaft and arm rotate (in the direction of the arrow on the shaft) the spring attempts to transmit this motion to the stop-wheel. However, this member is prevented from turning by the blocking of one of the lobes *F* on the wheel by a stop *G*. Instead, the wheel moves axially to the right (as indicated by the solid arrow) because of its threaded engagement with the arm. Then, when the stop-wheel has moved out sufficiently to clear the stop, the tension built up by the spring causes the wheel to rotate faster than the shaft and thus reverse its axial direction (the broken-line arrow).

Since this particular stop-wheel has six equally spaced lobes, it will rotate 60 degrees before the next lobe abuts the stop. Double frequency can be obtained by adding a second stop, *G′*.

The mechanism can be utilized in applications where the rotation of the input shaft itself is intermittent rather than continuous, since there will be a corresponding effect on the stop-wheel. For simplicity, the pointer is shown carried directly by the stop-wheel. However, if the angular movement of the pointer has to be increased or decreased, a gear transmission can be added.

Simple Device Allows Manual Overriding of Remotely Controlled Lever

Rod C (see Fig. 5), is reciprocated by extension B of split bushing A which is clamped to oscillating shaft D. To disengage A from D, arm F, which is attached to bearings G located in a recess in A, is rotated manually. Cam E then forces split bushing A apart, causing disengagement from shaft D. Further rotation of F will cause rotation of A.

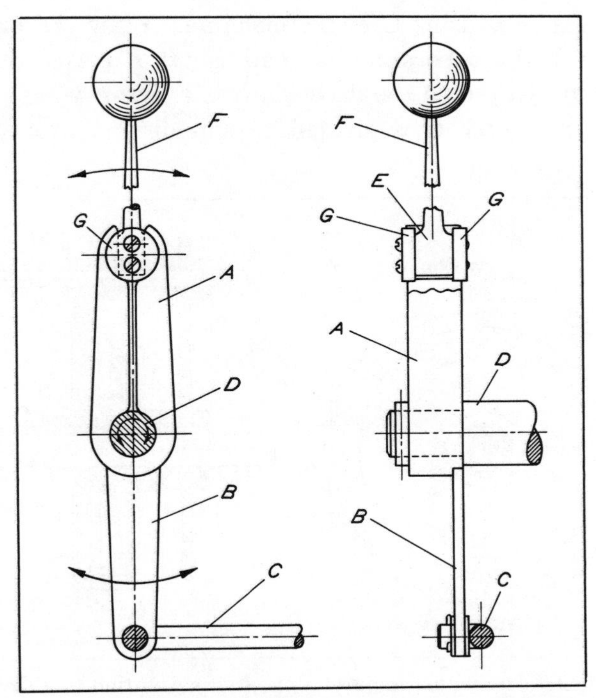

FIG. 5. Split bushing arrangement permits lever actuation by either manual or remote control.

One-Half Revolution — One-Half Pause Mechanism

A one-half turn trip clutch designed to operate two machine heads used for simultaneously twisting eyelets on wires is shown in Fig. 6. Fundamentally, this mechanism is based on a stationary cam and a pivoted latch which control the action of a pair of gears. The mechanism has proved successful in the application mentioned. The capability of the mechanism in driving several units is a decided advantage, and the arrangement has long-wearing characteristics.

The driven shaft makes a one-half revolution with the drive-shaft and then remains idle during the next one-half revolution of the drive-shaft. The mechanism indexes accurately and is suitable for operation at speeds of up to 30 rpm.

The stationary cam *A* is mounted on a short tube *B* which is held tight in housing *C* of the machine. A key prevents rotation of both the tube and the cam. Shaft *E* drives the entire mechanism. Keyed to the drive-shaft is a disc *F* which revolves at all times. Latch *G* is pivoted on this disc. Fastened to the

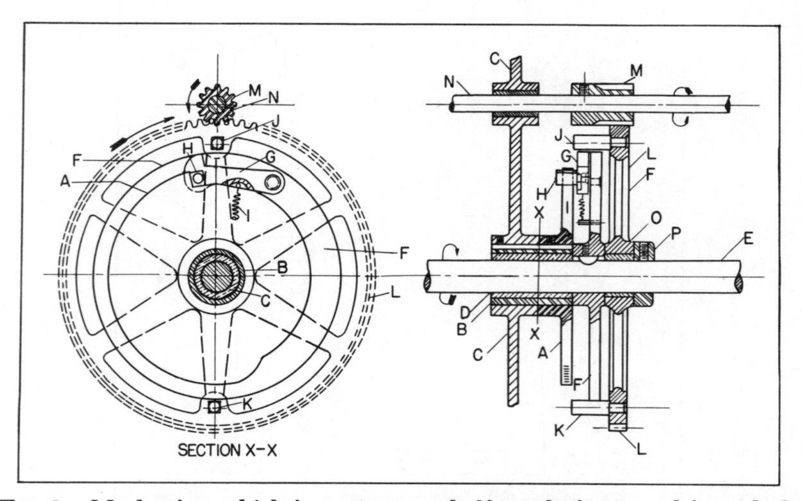

FIG. 6. Mechanism which imparts a one-half revolution to a driven shaft for a one-half revolution of the drive-shaft and then holds the driven shaft stationary as the drive-shaft completes its revolution.

latch is a spring *I* which is also secured to a pin on disc *F*. This spring tends to draw latch *G* toward the center of the mechanism.

Two pins *J* and *K*, respectively, are mounted diametrically opposite to each other on gear *L*. This gear turns one-half revolution during a one-half revolution of shaft *E* and then remains stationary during the remaining one-half revolution of the drive-shaft. Pinion *M* meshes with gear *L*, the gear having 120 teeth and the pinion having 12 teeth, so that in the one-half turn of the gear, the pinion makes five complete turns. The pinion is mounted on the driven shaft and drives it whenever gear *L* is in motion. Bronze bushing *O* is free to rotate on shaft *E*.

The operation of this mechanism is as follows. Shaft *E* drives disc *F* through a key. When latch *G* reaches the bottom of the cam, dog *H* forces the latch to swing radially outward so that the end of the dog engages pin *K* and carries this pin to the position shown occupied by pin *J*. At this point dog *H* rides off the high section of cam *A* and spring *I* pulls the latch inward to clear the pin that has just been indexed to the position shown occupied by pin *J*. The gears are then stationary until the latch is again carried to the bottom of the cam and the dog *H* contacts pin *J*, which has previously been indexed to that position. Gear *L* is then again given a one-half revolution.

Automatic Half-Nut Release for Thread Cutting

A mechanism that permits easy manual engagement and automatic disengagement between the half-nut and the lead-screw of a lathe is illustrated in Fig. 7. With this arrangement incorporated in the lathe apron, screw threads can be cut close to shoulders at a rapid rate.

A single half-nut is carried on the unrestrained end of arm *A*, and is swung downward to engage the lead-screw by an upward movement of hand lever *B*. Any positional change of lever *B*, which is conveniently mounted on the front of the apron, is transmitted to the half-nut by pin *C*. This pin is attached to

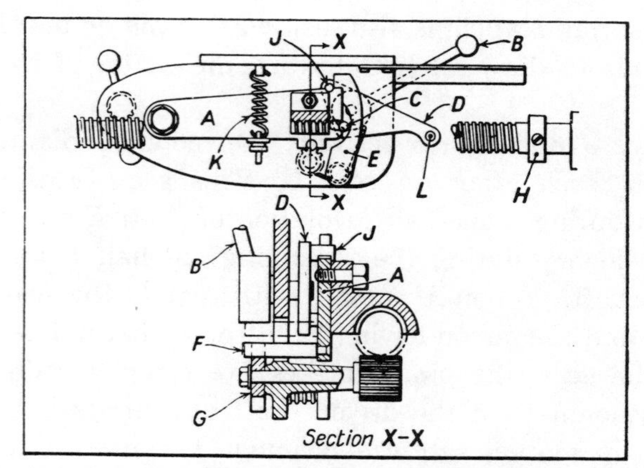

FIG. 7. Lever B is raised manually to engage the half-nut with the lead-screw, and adjustable collar H trips extension L for automatic release.

plate D and engages with a slot in the end of arm A. Plate D is connected to lever B by a shaft that passes through the apron. When the half-nut is engaged with the lead-screw, pawl E is turned counterclockwise by a spring to interlock with the upper end of arm A. In this manner, the half-nut is held in positive engagement with the lead-screw.

Referring to section X-X in the illustration, engagement of the half-nut is controlled by disc G which operates in conjunction with pin F on arm A. Disc G has two diametrically opposite slots, and, when screw-cutting is not actually in progress, is driven by a gear in mesh with the lead-screw. The disc replaces the screw-cutting dial usually provided on the carriage.

With this arrangement, upward movement of lever B causes pin F to make contact with the disc, so that the half-nut is temporarily held out of engagement with the lead-screw. The engagement is made when one of the slots in the disc is presented to pin F. In this way, the engagement of the half-nut is synchronized with the angular position of the work-piece in relation to the cutting tool.

At the end of the cutting stroke, an extension L on plate D makes contact with adjustable collar H mounted on the lead-screw. In consequence, plate D is pivoted in a clockwise direction. A second pin *J*, attached to this plate, causes pawl *E* to release arm *A*. At the same time, this arm is swung upward by the action of pin *C*, assisted by tension spring *K*, so that the half-nut is automatically disengaged. The cross-slide is then moved by hand so that the cutting tool is brought clear of the work, and the carriage is traversed toward the tailstock in preparation for taking the next threading cut.

An alternative method of insuring correct timing for engagement of the half-nut depends upon a tripping device, located at the left-hand end of the headstock which is coupled to the apron mechanism by a cable.

One-Revolution Clutch with Positive Action

For a special type of cutting-off machine it was necessary to supply a one-revolution clutch that would be positive in action and trouble-free. Figures 8 and 9 show a clutch designed to meet these requirements. The design is based on the action of a cog wheel and a driving pawl.

The driving member, pulley *A*, is keyed to drive cog wheel *B*. The cog wheel has six ratchet-shaped teeth and rotates on shaft *C*. Pawl *D* is spring-loaded and free to pivot in assembly *E*

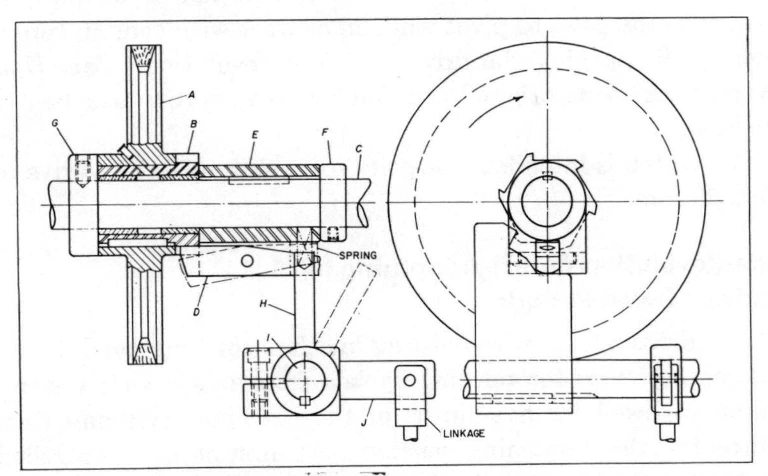

FIG. 8. One-revolution clutch which incorporates a cog wheel and pawl.

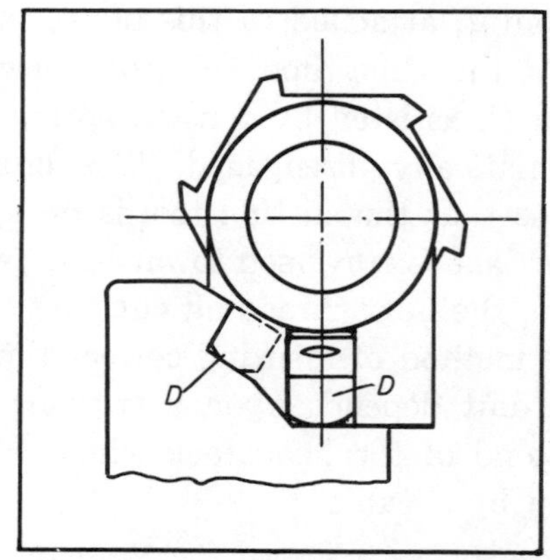

FIG. 9. Diagram showing pawl in entering and stopped positions.

which, in turn, is keyed on shaft *C*. The clutch is held in position on the shaft by means of collars *F* and *G*.

Cam *H* is keyed to the clutch operating shaft *I*. Also keyed to shaft *I* is lever *J*. This lever is used for operating the clutch. With proper linkage it can be manipulated by a foot-pedal.

In operating the clutch, cam *H* is rotated out of position to allow pawl *D* to engage a tooth of cog wheel *B*, as seen in Fig. 9. Cam *H* is then returned to its original position as the pawl and assembly *E* rotate shaft *C*. Continued rotation of assembly *E* will cause the pawl to pivot while in contact with cam *H*, consequently disengaging the drive after one revolution. Cam *H* is swung back to its original position by a coil spring attached to the linkage.

The clutch is intended primarily for a slow operating drive of 10 to 60 rpm.

One-Revolution Clutch Adaptable to Various Dwell Periods

The drive for a wire-twisting mechanism employed in an automatic device for forming eyelets from music wire was required to dwell for five-ninths of the machine cycle and then rotate for the remaining portion. At first a cam-controlled twisting arrangement was tried, but this failed to perform efficiently and was replaced by the one-revolution clutch shown in Fig. 10. The clutch not only provided a smoothly operating means of obtaining interrupted motion but also a method of accurately dividing the constant rotation of the driving member into the proper periods of dwell and movement. One feature of this clutch is that it is versatile and can easily be redesigned for other cycle timing.

Shaft *A*, which is the driven member, is supported in a sleeve bearing pressed in frame *B*. The driving member, sprocket *C*, rotates on shaft *A* and has four ratchet-shaped teeth *D* in the hub of the sprocket. Two pawls *E* pivot on pins *F* secured in disc *G* which, in turn, is keyed to shaft *A*. During periods of drive, each pawl is held in contact with a tooth on sprocket *C* by means of leaf springs *H* and stops *J*. These stops are carried in the bore of a shell *K* which is free to turn on shaft *A*. A collar *L* holds the assembly in place on the shaft.

To disengage shaft *A*, a latch *M*, operated by a tripping cam (not shown), is allowed to contact a tooth *N* on member *K*. (In Fig. 10, tooth *N* has just come into contact with the latch.) The latch stops the rotation of member *K*, but the continued rotation of disc *G* causes the pawls to pivot out of contact with teeth *D* disengaging the drive, and to move clockwise past stops

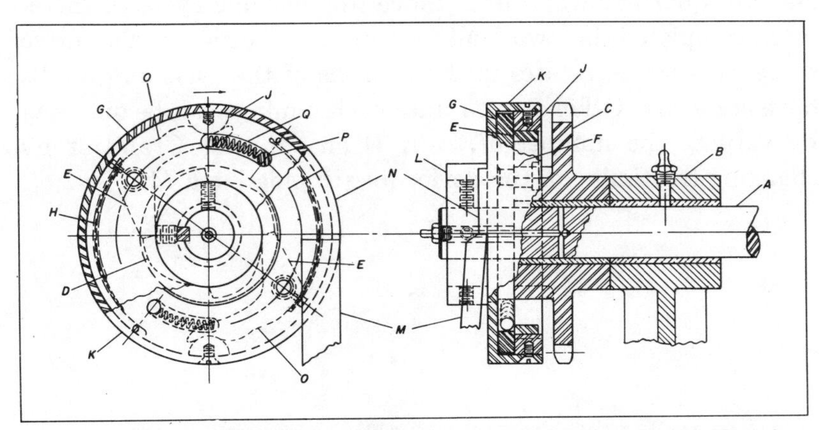

FIG. 10. One-revolution clutch that can be arranged to drive shaft *A* in a dwell-rotate cycle timed in various ratios.

J so that the stops contact the pawls at points *O*. As disc *G* rotates clockwise in relation to shell *K*, due to the friction between the rotating drive and disc *G*, two pins *P* secured in member *K* each compress a spring *Q* carried in an elongated radial slot in member *G*. These springs are heavy enough to rotate shell *K* counterclockwise and return the pawls into contact with the teeth on the drive sprocket when latch *M* releases tooth *N*. They are not heavy enough, however, to rotate disc *G* and shaft *A*.

When the pawls have rotated until they are out of contact with the teeth on the driving sprocket, shaft *A*, disc *G*, shell *K*, and the pawls remain stationary until the latch is tripped to release shell *K*. The release of shell *K* allows shaft *A* to rotate one full revolution before again being stopped by the action of latch *M*. To obtain the proper dwell period, latch *M*, however, is retained in contact with tooth *N* until one and one-quarter turns of the drive sprocket have been completed. The cam then trips latch *M*, and each pawl returns into contact with the adjacent tooth *D* (counterclockwise) on sprocket *C*. Timing of latch *M*, in this case, is accomplished by employing a single-lobe tripping cam geared to the constant speed drive to rotate once in each two and one-quarter turns of sprocket *C*.

Thus the shaft rotates one revolution and then dwells for one and one-quarter revolutions. Since the machine cycle is, therefore, completed in two and one-quarter turns of the drive sprocket, shaft *A* rotates for four-ninths of the cycle and dwells for five-ninths. Other dwell-rotate cycle timings can be obtained by varying the number of teeth *D* on sprocket *C* and/or by changing the timing of the device for tripping latch *M*.

CHAPTER 20

Miscellaneous Mechanisms

The mechanisms described in this chapter are those which were not readily classified in the general groups covered by the preceding chapters. They are included because of some interesting features or ingenious design.

Instantaneous Reversing Shaft-Traversing Device Replaces Troublesome Cam

A "cheeser-twister" in the textile industry is a yarn-winding machine. The machine puts the twist in yarn as it comes from the spinning machine, then winds it (zigzag, like twine) on a cylindrical cardboard core about 5 inches long, making a rough ball about 5 inches in diameter. The balled yarn is known as cheese.

It is important that tension of the yarn in the cheese be uniform and without slack because slackness results in a loose warp thread when the yarn has been woven into cloth. Wear on the drum cam causes the follower and, through it, the thread-guide traverse to dwell at the end of a stroke instead of reversing direction immediately. During the dwell the yarn builds up unevenly on the edges as undesirable slack, that causes loose warp threads.

The described mechanism eliminates the drum cam and its problem of wear. The new traverse therefore provides reliable yarn-tension control, and reduces machine down time and the cost of repairing and replacing worn cams.

In Fig. 1 is shown a cross-section assembly of the new traverse of the cheeser-twister. The cam-shaft has been shortened

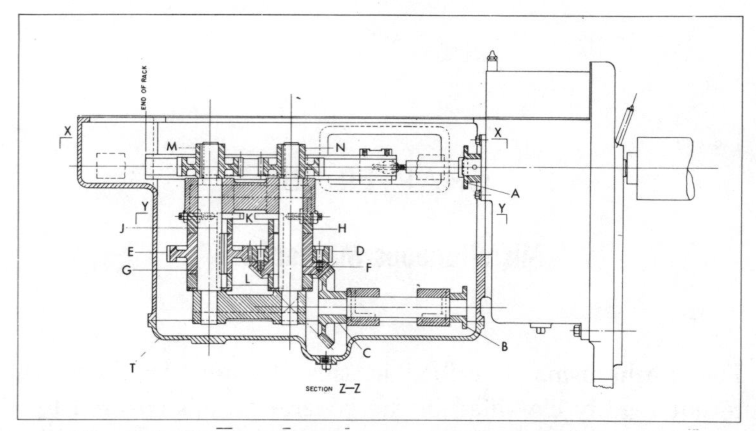

FIG. 1. View of thread-guide traversing mechanism shows design of the pinion-clutching arrangement on shafts *H* and *J*.

to provide for a driving sprocket *A* which transmits the power to the mechanism through the sprocket *B* by a roller chain.

The principle of the design is a parallel rack movement with alternative driving pinions mounted between the racks. The rotation of each pinion causes one rack to advance and the other to retract, thus creating a balanced motion in the machine.

Bevel gear *C* drives gears *D* and *E* which are free to rotate on shafts *F* and *G*. On the ends of the shafts, rack pinions *M* and *N* are keyed and always in mesh with the racks *O* and *P* (section X-X, Fig. 3).

Each of these shafts is fitted with round clutch keys *H* and *J*, each having half the diameter cut away as shown in Fig. 2, section Y-Y, and full-diameter bearings in the collars *K* and *L*, shown in section Z-Z.

The length cut away from the clutch keys is the length of the hub on gears *D* and *E*. The keys each have one end bent up at right angles to the diameter, providing a means for lever action to rotate them, when contacted by adjustable stop-screws *V*-1, *V*-2, *V*-3 and *V*-4. Thus, the rotating motion of the key serves as a clutch.

Each rack has a bracket fastened to it for the purpose of moving the traverse bars *T* and *U*, section X-X, Fig. 3.

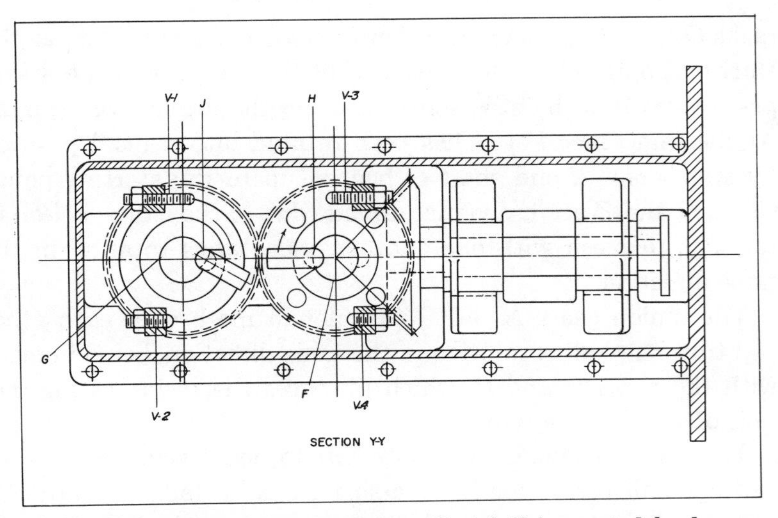

FIG. 2. Alternate clutching of pinions *G* and *F* is created by levers on pins *J* and *H* striking stops *V*-1 and *V*-2, and *V*-3 and *V*-4, respectively. (Note section Y-Y, Fig. 1.)

In the operation of this mechanism, as shown in section Y-Y, (Fig. 1) one key *J* is positive, while the other *H* is neutral. In this case, shaft *G* is a positive drive through gear *E*. At the same time shaft *F* is driven through the rack pinion *N* in mesh with

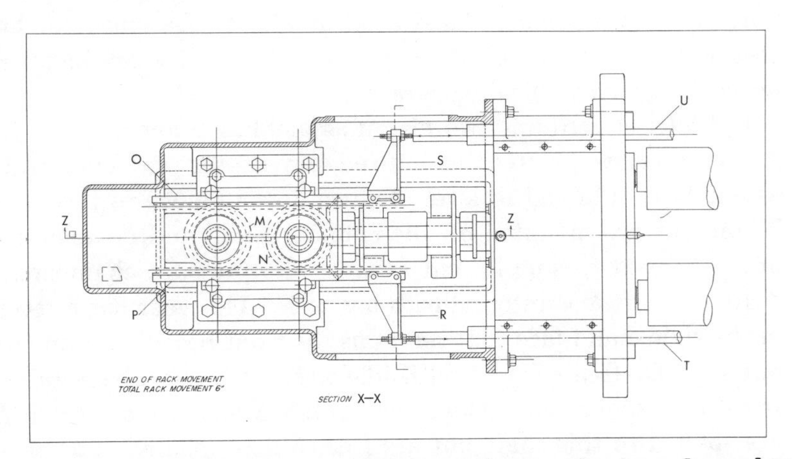

FIG. 3. Section through the shaft-traversing mechanism shows how brackets *R* and *S* transmit motion from racks *O* and *P* to the yarn guide-shaft traverse bars *T* and *U*, respectively.

racks *O* and *P*. Reverse rotation of shaft *F* turns it back to the starting point, when the lever end of the key *H* is rotated into positive position by stop-screw 3, to do the driving via gear *D*. At the same time key *J* has been rotated into neutral position by stop-screw 2, and shaft *G* backed up to the starting point. Gears *D* and *E* are individual drivers, due to the clutch action of the key, in mesh with one another and always rotating in the same direction.

The pinion gears *M* and *N* are not in mesh with each other and turn one-half revolution in the same direction, being in mesh with the racks *O* and *P*, which produces the reciprocating motion of the traverse bar.

The gears *D* and *E* make one-half to one revolution forward and one-half revolution in reversing gears *M* and *N*. Hardened chrome-alloy steel is used for the racks, gears, shafts, and keys.

The top of the case is covered with transparent plastic to permit inspection. The bottom has a removable castiron cover which supports the shaft bearings.

Transmission Remotely Controlled by Triple-Action Cylinder

A triple-acting cylinder arranged to remotely actuate a special purpose transmission is illustrated in Fig. 4. Designed for low cost and made of standard components, the device can be pneumatically or hydraulically operated.

Cylinder *A* contains two piston assemblies *B* and *C* (Fig. 5). Assembly *B* is pivoted on a fixed pin at the rod end, while assembly *C* is linked to level *D* but is free to move to the right. Trunnions are mounted on the cylinder shell, and flexible lines are employed to supply and exhaust the pressure chambers.

Input shaft *E* is integral with a gear *F*. This gear has a recess in the right-hand face that serves as the front bearing for an output shaft *G*. Gears *H*, *J*, and *K* idle on shaft *G* but are restrained from axial movement. Dual face clutch *L* and a rear clutch *M* are splined to this shaft but are free to float axially. The gears on shaft *N* are secured to it and are in positive drive with those on shafts *E* and *G*.

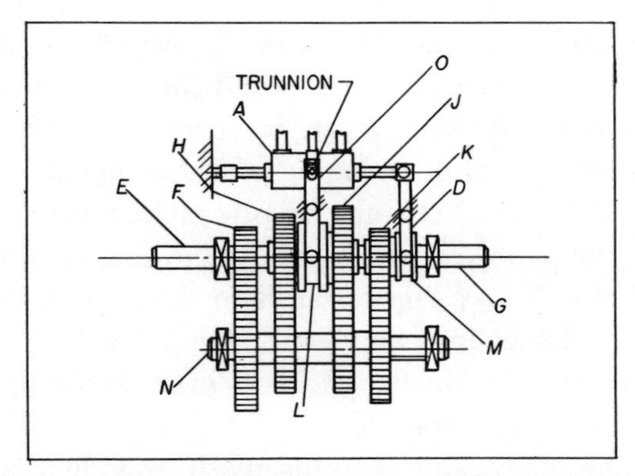

FIG. 4. Triple-acting cylinder *A* permits remote control of this gear transmission. Either pneumatic or hydraulic operation is possible.

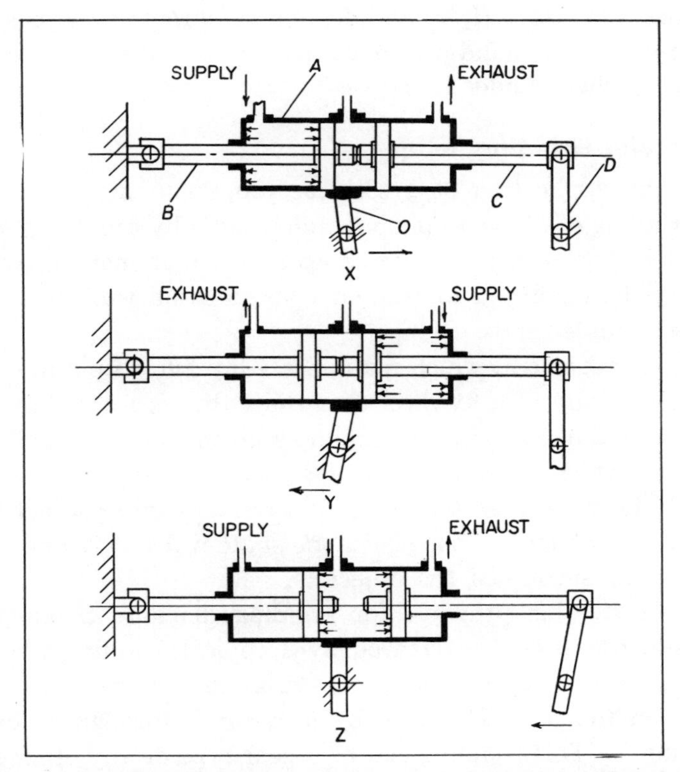

FIG. 5. Lever movements that result when a fluid is introduced under pressure into each section of the cylinder. A different section is vented in each case.

When the left-hand chamber (view X, Fig. 5) is pressurized, the cylinder body moves to the left, and the trunnions tilt a pivoting forked lever *O*. This lever, in turn, slides the dual face clutch *L* to the right, thus engaging gear *J* to output shaft *G* for drive. If the right-hand chamber is pressurized (view Y, Fig. 5) the cylinder body moves in the reverse direction and the left-hand face of clutch *L* engages gear *H* to the output shaft. Pressurizing the center chamber (view Z, Fig. 5) will force the rear piston-rod to the right, pivoting forked lever *D* so that it engages the rear clutch *M* with gear *K*.

Cone type clutches with meshing teeth provide a positive engagement with the gears, thus allowing a low pressure to actuate the system. Coil springs are mounted on shaft *G* between the clutches and gears *H*, *J*, and *K*. These springs overcome any piston-seal friction and disengage the clutches when the pressure actuating the cylinder is released.

Gyroscopic Grinding Setup

Unique application of gyroscopic procession makes possible the grinding of diamond phonograph needles by the device illustrated in Fig. 6. In use, the setup needs little manual control and can be employed for similar operations on work-pieces requiring rounded ends.

The arrangement, which resembles a toy gyroscopic top, consists essentially of a flywheel on a shaft that is rotated at high speed and allowed to precess around a given point. If, while the shaft is revolving, an object having an arbitrary profile is positioned along the precession axis and brought into contact with the precessing end of the shaft, the shaft will follow and press against the surface of the object.

To utilize this principle for grinding diamond phonograph styluses, the object is replaced by a conical grinding wheel *A* driven by a motor *B*, and the work-piece *C* is mounted in a chuck on the shaft *D* driven by a motor *E* through a flexible coupling *F*. This coupling can be a gear type or any other kind which will transmit a constant velocity. Motor *B* and the grinding wheel are mounted so that their axis is at an angle to that of motor *E*.

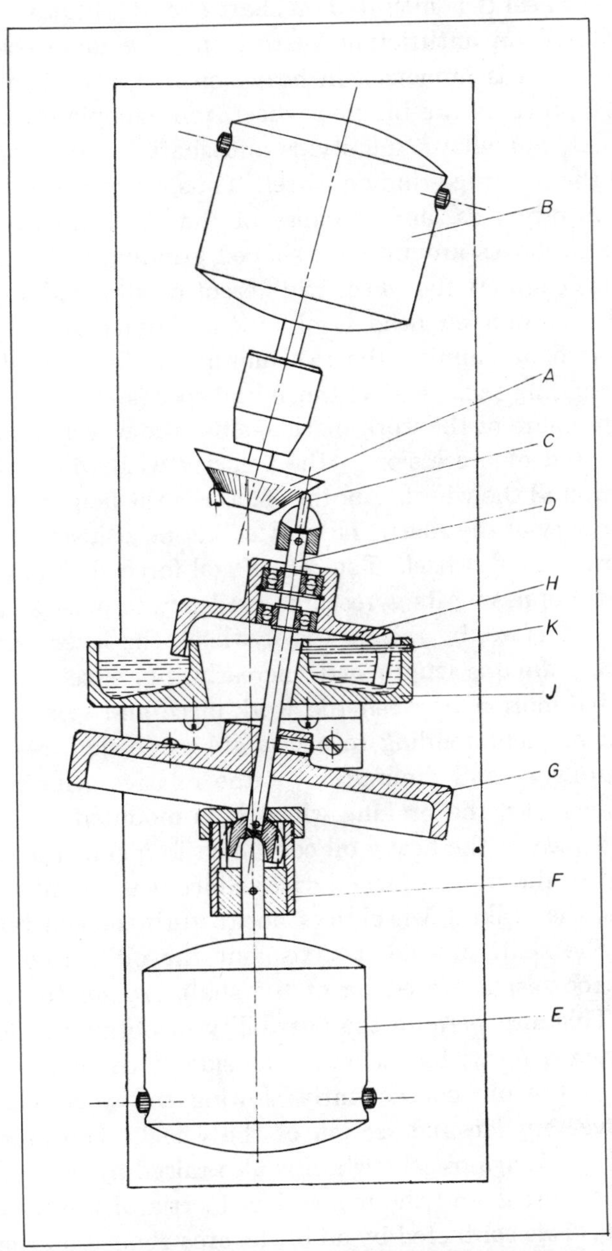

Fig. 6. A device for grinding diamond phonograph styluses. Gyroscopic precession causes the rotating stylus to press against and travel around the rotating conical grinding wheel *A*.

The flywheel G is mounted on shaft D and a braking wheel H is supported on antifriction bearings on the same shaft. The braking wheel is immersed in heavy oil in a fixed container J. A pin K prevents the braking wheel from completely revolving on its axis but allows the stylus and shaft to circle or precess around the rotating grinding wheel. Thus, work-piece C revolves about its own axis, and, because of the gyroscopic precession effect, also moves around cone-shaped grinding wheel A. Since the stylus contacts the wheel at different heights and angles successively, a rounded head is obtained. The speed of grinding is increased by running the two motors in the same direction, thereby adding their relative tangential speeds.

The pressure of the work-piece against the wheel is a function of the speed of precession — the velocity with which the shaft moves around the wheel. Another factor is the height of the center of gravity of the shaft. This latter can be adjusted by raising or lowering the flywheel. The centrifugal force of the flywheel is constantly opposing its gyroscopic tendency to press against the wheel. Therefore, by raising the flywheel, the lever arm of the centrifugal force is lengthened, increasing its effect.

Relative motion between the work-piece and grinding wheel causes a moment tending to accelerate precession, resulting in higher pressure and diminution of the relative grinding speed. To prevent this, the braking wheel H is mounted on the shaft and immersed in the heavy oil contained in member J, which is attached to the frame of the device. Since it is mounted on ball bearings, the braking wheel does not disturb the constant rotation of the shaft around its axis, but the oil's viscosity does prevent too fast a precession of the shaft around the grinding wheel. This also permits the possibility of obtaining elongated and flattened forms by inclining one side of the braking wheel more into the oil, consequently slowing the work-piece as it passes over any desired section of the wheel. The pressure of the work-piece against the wheel is also raised by increasing the speed of motor E and the moment of inertia of the flywheel.

The stylus is contacted by all of the area shaded on the grinding wheel as seen in Fig. 6. This is important for even wear

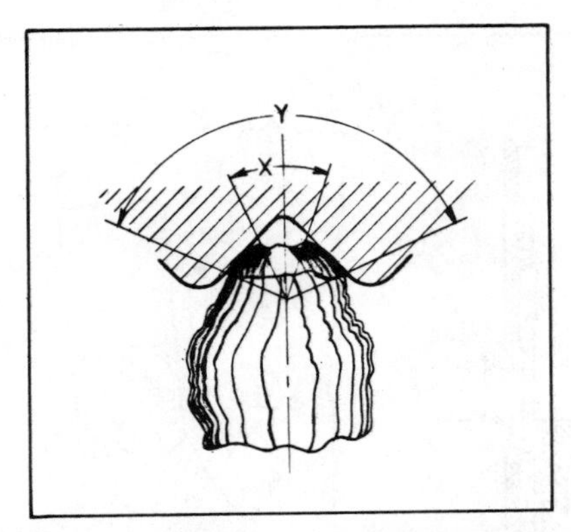

FIG. 7. Each stylus is required to be ground between the angles X (=60 degrees) and Y(=120 degrees) or for a spherical zone equivalent to 90 degrees. Included angle of grinding wheel cone (90 degrees) and axis inclination (15 degrees) is determined by this relationship.

of the wheel. Motor B can also be moved in the direction of its axis, thus changing the portion of the grinding wheel being used without affecting the adjustment of the whole assembly.

In the case under discussion, the stylus has to be ground only in the area between the angles X (= 60 degrees) and Y (= 120 degrees) as illustrated in Fig. 7, because the actual working angle of the diamond point in the record groove is equal to only 90 degrees. This means that one flank of the grinding wheel has to be inclined at an angle of 30 degrees from the vertical, and the other side inclined by $\frac{X}{2}$ or 30 degrees from the horizontal. Consequently, the required cone angle of the grinding wheel is 90 degrees and the inclination of motor B from the vertical is 15 degrees. The actual setup was built for grinding a diamond-stylus radius of 0.0008 inch.

Dial's Moving Witness Mark Compensates for Backlash

A moving witness mark on a dial controlling the angular position of gears, feed-screws, or levers solves the problem of discounting the influence of any backlash. In Fig. 8, the device is shown from the front and side. Dial A is fixed to shaft B which

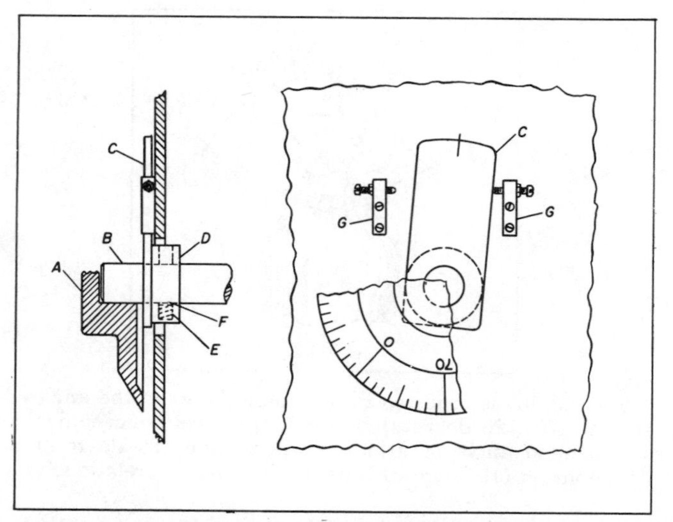

FIG. 8. The free movement of plate *C* between the dogs *G* equals the amount of backlash in the system that has to be taken up.

feeds into the machine or mechanism involved. The witness mark is scribed on plate *C*, which in turn is attached to hub *D* mounted on, but not fixed to, shaft *B*. A blind hole cross-drilled in the hub contains spring *E* and friction pad *F*. The pad drags against the shaft, transmitting the torque from the shaft to the hub. Adjustable stops *G* limit plate travel.

In each view, Fig. 9, the dial is set at 30. In view A, the setting was made by rotating the dial counterclockwise, while in view B, the dial has been rotated clockwise. Although the set-

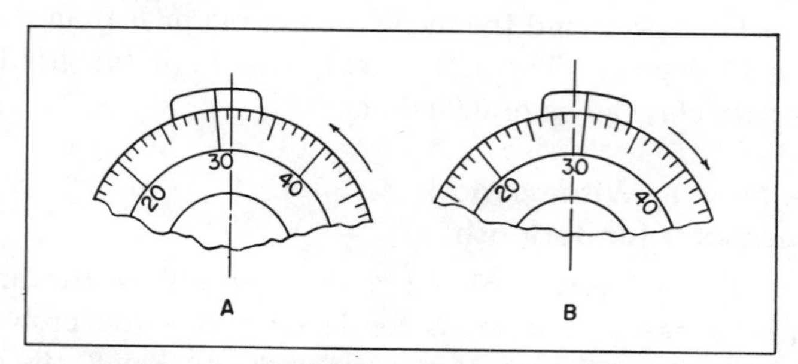

FIG. 9. Backlash in the system accounts for the slight difference in the angular disposition of the two dials and witness marks.

tings are the same, the dial in each instance is in a slightly different position, as is the witness mark. This difference is due to the backlash in the system between the dial (the input) and the gears, feed-screws, or levers (the output).

If the dial in view A is further rotated counterclockwise, the output motion will follow the input motion directly and immediately because the backlash has been taken up in this direction. But if the same dial is rotated clockwise, the input motion is not transmitted until the backlash has been taken up. As the dial moves clockwise, the witness mark moves with it until plate *C* abuts the right-hand stop *G*. Backlash has now been taken up, the output starts to move, and the dial reading changes accordingly.

Pneumatically Actuated Mechanical Oscillator

With the device shown in Fig. 10, a reciprocating mechanical motion can be generated from a constant gas pressure source. The unit is capable of functioning at frequencies as high as 100 cps and its oscillating rate for most operating conditions is independent of fluctuations in the applied pressure. Its inherent reliability is enhanced by the fact that it contains only one moving element.

Essentially, the mechanism consists of an oscillating piston and a cylinder. One end of the oscillating element is turned down to form an output shaft and the piston end is drilled with a number of air passages. The stationary member has a cylinder bore and three annular grooves with threaded external connections.

The two outermost cylinder connections are inputs fed by an air pressure source. The center connection is vented to the atmosphere. With the piston at the left, air from the left input enters the left end of the cylinder through the piston passages. Meanwhile, air previously under pressure in the right end of the cylinder is vented through other piston passages to the atmosphere. The right-hand input is closed by the piston. As a result the net force on the piston causes it to move to the right.

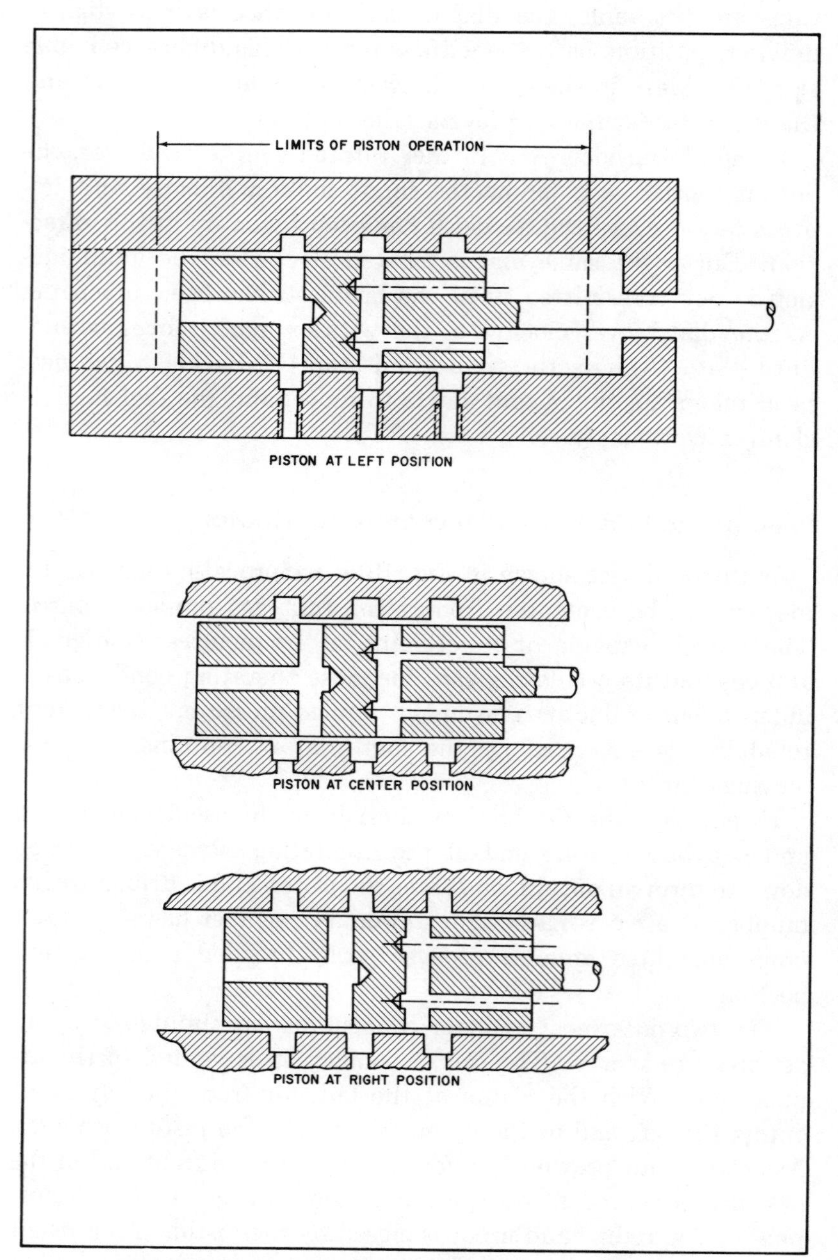

FIG. 10. Piston caused to vibrate by air pressure.

When the piston reaches the center position, it closes the left input and the vent connection to the right cylinder chamber. Further motion to the right allows the air in the left chamber to expand as the residual air in the right chamber is compressed by the piston.

With the piston at the right, the air in the left chamber starts escaping to the atmosphere and the right chamber is opened to the right-hand air supply. As the pressure in the right chamber rises, the piston first decelerates and then reverses its direction. On return of the piston to the left, the cycle repeats.

Reciprocating piston motion can be started from any initial position between the limits shown. The amount of pressure required for operating the mechanism depends upon the weight of the piston and the load attached to the output shaft. Initial pressure also has to overcome the initial friction of the piston in the bore.

Depending on the application, the mechanism can be built with one or two output shafts, one at either end. If it is desired to simply sense the motion of the piston electromagnetically, it can be built without an output shaft. Another application is to use the pulsating output air flow from the vent for pneumatic timing.

Plotting Circular Arcs with Inaccessible Centers

Plotting of circular arcs with inaccessible centers is possible, using a theorem of geometry that states that the locus of the vertex of a constant angle whose legs pass through two fixed points is a circular arc.

If points A and B in Fig. 11 are fixed and angle ACB is held constant, any position of point C, such as at C_1, C_2, or C_3, lies on a circular arc. Also, if angle ABF equals angle ACB, line EBF is tangent to the arc at point B.

This geometry principle provides a means of laying out a circular arc when access to the center point is inconvenient or impossible. Where the arc is to be drawn through three points, or to be drawn through one point and tangent to the line at an-

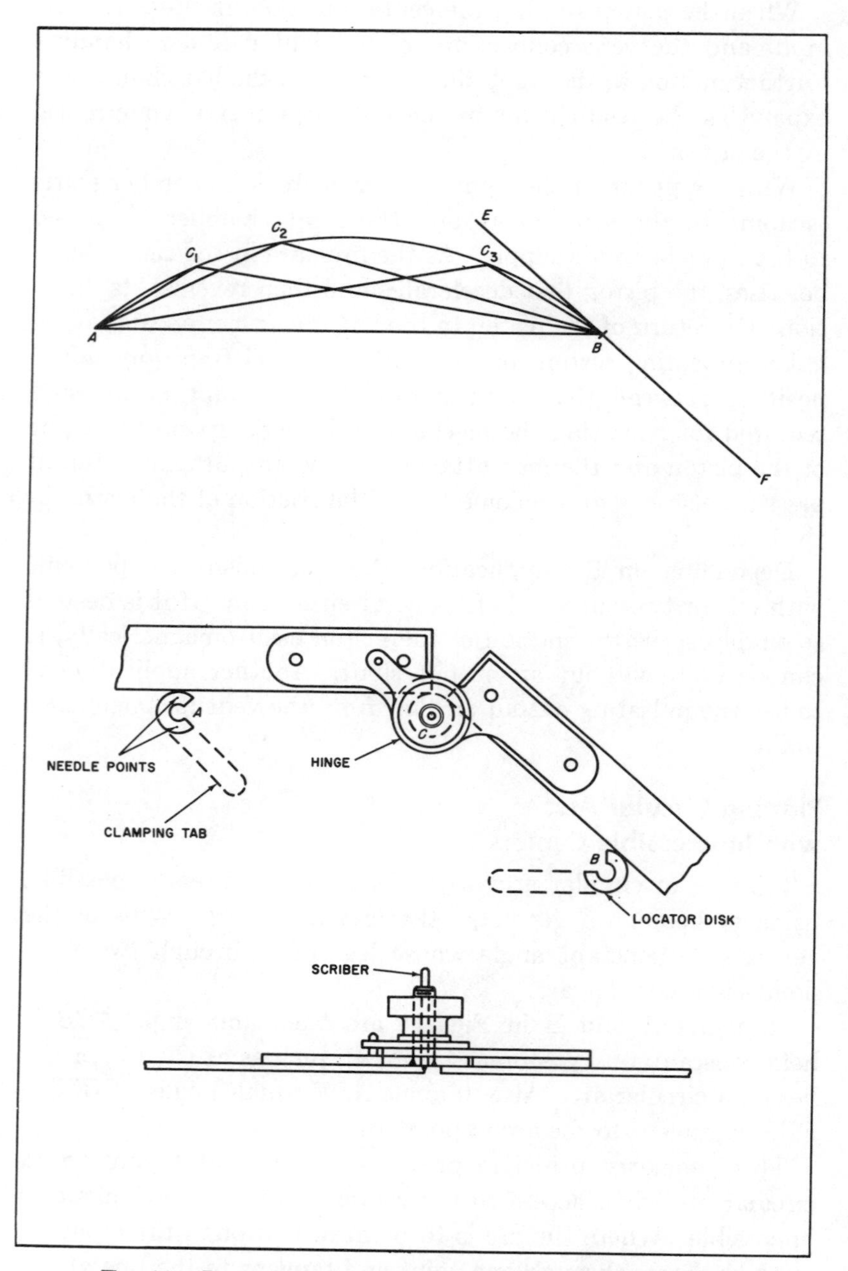

FIG. 11. **Instrument to draw arcs with inaccessible centers.**

other point, the angle involved can be reproduced on tracing paper and the resulting layout used as a template to mark off successive points of the required arc.

Where the problem is met frequently, construction of an instrument of the type illustrated may be justified. It affords additional accuracy by permitting a continuous drawing of the arc instead of the approximation achieved from plotted points.

The device consists of a pair of straight-edge blades connected by a clamp type hinge with its center offset from the blade edges. A hole is provided through the hinge center for the purpose of mounting a pencil lead or a steel scriber. Two discs with radii equal to the hinged offset are used as locators for the fixed points.

Center holes and cutaway segments in the discs allow the arc to be drawn through the fixed points. The locator discs can carry needle points for use on a drafting board, or clamping tabs (shown in dash lines) for part layout.

High-Speed Cutoff Mechanism

Length of the cut-off piece can be varied when using the simple but improved high-speed cutoff mechanism illustrated in Fig. 12. The device consists of a cutter blade attached to a circular disc on a shaft supported by two bearing brackets. The cutter blade is oriented so that when it is in its lowest position, as shown, it is perpendicular to the center line of the material being cut. The strip is supported by a work guide.

Angle α between the center line of the material and that of the cutter shaft can be determined by the equation:

$$\frac{2\pi RN}{60} \sin \alpha = V, \text{ or } \sin \alpha = \frac{30V}{\pi RN}$$

where R is the active cutting radius of the blade in inches, N is the rotary speed of the disc in rpm, and V the velocity of the material in inches per second.

In order to cut various lengths of strip material at different strip velocities V, the following principle is applied: The knife is rotated with a speed N equal to the number of pieces to be

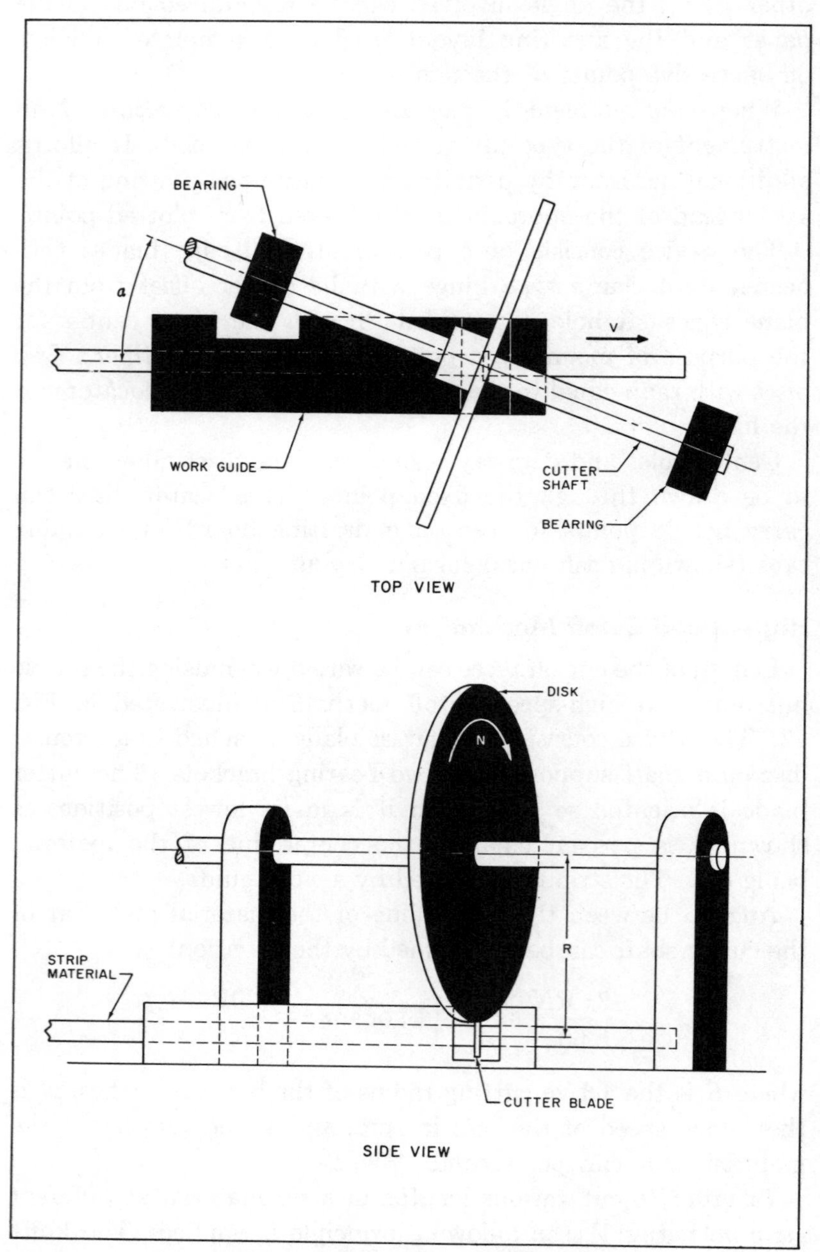

FIG. 12. Stock cut on the fly by shear on a disc.

cut off per minute. Angle α is then adjusted according to the equation and the blade angle is reset to cut perpendicular to the moving strip.

This device has been successful in cutting preslit tubes for can manufacture at a production rate of 1000-2000 cans per min. Actual deviation from an exact perpendicular cut is about 0.002 inch for an active cutting radius R of 2.5 inches.

Mechanism Compensates for Lead-Screw Pitch Errors

Pitch errors in the lead-screws of a thread-cutting or a thread-grinding machine can be compensated for by the mechanism shown in Fig. 13. The drawing shows, in view X, a sectional view of the lead-screw drive in which motion is transmitted through gearing and a differential unit. If the outer casing of the differential is held stationary, the differential acts as a coupling, and drive from the spur gear A is transmitted to the lead-screw in a ratio of 1 to 1, but in the opposite direction of rotation.

A non-self-locking worm B meshes with teeth cut on the periphery of the differential casing. On the outer end of this worm is keyed a plain spur gear C, as can be seen in view Y. Meshing with gear C are two rack-toothed plungers, each of which has a roller at the lower end making contact with one of a pair of templates D and E.

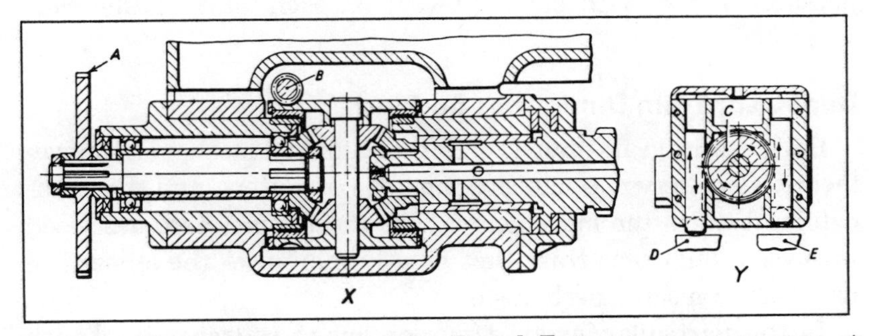

FIG. 13. The surfaces of templates D and E are convex or concave at different points to correspond to pitch errors in the lead-screw.

As gear A starts to rotate, the outer casing of the differential will turn in the same direction, rotating worm B and spur gear C until one of the plungers makes contact with a template. When this occurs, worm B can no longer rotate, and the outer casing of the differential is held stationary. Drive to the lead-screw is now transmitted in the manner described in the first paragraph.

As the lead-screw moves axially, the plunger travels along the upper edge of the template, and as long as the template is flat and parallel with the machine guide ways, the differential casing will be held stationary. However, the edge of the template is arranged with concave or convex curves, which correspond to the known pitch errors in the lead-screw, and when the plunger encounters one of these curves it is raised or lowered accordingly. These movements rotate gear C, which, in turn, rotates worm B, and thus the outer casing of the differential is also caused to rotate.

This rotation has the effect of adding to, or subtracting from, the rotary movement of the lead-screw, and correspondingly reducing or increasing the axial traverse of the carriage, thus compensating for pitch errors. In view Y, one of the plungers has contacted a concave area on template D, which has resulted in a counterclockwise rotation of gear C.

Because of the backlash between the lead-screw and its nut, only one flank of the thread is critical for each direction of carriage traverse. Two templates are necessary, therefore, to compensate for the pitch errors present on each flank of the lead-screw.

Dual Gear Train Diminishes Backlash

Backlash in gears that function to amplify mechanical movements, as in servomechanisms, analog computers, and dial indicators, distorts the movements. The mechanism here described involves a dual gear train that greatly minimizes the amount of backlash in a servomechanism.

In this particular instance, motion has to be transmitted with a ratio of 160:1. Backlash permissible for the slower pinion must not exceed 0.0025 degree. This accuracy is particularly important when the mechanism is in motion. The servo has worked for over one thousand hours and has backlash too small to be measured by normally available means.

A diagrammatic representation of the dual gear train principle appears in Fig. 14. The line of transmission from driving gear A to driven gear B extends both through intermediate gears C and D and through intermediate gears E and F. Through gears C and D, the ratio is 160:1, as required; but through gears E and F, the ratio is stepped up to 190:1.

For all practical purposes, gears C and D may be considered as a single compound gear. Gears E and F, on the other hand, are entirely separate. Gear E, free on shaft G, carries a pawl H. This pawl, in conjunction with a ratchet J pinned to the shaft, represents a one-direction clutch. On the other end of the shaft

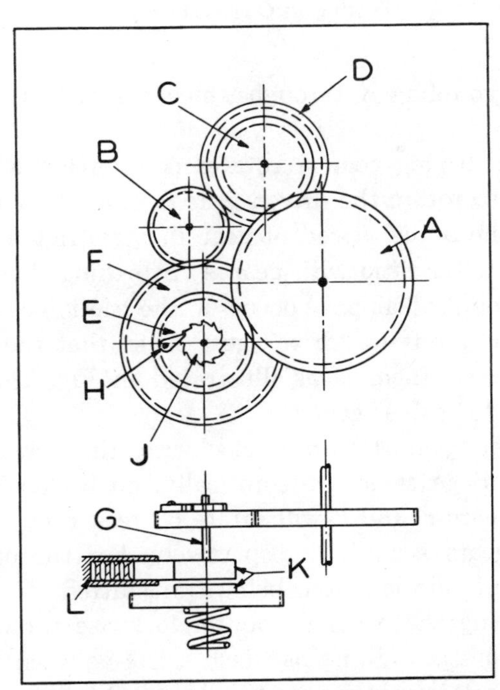

FIG. 14. Intentional slippage in the friction coupling K compensates for the differences in the ratios of the two gear trains.

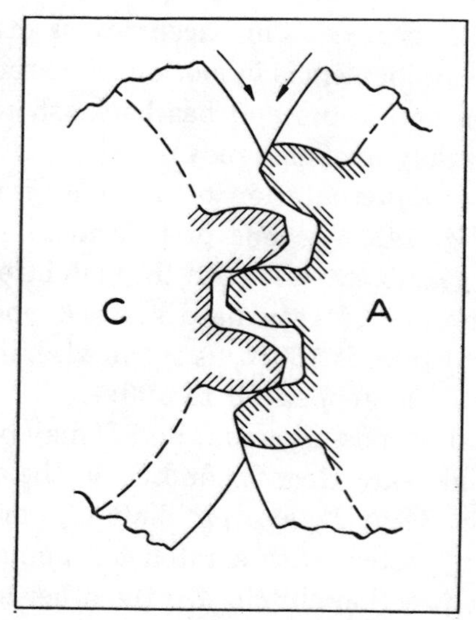

FIG. 15. When gear A rotates counterclockwise, the tooth-flank contact with gear C is as shown.

is a friction coupling K through which the drive is transmitted to gear F.

When the driving gear A rotates counterclockwise, gears E and F tend to rotate the driven gear B at a 190:1 ratio. However, since the actual speed of gear B is restricted by gears C and D to a 160:1 ratio with gear A, two things happen. First, a certain amount of slippage occurs in the friction coupling. Secondly, gear C tends to "drive" gear A, so that the tooth-flank contact between them is as illustrated in Fig. 15. Similarly, gear B tends to "drive" gear D.

Then, when gear A rotates clockwise, the drive is through the compound gears, and theoretically, no backlash has to be absorbed. Gears E and F will rotate by reason of their engagement with gears A and B, respectively, but the one-direction clutch prevents motion from being transmitted.

Since, during the original counterclockwise rotation of gear A, gear F drove gear B, no backlash exists between gears F and B when the rotation is reversed and gear B becomes the driver of gear F which now idles.

Also, the mild drag of a brake *L*, Fig. 14, on shaft *G* prevents gear *C* from running ahead of gear *A* when the latter is rotating clockwise. Since the brake exerts pressure on the shaft at all times, the friction moment of the brake must be stronger than the remaining moment transmitted by the one-direction clutch when rotating in its free direction. In order to attain the greatest accuracy in the mechanism, the friction moment of the brake should correspond to the following equation:

$$M_{br} + M_{set} - M_{odc} = M_{coup} - M_{set}$$

where,

M_{br} = friction moment of brake

M_{set} = moment needed to rotate output shaft and overcome friction in mechanism

M_{odc} = remaining moment of one-direction clutch in free direction

M_{coup} = moment transmitted by coupling

(All of these moments have to be measured on the same axis.)

By thus maintaining a pressure between the teeth of the gears that is equal in both directions, the moment of the mechanism

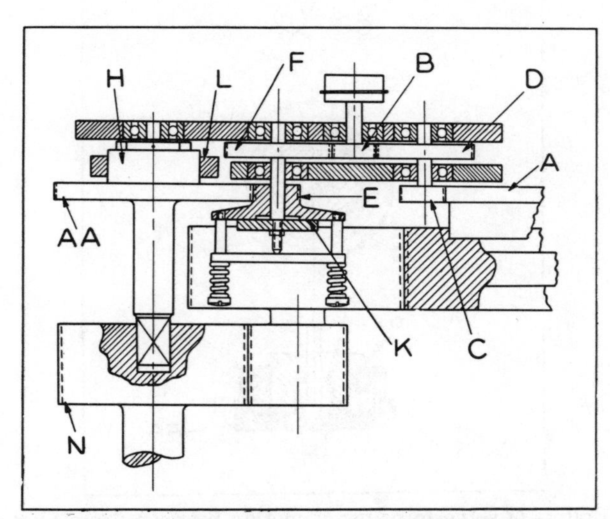

FIG. 16. Cross-sectional view of the servomechanism built around the principle in Fig. 14.

is assured of being constant. This, incidentally, is a feature not found in spring type backlash eliminators.

The actual mechanism that was built is shown in cross section in Fig. 16. Identifying letters on the drawing correspond to those for the counterparts in the diagrammatic representation, Fig. 14. In arranging the transmission, it was found practical to include two driving gears, *A* and *AA*. Both are powered by the same motor pinion *N*. From gear *A* to gear *B*, the train runs through gears *C* and *D*. And from gear *AA* to gear *B*, the train runs through gears *E* and *F*.

The one-direction clutch *H*, the brake *L*, and the friction coupling *K* function as previously described. Details of the

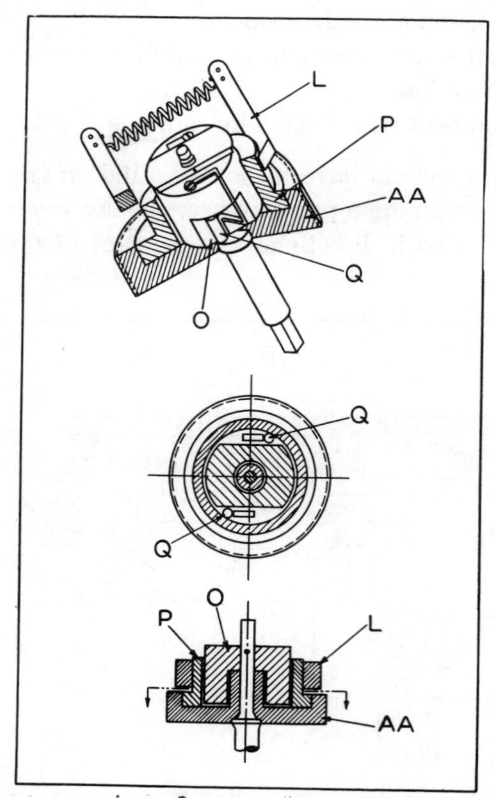

FIG. 17. Rollers *Q* serve to transmit motion between drum *O* and sleeve *P* in one direction only.

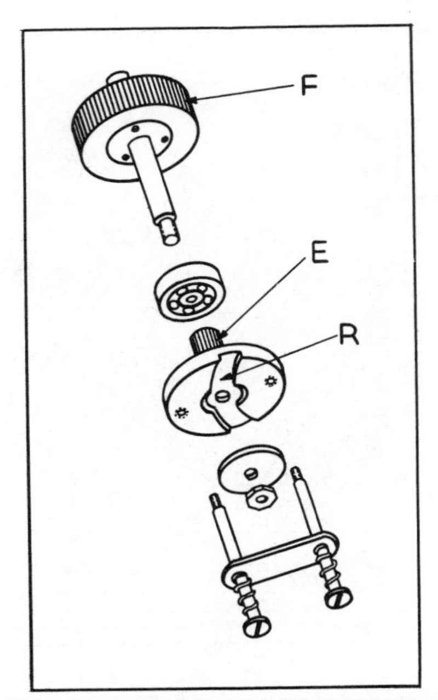

FIG. 18. Exploded view shows the friction coupling between gears *E* and *F*.

clutch and brake appear in Fig. 17. The clutch consists of a drum *O* and sleeve *P*. A pair of rollers *Q* transmit motion in but one direction.

An exploded view of the friction coupling is shown in Fig. 18. Although the relative motion between the two discs of the coupling is small, together they revolve rapidly. A groove *R* in the larger disc produces a flow of air that serves to cool the coupling.

Hydraulically Operated Cross-Slide for a Copying Lathe

A hydraulically operated cross-slide is shown in Fig. 19. Two templates of the part to be turned, as indicated by the broken lines at *A* and *B*, are secured to a fixed member of the lathe. This arrangement enables shoulders up to an angle of 90 degrees to be reproduced on both the headstock and tailstock ends of the work.

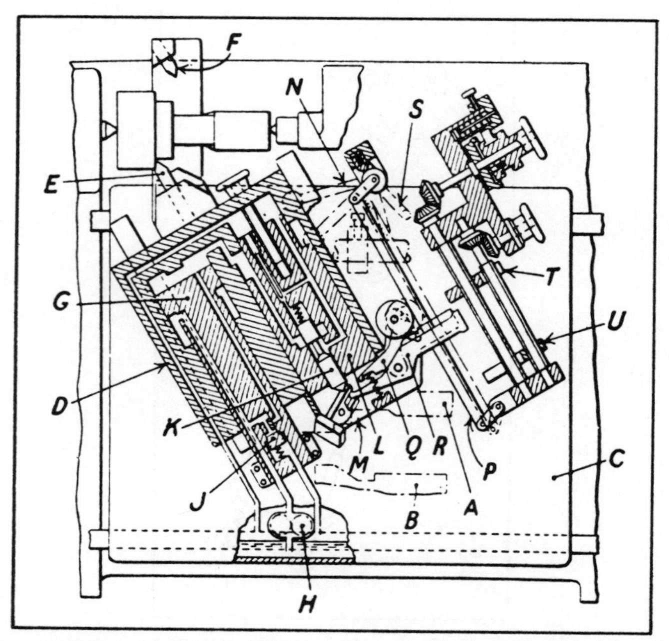

FIG. 19. Sectional view of a hydraulically operated cross-slide for a copying lathe, employing two templates, *A* and *B*, of the part to be turned.

Saddle *C*, mounted on ways beneath the lathe centers, has angular guide ways on which the cross-slide *D* can be moved hydraulically toward and away from the work-piece. A toolholder of the upper end of cross-slide *D* carries the cutting tools *E* and *F*. Tool *E* is used in conjunction with template *A* for turning stepped portions on the tailstock end of the piece. Tool *F* is employed, together with template *B*, when profiles are to be reproduced on the headstock end of the part. A bore is provided in cross-slide *D* to form a hydraulic cylinder which surrounds the fixed differential piston *G* carried on saddle *C*.

In operation, saddle *C* is traversed on the ways, and oil is delivered to both sides of piston *G* by the pump *H*. The pressure in the lower end of the cylinder is maintained constant by a spring-loaded relief valve *J*. When cross-slide *D* is to be held stationary or moved away from the work axis, oil is permitted to escape from the upper end of the cylinder and return to the reservoir by way of the spring-loaded piston type valve K (fitted to body L), and through passages formed in the body and in cross-slide D.

Alternatively, when the cross-slide is to be traversed toward the work axis, valve K is moved downward by spring action so that the oil flow is interrupted and the pressure in the upper end of the cylinder is increased. Movement of valve K and, consequently, of cross-slide D toward and away from the work, is controlled by a stylus which engages template A or B. The stylus is pivoted on a pin fitted to the bracket M, the latter being attached to the lower end of valve body L.

Cutting tools E and F are brought into use, as required, by movement of a lever N, which together with its associated control mechanism, is carried on brackets on the cross-slide. When this lever is turned to its lowest position, bar P (connected to it and to a bracket at its lower end by swivel links) is caused to move in a curved path to the right. The arm Q, carried on a cross-pin fitted to the bracket M, is then swiveled by spring action so that a pin attached to it makes contact with the lever R. Lever R is carried on the same pin as arm Q and its left-hand end engages a slot in the stylus.

Due to the swivel action of arm Q, and consequently of the lever R, valve K is moved upward causing the cross-slide D to be moved away from the work. Thus, cutting tool F is brought into use, and the stylus makes contact with the template B.

Conversely, when lever N is set in its upper position, bar P moves to the left to engage the roller mounted on arm Q, thereby causing the pin fitted to the latter to be brought clear of lever R. Thereupon, valve K is moved downward by spring action with the result that cross-slide D is moved toward the work. In this way, cutting tool E is brought into use and the stylus engages template A.

If desired, cutting tool F can be automatically brought into use when the saddle reaches the extreme left-hand position by means of the lever S. This lever is mounted on the same spindle as lever N and is arranged to make contact with a horizontal adjustable stop (indicated by broken lines), whereupon bar P

and the stylus mechanism are actuated. The sliding motion of saddle *C* is then reversed.

The movement of cross-slide *D* in each direction can be limited independently of templates *A* and *B* by the stops *T* and *U*. These stops are engaged by the right-hand end of lever *R* and can be adjusted toward and away from each other by handwheels, through bevel gears and screw systems. The handwheel controlling stop *T* is provided with a circular scale to enable parts to be turned by tool *E* without the need for a template.

Valve body *L* can be adjusted axially in a second bore provided in cross-slide *D* so that the distance between one of the cutting tools and the stylus can be set to produce stepped parts of the required diameters.

Tension Regulating Device for Coiling Machines

One of the final phases in the fabrication of wire is the winding of the product around a reel as it leaves the annealing furnace. Winding is done by rotating the reel and guiding the wire back and forth along the axis of the reel until layers of wire have been built up.

If the reel rotates at a fixed speed, the tension on the wire increases for each layer, since the wire is forced to feed faster as the diameter of the reel body is built up. Eventually, the wire may snap, or some other failure take place. The accompanying diagram shows the principle of a mechanism designed to maintain a state of equilibrium in which the tension on the wire is constant.

In advancing from the annealing furnace to the reel *A* (see Fig. 20), the wire is directed through a control unit *B*. The reel is driven by a motor *C* through a drive-shaft *D*. A weight *E* produces sufficient drag in the control unit to keep the wire taut as it is wound on the reel. The tendency for the wire to speed up, and for the tension on the wire to increase, is countered by a movement of the control unit. This unit rotates through a small angle after each layer has been wound, to a total of a degrees.

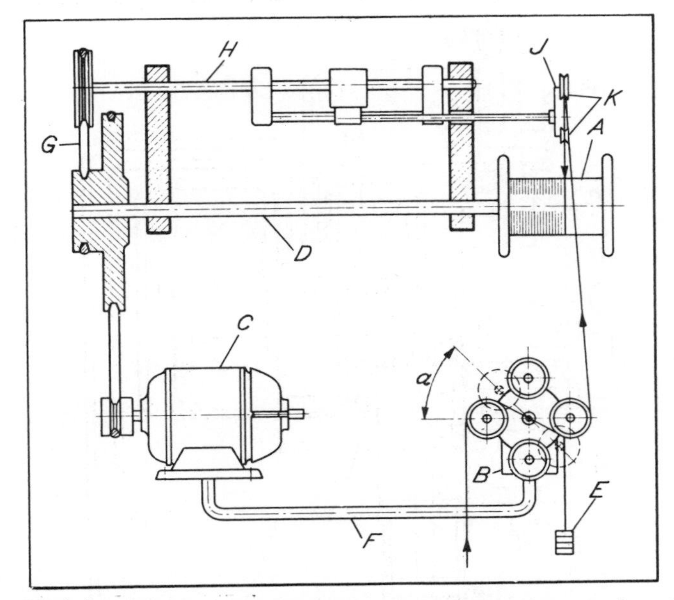

FIG. 20. The control unit B maintains constant tension on the wire by decelerating the motor C.

In pivoting, the control unit actuates a rheostat which, through a cable F, decelerates the motor. Thus, by driving the reel at a progressively slower rate, the feed and tension of the wire are held constant. For any particular wire size, the tension can be set by varying the weight E in the control unit.

A V-belt G causes the wire-guide shaft H to turn with the drive-shaft so that the lineal movement of the wire-guide roller J is synchronized. This roller has a friction contact with two sheaves K, the axes of which are able to float out from a horizontal position to permit the wire to be properly wound around the reel.

Hydraulic Copying System Controls Two Lathe Tools

A patented hydraulic copying system for a railroad car wheel lathe is shown in Figs. 21 and 22. This device permits simultaneous turning on the tread and flange portions of the car wheel

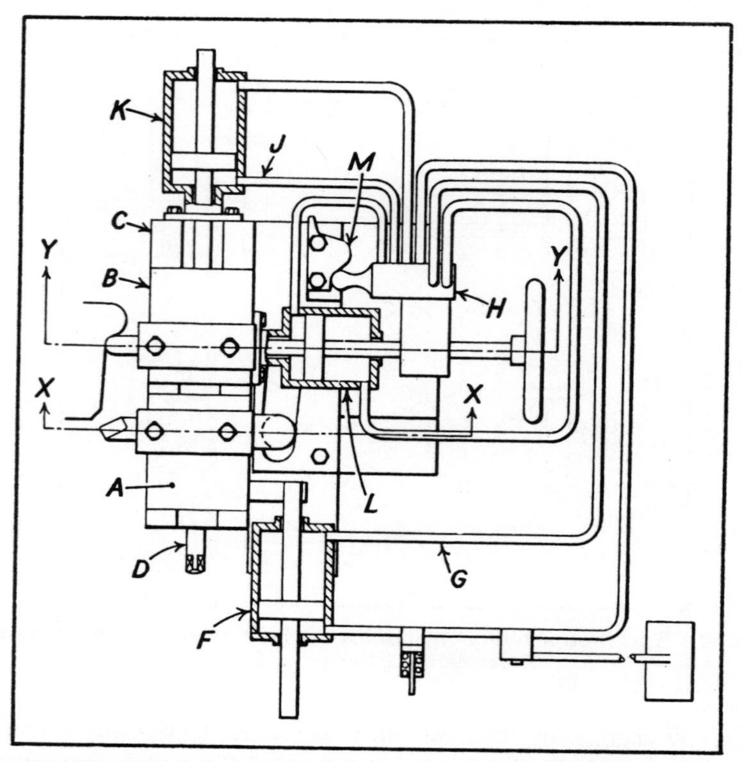

FIG. 21. Plan view of a hydraulic copying system for a railroad car wheel lathe that permits turning tread and flange portions simultaneously.

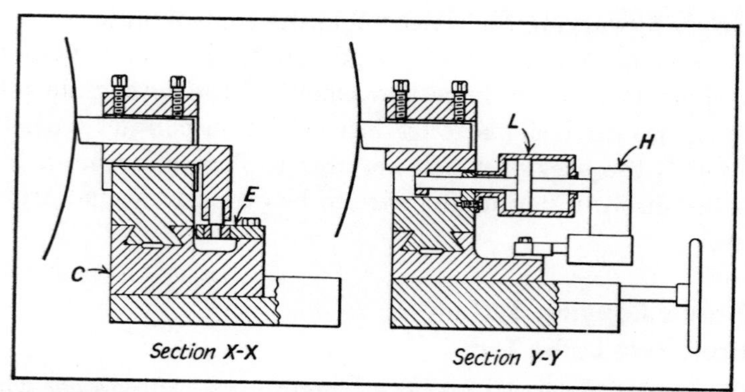

FIG. 22. Sectional views of copying system illustrated in Fig. 21 show template E for tread form, and spoolvalve H attached to piston-rod of cylinder L.

by separate, single-point cutting tools. Thus, cycle time can be reduced and production increased.

As seen in Fig. 21, the cutting tools are mounted on compound slides *A* and *B*, which can be traversed parallel with the work axis on way *C*. The entire assembly can be adjusted toward or away from the work on ways provided on the base.

During the machining cycle, rotation of shaft *D* causes slide *A* only to be traversed parallel with the work axis for turning the tread portion of the wheel. The required tapered form of the tread is produced by the action between a follower roll attached to the slide and the template *E* (section X-X in Fig. 22) which is secured to way *C*.

This traversing motion causes a piston-rod attached to slide *A* to be moved axially in cylinder *F*. As a result, hydraulic fluid is discharged from the inner end of this cylinder, and through pipe *G* to a spool-valve *H*. When profile-turning is not being performed on the wheel flange, the fluid is directed from the valve through pipe *J* to the inner end of cylinder *K*. The piston-rod of this cylinder is attached to slide *B*. Thus, slides *A* and *B* are traversed in the same direction at similar speeds.

As shown in section Y-Y of Fig. 22, valve *H* is attached to one end of a piston-rod. The opposite end of this rod is coupled to the slide carrying the cutting tool for profile-turning the wheel flange. Cylinder *L* is fixed to slide *B*. When profile-turning the flange, a spring-loaded follower (mounted on one end of the valve-spool) is held in contact with template *M*, which is fixed to way *C*.

When the valve-spool is moved outward, due to the action of the template on the follower, the hydraulic fluid from cylinder *F* is directed simultaneously to the inner ends of cylinders *K* and *L*. As a result, the traverse rate of slide *B* is reduced, and the profiling tool is moved away from the work axis to turn half the wheel flange. When the largest diameter of the flange has been turned, the valve-spool is moved in the opposite direction by spring action. This keeps the follower roll in contact with template *M* as slide *B* continues to move in the same direction, as can be seen from Fig. 21.

The hydraulic fluid is then directed to the outer end of cylinder *L*, as well as to the inner end of cylinder *K*. In this way, the profiling tool is moved toward the work to turn the other half of the wheel flange, reproducing the shape of template *M* on the work-piece.

Pitch-Error Compensating Device for Height Gage

An unusual height gage is equipped with an adjustable scale for direct reading and a measuring head that is positioned on the guide column by means of a lead-screw. A vernier, with a dial that is coupled to the lead-screw, permits setting of the instrument to within 0.0001 inch, normally without the use of an optical aid. This accuracy, however, is only made possible by an arrangement that automatically compensates for small pitch errors along the full length of the lead-screw.

The height gage has a tubular column *A* (Fig. 23) mounted on a hollow base *B*. A 10-pitch lead-screw *C*, 8½ inches in length, is supported upright inside of the column by a special anti-friction bearing *D* (Fig. 24) attached to the top of the base. The screw thread is of modified Acme form with a root diameter of only 0.190 inch. A plain bushing, lightly pressed in a suitable cap that is screwed on to the top of the column, retains the lead-screw and keeps it in axial alignment.

A measuring head *E* has a slide fit over the column and is fastened to a tongue integral with a nut *F* that has a close fit on the lead-screw. For nearly its entire length, the column is provided with a slot 0.200 inch wide to permit passage of the tongue.

Drive for the lead-screw is simply a chain of 50 diametral pitch gears (not shown). There are four gears in all: one is attached to the bottom of the lead-screw; one is an intermediate; and the other two are drivers chosen to give a 2-to-1 speed ratio for rapid traverse, and a 3-to-1 speed reduction for fine adjustment. The setting of the measuring head is accomplished by two knurled knobs protruding from the base of the height gage.

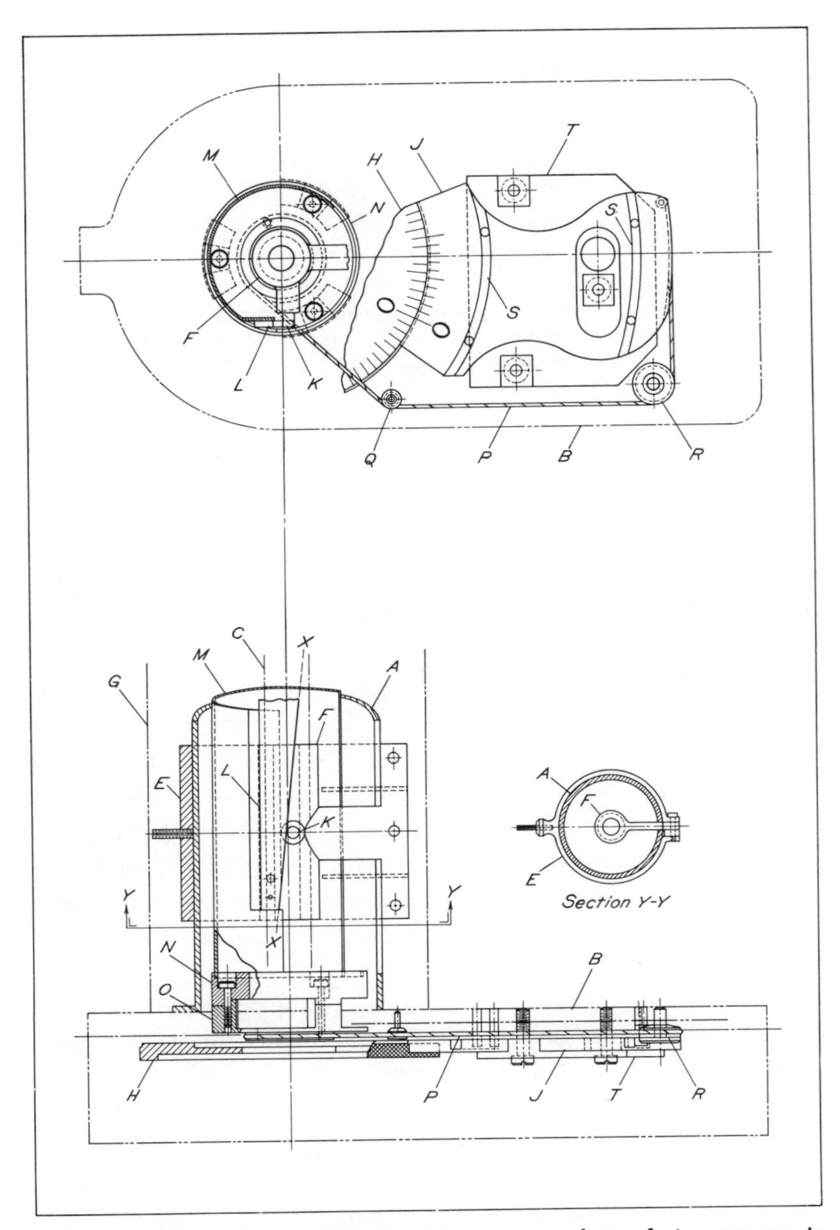

FIG. 23. This mechanism automatically resets vernier scale to compensate for pitch error in the lead-screw of special height gage. For clarity, components not used for this function have been eliminated.

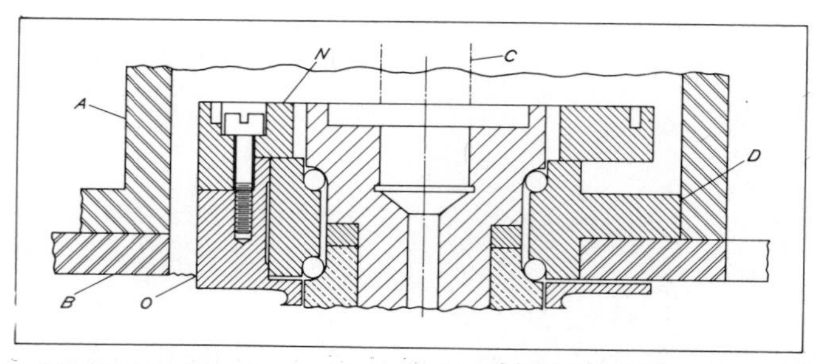

FIG. 24. Enlarged view shows details of main bearing *D* and parts *N* and *O*. Outer race of the bearing has a flange that is cut away at three points to permit passage and movement of part *O*.

A sheet-metal cover *G* encloses the column except for the scale and the slot that provides clearance for the measuring head.

A dial *H*, coupled to the lead-screw, is visible through a window in the base of the gage and is used in conjunction with a vernier scale *J* to set the measuring head. The main scale is vertical. Also, exerting light pressure on a button on top of the base frees the dial for readjusting to zero and simultaneously locks the lead-screw in place. This arrangement permits the scale and the dial to be quickly and easily set to zero, regardless of the position of the head within its full range of 7 inches. The height gage is, therefore, direct reading, and consequently the usual paper work required for setting the gage is greatly reduced. Downward measurements, although involving subtraction, present no problem since the dial is simply read backwards, as is the micrometer dial on the feed-screws of many machine tools.

Construction of the measuring head is shown in section Y-Y, Fig. 23. The center hole in the tongue of the nut is so accommodate a knurled knob for clamping the measuring head to the column. Scriber clamping arrangement (not shown) is attached to the measuring head by means of the stud located opposite the tongue. The accuracy of measurements made with this height gage would normally depend on the accuracy of the lead-screw were it not that the arrangement incorporates an automatic pitch error compensating device. This device advances or retracts the vernier-plate according to the position of the measuring head. Amount and direction of vernier movement are controlled by the contour of a compensating bar. The pitch-error compensating arrangement and associated parts are illustrated in detail in Fig. 23.

Actuating pin *K* is lightly pressed in a boss on nut *F*, and is always in contact with the working edge of compensating bar *L*. The bar is attached to a member *M*, which is made of light tubing. Member *M* stands upright between the inner wall of the column and the screw. Sufficient clearance prevents interference between these parts. The bushing forming the top bearing for the screw also serves as the top pivot for member *M*, the bottom end of which fits in the annular groove in a part *N*. A small screw, not shown, restrains member *M* from turning in the groove. Three screws fasten member *N* to three bosses milled on the upper face of a member *O*. The shape of these bosses may be seen in Figs. 23 and 24. Before assembly, the bosses on member *O* are inserted into openings in the top of the base. These openings surround main bearing *D*, and are shaped as shown to permit members *N*, *O*, and attached parts to have a limited rotational movement.

The body of the main bearing fits snugly in a reamed hole in the top of the base and is flanged. Three screws secure the flange to the top of the base. In addition, the flange is cut away the same as the base to allow clearance for the bosses on part *O*. Member *N* rests on the top face of the bearing body which is somewhat higher than the top of the flange. The bosses on member *O* are a free fit on the periphery of the outer bearing race. This arrangement permits members *N* and *O* to straddle the bearing support flange and to rotate a few degrees on the bearing.

A groove of semicircular cross-section is machined into the periphery of member *O* to accommodate a cord *P*. One end of this cord is attached to member *O*, and the other end is attached to vernier-plate *J*. In addition, cord *P* passes over pulleys *Q* and *R*. Two grooves of rectangular cross-section are cut in the vernier-plate along arcs centered on the axis of the main bearing.

These grooves accommodate two phospher bronze guides *S* of the same radii. Each guide is attached to the base with two 0.063-inch diameter pins that are a light press fit in locating holes. The guides allow the vernier-plate to move in an arc. Retaining plate *T*, held in place with three screws, supports the vernier-plate. A suitable tension spring (not shown) is attached to the vernier-plate to exert a constant pull on the related parts and keep the operating edge of compensating bar *L* in contact with actuating pin *K*. Dial *H* rotates with the lead-screw.

As an illustration, let X-X represent the operating edge of compensating bar *L*. Assuming the measuring head is being raised for a distance of 0.5000 inch, actuating pin *K*, moving with the nut, slides along the bar. Since the bar is inclined in relation to the axis of the column, the sliding action of pin *K* imparts a clockwise movement to members *M*, *N*, and *O*, and the pull of the cord advances the vernier accordingly. Advancement of the vernier is necessitated by the fact, previously established, that a little more than five turns of the 10-pitch screw is required to raise the measuring head exactly 0.5000 inch. If the vernier remained stationary, the zero on the dial would be slightly ahead of the zero on the vernier. Disregarding the pitch error and coinciding the zeros of the dial would, of course, result in a false setting. This has been avoided by the automatic adjustment of the vernier-plate setting.

Finding the pitch errors of the lead-screw and bringing the compensating bar to the desired shape are simple undertakings carried out at the time of assembly. Pitch errors are first ascertained by recording the dial readings of the gage when measuring the height of gage-blocks. Then the compensating bar is detached from the easily removed member *M* and is filed to a shape that will automatically position the vernier scale for accurate readings. Since the back edge of the bar is straight, the contour may be easily measured with a 1-inch micrometer. A contour filed to within plus or minus 0.001 inch of the desired shape will meet practical requirements, for movement of the pin *K* in relation to the change in the reading on the vernier scale of the height gage is in the ratio of over 50 to 1.

Mechanism for Checking the Torque of Tappet Screws

Tappet screws have to fit a tapped hole properly for an assembly to pass inspection. If the fit is too loose (30 inch-pounds is the minimum torque specified), the screws may become slack during operation; and if too tight (120 inch-pounds is the maximum allowed), adjustment for making service repairs difficult.

A tapping machine — equipped with a socket wrench instead of a tap — is used to drive the screws after starting them by hand.

To check the torque applied, the machine is equipped with a special mechanism so designed that if the torque needed at any point in driving the screw to a specified depth is outside the limits set, the assembly is rejected. Assemblies coming within the limits are automatically dropped into a tote box below the machine.

To check the torque applied to the screw *A*, seen in Fig. 25, rocker arm *B* is mounted on a plunger *C*. The cylinder *D* holding this plunger is mounted on a turntable *E*. The turntable is thus free to rock horizontally through a certain angle without significant friction when torque is applied to the screw. The screw is located directly under the socket wrench held in the vertical spindle of the machine.

Before loading the rocker arm over the plunger, the operator starts the screw by hand. As the rocker is mounted on the plunger, it pushes pin *F* inward to actuate limit switch *O* and automatically start the machine. The wrench then grips the screw and feeds it downward into the arm, applying whatever torque is required.

Projecting radially from the turntable is an arm *G* that carries a roller *H* in contact with a piston *J*, which is pressed outward by a spring. The spring presses the roller against a fixed stop-screw *K* with sufficient pressure to resist a torque of 30 inch-pounds. If the torque applied to the screw does not exceed this amount, the roller will not leave the stop and the plunger *C*

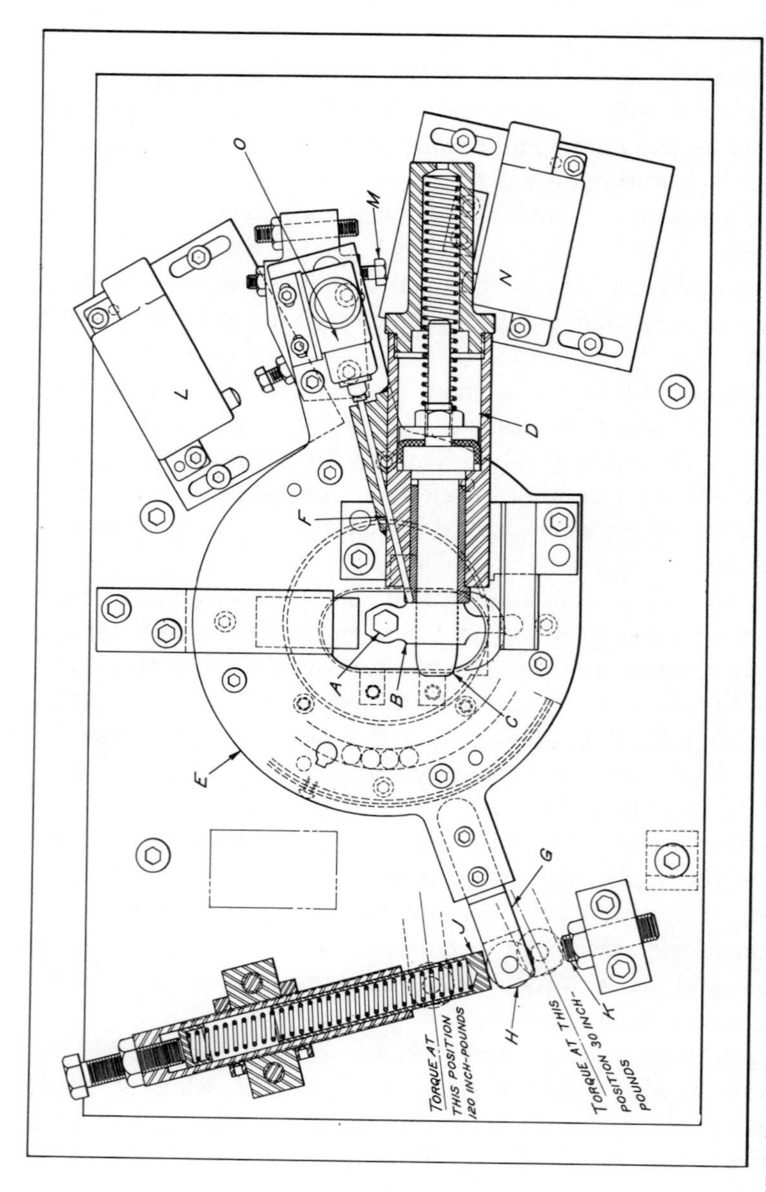

FIG. 25. Assembly drawing of the torque inspection mechanism. The rocker arm *B* is applied over plunger *C* in a cylinder mounted on turntable *E*. Torque is measured, in effect, by the resistance of a spring in back of piston *J*.

will not retract at the end of the cycle because switch *L* remains open. In this event, the operator knows that the screw fits too loosely. Consequently, he must remove the rocker arm from the plunger by hand and place it in a reject box.

If the torque exceeds 30 inch-pounds, roller *H* will be swung clockwise against steadily increasing pressure of the spring behind piston *J*. If the torque reaches 120 inch-pounds, the roller will reach the position indicated in Fig. 25. When this occurs, a stop-screw *M* on the opposite side of the turntable contacts switch *N*, and by tripping this switch prevents plunger *C* from retracting at the end of the cycle. This occurrence warns the operator that the torque is above the limit set. He consequently removes the assembly by hand and puts it in a reject box.

In case the torque applied is not outside the specified limits, the tapping machine spindle feeds down a specified amount in driving the screw and then automatically retracts. As the spindle reaches its upper limit, it trips a switch that operates a solenoid. This, in turn, opens an air valve, admitting air to the cylinder whose piston retracts plunger *C*. This retraction does not occur, however, if roller *H* does not leave the stop-screw *K*, or if the roller moves far enough to trip the switch *N*.

When the applied torque is within the limits set, retraction of the plunger *C* releases the assembly and it drops through a hole in the base of the machine and into a box containing acceptable parts. A spring then returns the plunger, ready to repeat the cycle. With this set-up, it is possible to apply screws to 362 rocker arms per hour on each machine.

All the operator does is to start each screw and then load the rocker arm on the plunger. The remainder of the cycle is automatic and the operator does not even have to remove the workpiece unless it is unacceptable.

As the cylinder carrying plunger *C* is mounted on a turntable that must move freely, the air lines to the cylinders are made from flexible rubber tires. The resistance of these tubes to bending as the turntable rocks has a negligible effect on the torque applied.

INDEX